18th and 19th Century Porcelain Analysis

Howell G. M. Edwards

18th and 19th Century Porcelain Analysis

A Forensic Provenancing Assessment

Springer

Howell G. M. Edwards
Chemical & Forensic Sciences
University of Bradford
BRADFORD, UK

ISBN 978-3-030-42191-5 ISBN 978-3-030-42192-2 (eBook)
https://doi.org/10.1007/978-3-030-42192-2

Cover picture : Spill vase, of trumpet shape, photographed in transmitted light to illustrate its superb and clear white translucency. Decorated with a geometric lattice of green pigment containing single pink roses, and finely gilded with circlets of leaves and lozenges. Unmarked, but non-destructive analysis indicates that it is soft paste porcelain from the Nantgarw China Works, ca. 1817-1820.

This Springer imprint is published by the registered company Springer Nature Switzerland AG.
The registered company address is: Gewerbestrasse 11, 6330 Cham, Switzerland

This book is dedicated to the memory of my beloved late wife, Gill, whose constant encouragement was so material to its completion, and also to our daughter, Kate, whose support was ever present. Without their participation in this enterprise, the outcome of many years' research would not have been realised.*

Howell G.M. Edwards
January, 2020

**Gillian Patricia Edwards: 21st November, 1944–7th December, 2019*

Preface

This book addresses the contributions made by analytical chemistry to the characterisation of eighteenth- and early nineteenth-century English and Welsh porcelains commencing with the earliest reports of Sir Arthur Church and of Herbert Eccles and Bernard Rackham using chemical digestion techniques and concluding with the most recent instrumental analytical experiments, which together span more than a hundred years of study, from the earliest experiments which required necessarily the sacrifice of significant portions of each specimen, which may already have been damaged, to the latest experiments which needed only microsampling or the non-destructive interrogation of valuable perfect specimens. A comprehensive survey is undertaken of more than 20 manufactories of quality porcelains. The correlation is made between the quantitative elemental oxide determinations of the scanning electron microscopic diffraction and X-ray fluorescence data and the qualitative molecular spectroscopic Raman data to demonstrate their complementarity and use in the holistic forensic assessment of the origin of the fired porcelains; this will form the groundwork for the adoption of combined analytical techniques for the attribution of unknown or questionable porcelains to their potential source factories. The book will also examine the perception of what constitutes a porcelain, its various definitions and the assignment of porcelains to types which currently employs the definitions of hard paste, soft paste, hybrid, magnesian and bone china from the conclusions derived from the analytical data and a consideration of the raw materials employed in their manufacturing processes. During the discussion of this analytical evidence, several themes and protocols have been established for its utilisation in the potential identification of porcelains, and several case studies undertaken for this purpose are cited. The book will be of interest to analytical scientists, to museum ceramic curators and to ceramic historians.

Emeritus Professor of Molecular Spectroscopy, Chemical & Forensic Sciences, School of Chemistry and Biological Sciences, Faculty of Life Sciences, University of Bradford, Bradford, West Yorkshire, United Kingdom

Professor Howell G. M. Edwards

Acknowledgements

The author is deeply indebted to the many colleagues and friends who have contributed to the ideas, practicalities, experiments, discussions and concepts behind this book. In particular, he would like to recognise the special assistance of the following:

Charles Fountain, Director of the Nantgarw China Works Trust, Nantgarw, South Wales, UK
Kate Edwards, MSc (Dunelm), CGeol, FGS, Sheffield, UK
Dr. Peter Bradshaw, Sheffield, UK
Guy Fawkes, Farnley Hall, Otley, North Yorkshire, UK
Dr. Morgan Denyer, BSc, PhD (UCNW), Penrose Antiques, Bradford, UK
Rachel Denyer, BA (UCNW), MA (OU), Penrose Antiques, Bradford, UK
Peter Frost-Pennington, Muncaster Castle, Cumbria, UK
Bryan Bowden, Harrogate, North Yorkshire, UK
Dr. Danita de Waal, Pretoria, South Africa
Dr. Elizabeth Carter, BSc, PhD (QUT), University of Sydney, Australia
Professor Philippe Colomban, Sorbonne University, Paris, France
Professor Peter Vandenabeele, University of Ghent, Ghent, Belgium
Professor Jan Jehlicka, Charles University, Prague, Czech Republic
Rev James Dickinson, Chesterfield, UK
Trustees of the Estate of Dr. John Twitchett, Somerset, UK
Dr. Ross Ramsay, Kerikeri, New Zealand
Dr. William Jay, Monash University, Melbourne, Australia
Andrew Renton, BA (Oxon) Keeper of Art, National Museum of Wales, Cardiff
Dr. Alexander Surtees, BSc, PhD (Bradford), University of Bradford, UK

Contents

List of Figures

List of Tables

About the Author

Howell G. M. Edwards, MA, BSc, DPhil, CChem, FRSC Howell Edwards is Professor Emeritus of Molecular Spectroscopy at the University of Bradford. He studied Chemistry at Jesus College, University of Oxford, and after completing BA and BSc degrees, he completed his doctorate there and then took up a Research Fellowship at Jesus College, University of Cambridge. He joined the University of Bradford as a Lecturer in Structural and Inorganic Chemistry, becoming Head of the Department of Chemical and Forensic Sciences, and was awarded a Personal Chair in Molecular Spectroscopy in 1996. He has received several international awards (Sir Harold Thompson Memorial Award, Charles Mann Award, Emanuel Boricky Medal, Norman Sheppard Award) in a spectroscopic career which has resulted in the publication of over 1300 research papers in Raman spectroscopy and the characterisation of materials, along with six books on the application of this technique to art, archaeology and forensic analysis. He has had a lifelong interest in the porcelains of William Billingsley, especially those from the Derby, Nantgarw and Swansea china factories. He has authored three major books on Nantgarw and Swansea porcelains: *Swansea and Nantgarw Porcelains: A Scientific Reappraisal*, *Nantgarw and Swansea Porcelains: An Analytical Perspective* and *Porcelain to Silica Bricks: The Extreme Ceramics of William Weston Young, 1776–1847*, all published by Springer, Dordrecht. He has also produced several monographs on the personalities involved in these factories: *William Billingsley: The Enigmatic Porcelain Artist, Decorator and Manufacturer*, *Nantgarw Porcelain: The Pursuit of Perfection*, *Swansea Porcelain: The Duck-Egg Translucent Vision of Lewis Dillwyn* and *Derby Porcelain: The Golden Years, 1780–1830.* Howell Edwards is Honorary Scientific Adviser to the de Brecy Trust on the scientific evaluation of artworks and paintings.

Chapter 1
Porcelain and Its Composition

Abstract The definition of porcelain is followed by a survey of the different types of porcelain manufactured, including bone china, which forms the basis for their analytical assessment in terms of the raw materials used in their composition. The birth of porcelain in China, its growth in Europe from an accepted start in Meissen in 1709 and the emergence of the earliest English porcelain factories in the 1740s is summarised.

Keywords Hard paste porcelain · Soft paste porcelain · Celadons · Faience · Majolica · Meissen · Bow · Pomona · Chelsea · Raw materials for porcelain bodies · Biscuit porcelain

1.1 What Is Porcelain – The Birth of Porcelain in Europe and Its English Roots?

The definition of *porcelain* seems to be relatively straightforward in that it describes a synthetic ceramic material, generally white in appearance, which has been fired at a high temperature and has a *translucency* to visible light: this definition clearly differentiates porcelain from earthenware, faience, majolica and other non-translucent ceramic materials which may have also been fired in high-temperature kilns and which may have been treated with an applied, highly reflective glaze. The presence of a surface glaze or otherwise on ceramic porcelain materials is actually irrelevant in terms of this definition since unglazed porcelain has been recognised from early times and is referred to as "biscuit" porcelain: this is usually the product of a first time firing of an article comprising the porcelain paste, which is then subjected to further firing after any appropriate glazing and enamelling decoration have been applied. An example of the high esteem in which pure, unglazed biscuit porcelain was held in the eighteenth century is seen in Fig. 1.1, which shows a figure of *Flora* from the Derby China Works, *ca.* 1790–95, depicting an operatic Parisian flower girl portrayed by Angelika Kaufmann and modelled by the famed Jean-Jacques

H. G. M. Edwards, *18th and 19th Century Porcelain Analysis*,
https://doi.org/10.1007/978-3-030-42192-2_1

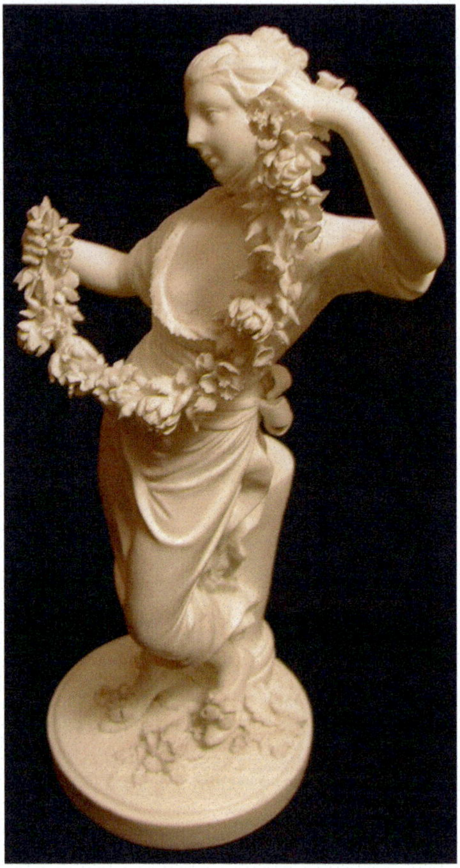

Fig. 1.1 Derby porcelain, *ca.* 1790–95, beautiful, unglazed biscuit porcelain figurine depicting "*Flora*" in a classical pose of "*An Opera Girl in Paris*" wearing a late Georgian period dress resting against a ruined Greek pillar and carrying a garland of flowers. Height 12″ (30 cm). This figure is beautifully crafted by Joseph "Jockey" Hill, whose rebus comprising a triangle is impressed on the base along with the crown and crossed batons of Derby and the number 390 in cursive script. This figure probably used models from Angelika Kaufmann as a basis of the design. It does not appear in the Derby figure records (Twitchett, *Derby Porcelain, 1748–1848: An Illustrated Guide*, 2002) and it is extremely rare if not unique, as researched by Dr Peter Bradshaw and illustrated in his book on *English Porcelain Figures of the 18th Century. In a Private Collection*

Spangler (Edwards, *Derby Porcelain: The Golden Years*, 1780–1830, 2018b; Bradshaw, *English Porcelain Figures of the 18th Century 1745–1795*, 1981; Bradshaw, *Derby Porcelain Figures of the 18th Century*, 1990); the absolute perfection required for the artistic production of such a figure was essential because blemishes could not be hidden or masked by subsequent glazing or enamelling (Twitchett, *Derby Porcelain 1748–1848*, 2002). It is of interest to record here that the term "porcelain" actually originates from the Italian "*porcelanna*", which highlighted the analogy between the glazed white ceramic artefact and a natural cowrie shell, accentuating thereby the clear white glaze and the hard texture – but as we now realise, of course, not all porcelain is glazed and hard!

What is much more difficult to determine more accurately, however, is the type of porcelain which occurs within this broad generic ceramic definition which has been produced from several manufacturing routes, and the real meaning of historic terms such as hard paste porcelain (*pate dure, porcelain francais*), soft paste porcelain (*pate tendre*), hybrid porcelain, true porcelain, artificial porcelain, phosphatic porcelain, siliceous porcelain, soapstone porcelain, magnesian porcelain, glassy porcelain and bone china, which are all categories encountered in the ceramic literature within this specific genre of porcelain. Although Sir Arthur Church is usually credited with performing the first analyses on English porcelains towards the end of the nineteenth century (Church, *English Porcelain*, 1894), Herbert Eccles and Bernard Rackham published the first comprehensive analytical and detailed study of English porcelains in 1922 (*Analysed Specimens of English Porcelain in the Victoria and Albert Museum Collection*, 1922) which cited their actual analytical data, and on the basis of their results they proposed the following categories of porcelain, namely, hard paste, soft paste, soaprock, glassy and hybrid. In later years, these have been further proscribed to *hard paste* and *soft paste porcelains;* the latter has then been further subdivided into “glassy”, “soapy” and “bony” categories, which describe porcelain which contains glass frit, soapstone and bone ash, respectively. Savage and Newman (*An Illustrated Dictionary of Ceramics*, 1976), after careful consideration, recognised only three categorical types of porcelain:

True porcelain: hard paste, *pate dure, porcelain royale* (Sevres);
Artificial porcelain: soft paste, *pate tendre, porcelain de France*, frit porcelain (containing powdered glass used as a substitute for the feldspar in true porcelain);
Bone china: containing calcined bone ash.

The association of the two terms “hard paste porcelain” and “true porcelain” was first proposed by Alexander Brongniart (Ramsay and Ramsay 2008, p.237), who appropriated the word *hard* from the high temperature, “hard firing” involved in the kiln process for its original Chinese manufacture. In contrast, Godden (*Godden's New Guide to English Porcelain*, 2004) identified just two types of porcelain: *hard paste* and *soft paste*. The former category included Chinese, Japanese, Meissen and some European porcelains, whereas the latter category comprised the production of the early French factories such as those at Sevres (1745–1772), Vincennes, St. Cloud, Mennecy, Chantilly and most English factories. Bone china was, therefore, not categorised separately. The main stipulation and criterion for the generic classification of porcelain seemed to be related to the actual firing temperature adopted in its production; hard paste porcelain (containing kaolinite plus a feldspathic flux) required a kiln firing temperature of at least 1300–1400 °C or possibly more, whereas soft paste porcelain needed kiln firing at the relatively lower temperature range of 1100–1200 °C. The practical differentiation between these hard paste and soft paste porcelains after firing was accomplished by only a visual and textural observation – hard paste porcelain was physically hard, had a white, glittery, hard glazed surface, was cold to the touch, was brittle and exhibited fractures on chipped surfaces which were conchoidal in appearance, unlike soft paste porcelain which was much more fragile, less robust in handling and texturally smoother to the touch.

The alternative and perhaps rather obvious and simplistic categorisation of a porcelain ceramic material which can be accomplished in terms of its composition and raw material components as hard paste, soft paste or glassy, for example, also fails remarkably in this respect and several authors have attempted to address this topic hitherto in the literature, generally with varying levels of success, which has resulted in them being then forced to assign several porcelains to rather indefinite categories such as a generic "hybrid" porcelain or even a "bone china", which necessarily avoided more accurate and specific descriptors. Indeed, the porcelain categories first adopted by the earliest writers on the subject now seem to be totally inadequate to effectively describe the real nature of their porcelains under discussion. For instance, the earliest categorisation of Chinese porcelain as "*hard paste*" undoubtedly referred to its ability to create a spark when struck with a flint or of the smoothing of a bodily fracture which could only be accomplished using a steel file! This is obviously neither a scientifically practical nor an acceptable method of evaluating the category of a precious piece of Chinese, Bow, Bristol or Chelsea porcelain. Likewise, the category of a "*soft paste*" porcelain is not necessarily deemed to contain ceramics as soft and pliable as the name implies and some soft paste porcelains containing soaprock as a component are indeed very robust and resistant to both thermal and mechanical shock, as evident from Lewis Dillwyn's experiments at Swansea carried out between 1815 and 1817 for the improvement of his much admired soft paste duck-egg porcelain by the addition of soaprock and the consequent reduction of its phosphatic calcined bone ash component (Dillwyn, *Notebooks*, see Eccles and Rackham, *Analysed Specimens of English Porcelain in the Victoria and Albert Museum Collection*, 1922).

There is no doubt that the first porcelain manufactured historically was created by the Chinese approximately one thousand years ago, it is believed in the Song Dynasty (960–1279 CE), when mixtures of kaolin (china clay; Chinese, *gaoling*) and *petuntse* (porcelain stone) were heated to a high temperature in a kiln to manufacture the characteristic Chinese "hard paste" porcelain, the export of which with some refinement soon took Europe by storm several centuries later in the early sixteenth century, and thereafter generated many unsuccessful attempts by others at simulating this wonderful imported ceramic locally. It is believed that the Emperor Zhenzong first perfected the manufacture of Chinese porcelain around 1004 CE in the kilns at Jingdezhen and some of the earliest porcelains emanated from the Ju kilns a few years later. Several authors have proposed that the precursor ceramics to this porcelain appeared some centuries earlier in the form of glazed earthenwares and a type of faience and glazed tile (Yanyi 1987; Rado, *An Introduction to the Technology of Pottery*, 1969). This wondrous material, first mentioned in the annals of the travels of Marco Polo (*Il Milione*, translated 1928), soon became the medium of choice for artistic and functional ownership by a discerning clientele in Europe (Edwards, *Nantgarw and Swansea Porcelain: An Analytical Perspective*, 2018a). Tite et al. (1984) have made a technical study of the early Chinese porcelains of the Yuan Dynasty (founded by Kublai Khan in 1272 CE, which lasted until it was overthrown by the Ming Dynasty in 1368 CE) and have concluded that the earliest Chinese porcelain manufactured in Jingdezhen (Ching-te-chen) in the south east of the country was the thickly potted and heavily glazed Yinqing (*qingbai*) ware in the early eleventh century, but by the late fourteenth century significant technical developments had been made in the selection of the raw materials and in the thermal tuning of the empirical kiln firing processes to create the finer Chinese porcelain that we know and recognise today as "eggshell" porcelain. The major step-change here seemed to be the replacement of the original kaolinized feldspathic stone with kaolin and a fine-grained quartz feldspar known as *petuntse* (Addis 1981; Sundius and Steger, *The Constitution and Manufacture of Chinese Ceramics from Sung and Earlier Times*, 1963; Wood 2000). Addis (1981) has proposed that the kaolin content in the recipe formulation was the most significant reason for the increased perceived quality of the much esteemed underglaze blue decoration in Chinese porcelains towards the end of the Yuan Dynasty in the mid-fourteenth century. The potential creators and simulators of Chinese porcelain in Europe in the late seventeenth century, however, suffered a major disadvantage in that they had no knowledge at all of the constituents used in porcelain manufacture in China, which were a very closely guarded secret, and chemical analytical science simply did not exist for the determination of the materials used in its construction which could be derived from the completed, fired and imported artefacts. Despite these detrimental factors, several early experimental examples of European porcelains appeared, such as the Medici porcelains, which were first reported in Florence in 1575 in a sponsored venture by Francesco de Medici, Grand Duke of Tuscany, and thereafter followed under the Royal patronage of King Louis XIV in France in 1698 at the St Cloud factory. It remained for Ehrenfried von Tschirnhaus and Johann Bottger to claim the first successful experimental manufacture of porcelain in Europe, under the patronage of King Augustus II, the Elector of Saxony, at Meissen in 1709, which then went into its initial commercial production in 1713. Their procedure and this date are usually hailed as the unequivocal start of the European challenge to the esteemed Chinese porcelain imports but later on we shall revisit this salient feature in the light of recent documentation that has surfaced which suggests that earlier versions of a quite acceptable porcelain body were actually manufactured in Europe, some 40 years hitherto, and that von Tschirnhaus and Bottger seemed to have had some knowledge of the procedures and materials involved in their manufacture. The Meissen factory used at first a rather impure and compositionally variable kaolin (china clay) which was sourced locally from Colditz in Saxony, and later a purer china clay from Aue, which was mixed with alabaster (gypsum, $CaSO_4.2H_2O$) as a high temperature flux from Nordhausen and fired as a mixture in a kiln at 1400 °C (Kingery, *Ceramics and Civilisation, High-Technology Ceramics, Past, Present and Future*, 1986). As needs to be highlighted here, this was not a primarily a hard paste porcelain, or even perhaps a "true porcelain" as made by the Chinese, but rather comprised a silica–alumina-calcareous (SiAlCa) fired vitreous ceramic body – which intriguingly corresponds closely to that of an earlier but perhaps more problematic seventeenth century venture in England at the Fulham Pottery led by John

Dwight, of which von Tschirnhaus was clearly aware before he commenced the manufacture of porcelain at Meissen, as will be discussed later.!

This discovery at Meissen undoubtedly generated the setting up and subsequent growth of several small European porcelain manufactories, notably those in France at Vincennes, Chantilly, Mennecy, Sceaux and Sevres. It was soon realised that the new porcelain from these manufactories was significantly different from the imported Chinese analogue because their bodies invariably contained powdered glass (then referred to simply as "*frit*", or "*cullet*"), which was initially a crown glass or soda glass, until the heavier and more refractory flint glass or lead glass was adopted, which was presumably added as an aid to improving the visual translucency of the fired ceramic body. This is not an unexpected procedural concept as for many years the establishment of a glass industry and the working of molten glass, especially low-melting alkaline soda-lime glasses, was quite well-developed in Western Europe, and had been especially refined by the Venetians in Murano. The practice of adding a glass frit to a porcelain body formulation recipe was still maintained throughout the eighteenth century and well into the nineteenth century by many factories until the negative toxic effects of the adoption of a heavy flint glass-containing lead oxide into the porcelain bodies and glazes, particularly upon the workforce in the manufacturing workshops, was fully appreciated and gave rise to its usage being reconsidered. It is recorded that some flint glasses actually contained up to 60% lead oxide additive to increase their brilliance and translucency, especially for cutting and etching purposes and the use of this flint glass cullet as a component in porcelains would therefore incur a significant level of lead toxicity in the resultant porcelain, especially if it was also used in the glazing procedures and would certainly have an ongoing cumulative and negative effect on the workforce through exposure and handling in the workshops. It is worth noting that earlier majolica and faiences used the more consumer friendly tin-glazing process, which was usurped by the more toxic lead glazing in the eighteenth century and that this move was not redressed until John Rose of the Coalport China Works introduced his award winning, patented lead-free glaze for his Coalport porcelains in 1820. It is interesting that in the 1720s and 1730s, there is no established record of any porcelain manufacture being undertaken anywhere in England – despite there being a well-established glass industry, as exemplified by George Ravenscroft, who patented his lead-crystal glass and set up his glass works in Vauxhall, London, in the latter part of the seventeenth century, and the presence of many locally-based small earthenware ceramics works. However, it is very likely that towards the end of the seventeenth century, some partially successful experiments in vitreous crucible technology had resulted in the production of specimens of porcellaneous materials which drew the appropriate comments from several observers who had seen the artefacts successfully produced therefrom, a theme to which we shall return later.

The commencement of English porcelain manufacture in the 1740s as officially recorded presaged a lively growth in factories in the 1750s, many of which were based upon earlier local earthenware sites and potworks, such as those at Liverpool and in Staffordshire (Table 1.1) (Bemrose 1975; Boney, *Liverpool Porcelain of the 18th Century and Its Makers*, 1957; Hillis, *Liverpool Porcelain*, 2011). Owen (1998)

Table 1.1 Early eighteenth century English porcelain factories, their founders and type of porcelain made

Factory	Start year	Founder(s)	Type of porcelain
Pomona	1744	William Steers	Glassy
Bow	1744	Edward Heylyn & Thomas Frye	Glassy
Limehouse	1745	Joseph Wilson	Soft Paste
Chelsea	1745	Nicholas Sprimont & Charles Gouyn	Soft Paste/Glassy
Derby	1748	Andrew Planche	Soft Paste
Bristol	1749	Richard Champion & Benjamin Lund	Hard Paste/Glassy
Longton Hall	1749	William Littler	Soft Paste
St James	1749	Charles Gouyn	Soft Paste
Worcester	1751	Dr John Wall & William Davis	Soaprock
	1791	Robert Chamberlain	Soapstone
Vauxhall	1751	Nicholas Crisp & John Sanders	Soft Paste
Liverpool	1754	Richard Chaffers	Soapstone
	1754	Samuel Gilbody	Soapstone
	1756	William Reid	Soapstone/Bone ash
	1765	Philip Christian	Soapstone
	1779	John & Seth Pennington	Soapstone
Isleworth	1756	Joseph Shore	Soft Paste
Lowestoft	1757	Philip Walker & Robert Browne	Soft Paste
Bovey Tracey	1750	William Ellis	Hard Paste
	1766	NIcholas Crisp	Hard Paste
Plymouth	1768	William Cookworthy	Hard Paste
Chelsea -Derby	1770	William Duesbury	Soft Paste
Caughley	1772	Ambrose Gallimore & Thomas Turner	Soapstone
Spode	1776	Josiah Spode	Soft Paste/Bone china
New Hall	1781	Jacob Warburton & Samuel Hollins	Hard paste
Minton	1793	Thomas Minton	Soft Paste/Bone china
Ridgway	1794	Job & George Ridgway	Soft Paste
Davenport	1794	John Davenport	Soaprock
Coalport	1795	John Rose	Soft Paste/Hybrid
Pinxton	1796	John Coke & William Billingsley	Soft Paste

has pointed out that, unlike their French and Saxon porcelain factory counterparts, the fledgling English factories did not receive any direct Royal patronage or funding support and that this would have inevitably forestalled their foundation and commercial production research experiments on economic grounds. In fact, an emotional plea for financial support for British porcelain manufacture generally was made unsuccessfully by William Billingsley, Samuel Walker and William Weston Young in their *Memorial* document of September, 1814, in which they sought the help of the British government for their new porcelain production venture at the Nantgarw China Works, citing the aid given by firstly French royalty and then the revolutionary French government for the manufacture of their porcelain at Sevres with which they would be in direct competition (Edwards, *Porcelain to Silica*

Bricks: The Extreme Ceramics of William Weston Young, 1776–1847, 2019). By this time, of course, other global factors were operating which made the economic climate even less conducive to governmental support in Great Britain, including the expensive ongoing Napoleonic war with France and her allies and the recent hostilities of 1812 and thereafter with the emergent United States of America.

Perhaps the most important and often quoted landmark in European porcelain manufacture, after the discovery of a commercially viable product by von Tschirnhaus and Bottger at Meissen in 1709, was provided in the second and third decades of the eighteenth century with the key information on Chinese porcelain technology released by the French Jesuit priest, Father Francois Xavier d'Entrecolles, who was assigned originally to the Peking Jesuit mission but then re-assigned to his residency in Jingdezhen, which was famous already for its porcelain manufacturing. D'Entrecolles managed to discover the formulation recipe and procedures for Chinese porcelain then being made and fired in the large dragon kilns at Jingdezhen (Ching-te-Chen) in the Jiangxi province in SE China, which had been a major centre for porcelain manufacture for many centuries, especially in the Ming Dynasty (1348–1640) and even earlier in the Song and Yuan Dynsaties, from two of his new Chinese converts to Catholicism at confession. D'Entrecolles revealed for the first time (Savage, *Dictionary of Antiques*, 1970) that the two main components of Chinese hard paste porcelain were kaolin and a porcelain-stone (*petuntse*):

> *In China it was made from kaolin (china clay) and a fusible feldspathic rock termed pai-tun-t'zu, the latter usually written in in the form employed by the French 18th Century missionaries, who called it petuntse*

He immediately realised the importance of his discovery, which he transmitted in two letters to his superiors in Paris, specifically the Jesuit friar Gorry, which were despatched in 1717 and 1722: these are reproduced fully in Jean-Baptiste du Halde's text (*The General History of China Containing a Geographical, Historical, Chronological, Political and Physical Description of the Empire of China, Chinese Tartary, Corea and Thibet*, English translation, 1736) which was released first in France. The appearance of these letters in France which contained details of the Chinese recipe for the production of porcelain in the late 1710s and early 1720s is an intriguing one and has been the subject of a swashbuckling mystery novel woven around the discovery of porcelain composition of subsequent national and international importance (Keating, *The Hunt for White Gold*, 2011). In addition, it is little realised that D'Entrecolles also sent some specimens of Chinese porcelain from the Jingdezhen kilns directly to France and that these were subjected to the first ever analyses by Rene Antoine Ferchault de Reaumur (1683–1757) between 1719 and 1735; perhaps best known for his expertise and discoveries in entomology, Reaumur actually made his own "opaque porcelain" in 1740, which he advocated for use in thermometry (he is still known for the Reaumur scale of temperature) . This prior knowledge of the details of Chinese porcelain composition gave the French porcelain manufacturers the leading edge over their European competitors for a few years in the production of a quality porcelain in direct competition with the Chinese imports. It was only a few years later, therefore, that the manufacture of porcelain formally commenced in England, with the oldest factories being established at Pomona, in Newcastle-under-Lyme in Staffordshire and at Bow, in London, both of which started up in 1744–1745 (Bemrose 1975; Ramsay and Ramsay 2008). A list of the earliest English porcelain factories in the eighteenth century has been given in Table 1.1: this comprehensive list is not designed to be exhaustive or discriminatory, as sometimes several small factories opened together in the larger cities and often exchanged their workforce personnel so that the positive unequivocal identification of their individual manufacturing output is frequently difficult for ceramic historians to assess and collate. In 1758, the Comte de Brancas-Lauraguais found deposits of kaolin in Alencon in France, and the Orleans manufactory there was already using a local French kaolin extensively from St. Yrieux-la-Peche by 1764. In England, William Cookworthy discovered a source of high-quality china clay in Cornwall near St Austell in the 1740s, calling it "*vitreous earth*", which over the next decade then gave rise directly to the early English ceramic establishments at Plymouth, Bristol, Chelsea and Bow. The Comte de Brancas-Lauraguais had meanwhile moved to England in the 1760s, possibly to escape ongoing political persecution in France, and he became involved in several English factories making porcelain, such as that at Chelsea; and he patented the use of Cornish china clay in his own English-based porcelain manufacturing enterprise in 1766. Very little of this Brancas-Lauraguais porcelain now remains and sadly, several key and documentary pieces were destroyed in the Alexandra Palace Exhibition fire in the 1880s (Edwards, *Nantgarw and Swansea Porcelain; An Analytical Perspective*, 2018a). Sir Arthur Church comments upon a specimen of this Brancas-Lauraguais porcelain which was badly damaged in the fire, only the handle of which then remained upon recovery, and which he subsequently analysed destructively (Church, *English Porcelain*, 1894).

No fewer than five factories were already established in London in the 1740s and 1750s (although it is quite possible that we should now add a few more smaller and lesser-known factories to this list), namely, Chelsea, Bow, Vauxhall, Limehouse and Isleworth. Porcelain was made at between five and seven distinct sites in Liverpool (Watney, *Liverpool Porcelain of the 18th Century*, 1997; Hillis, *Liverpool Porcelain, 1756–1804*, 2011) during the second half of the eighteenth century, including Brownlow Hill, Shaws Brow and Islington, which involved leading potters and "chinamen" such as William Reid, Samuel Gilbody, Thomas Wolfe, John Sadler, Philip Christian, William Ball, Richard Chaffers, John and Seth Pennington, and finally Samuel Worthington at the Herculaneum Pottery towards the end of the 1790s. From the two already cited very early English porcelain factories at Pomona and Bow, both of which were established in 1744, the list had grown in England to at least 30 by the end of the Century as can be seen in Table 1.1. The Pomona Potworks, founded by William Steers in Newcastle-under-Lyme, Staffordshire, in 1744 is believed to be one of the very first porcelain manufactories in England, but little now remains of its production of useful porcelain tea wares as it was probably primarily experimental in this respect: wasters and shards found on the factory site are now all that remain of this enterprise, some of which are found to have been decorated in a blue and white palette, but no finished, complete items of unbroken Pomona porcelain have been so far identified.

An analysis of the information listed in Table 1.1 indicates that many of these factories were set up as family concerns and their porcelain adopted the names of their local towns or villages; most English porcelains can be broadly described as *soft paste* in category, especially those manufactured earlier in the eighteenth century – only later did the addition of bone ash to the china clay become universally adopted, and then we have the growth of factories such as Spode, which specialised in the production of its characteristic bone china which set the scene and standard for the expansion of other English factories in this area. The degree of attrition of these very early porcelain manufactories was very significant and only one, Worcester, founded on the 16th May 1751 by Richard Holdship, Dr John Wall and William Davis with other investors, has survived through in unbroken manufacturing to the present day. The Derby manufactory also lays claim rightly to be a survivor of these earliest ventures in English porcelain manufacture with a foundation date in 1748 but the original works closed in 1848, although it did start up again a few years later at a new site and this has survived to the present day as the Royal Crown Derby China Works (Twitchett, *Derby Porcelain, 1748–1848*, 2002). There was still a variety of porcelain types being made by these early English factories: Pomona, Chelsea, Derby and Bow all started as soft paste factories, Plymouth was the first to make a hard paste porcelain in 1768, whereas Limehouse (1745–1748), Worcester (1751- present day), Bristol (1749–1752) all used steatite or soapstone in their raw materials, so these products can all be categorised correctly as magnesian soft paste porcelains. After the initial closure of Benjamin Lund and William Miller's soft paste Bristol factory in 1752, when it was taken over in the same year by the Chelsea China Works, a new Bristol factory started up again in 1770 to manufacture hard paste porcelain, subsuming the Plymouth factory enterprise.

What has been tacitly assumed by chroniclers and historians hitherto is that the Chinese made huge quantities of porcelain for home use and for export at several diverse sites using exactly the same kiln firing processes and according to the same compositional recipes and formulae, from the identical raw materials and without any changes being made to the formulation recipe at each site over many hundreds of years' production. Professor Victor Owen, among others, has rightly challenged this assumption (Owen 2002) and he has revealed several interesting facts about Chinese porcelain production which have remained literally buried historically. This has been taken up in a scholarly and comprehensive treatise by Ramsay and Ramsay (2008) on the origins of the earliest porcelains made at Bow and is summarised below: the clear inference is that Western thought and ideas about the manufacture and formulation of Chinese porcelain all originate with the information provided by Father Francois Xavier d'Entrecolles SJ to the Jesuits in France in the period 1717–1722. In his first letter in 1717, d'Entrecolles mentioned that there was a compositional variability in the Jingdezhen porcelain paste formulations depending upon the quality of the porcelain that was being manufactured there: the highest quality Jingdezhen porcelain had a compositional ratio of kaolin to *petuntse* (porcelain stone) of 1:1, then medium quality porcelain was made with the ratio of kaolin:

petuntse ingredients of 4:6, and finally the poorest grade porcelain was made with the ratio of kaolin: *petuntse* of 1:3. These ratios are still maintained today for the production in the china factories at Jingdezhen. Carswell (*Blue and White Chinese Porcelain Around the World*, 2000) and Wood (2000) have further fuelled this discussion about recipe variability in composition with the conclusion that the formulations were even more complex than the simple recipes indicated by d'Entrecolles and it has been suggested not only that the compositions varied with different periods of manufacture but also with the location of the factory sites, with a major divide occurring between north and south China; the northern sites made porcelains that were richer in clay, whereas porcelains from the southern sites were richer in quartz, and this has been confirmed in the analytical studies of Ramsay and Ramsay (2008). Wood (2000) has also proposed that kaolin for a long time was merely an optional extra in raw materials of the major southern factories exporting china to the West, whereas in most European porcelains it formed the basic ingredient of the paste composition for the majority of factories.

1.2 Raw Materials in Porcelain Manufacture and Their Geological Sourcing

It is appropriate at this point to consider in detail the properties and chemistry of some of the key raw materials that were involved in the manufacture of porcelain: in the earlier literature, some confusion has been created inevitably because of an imprecise description or incorrect labelling accorded to some of these materials, whose chemical composition is sometimes itself rather indefinite anyway, and ceramic authors and historians may unwittingly have compounded the situation by using generically and non-specifically applied material terms such as "clay", "stone" and "glass". Firstly, it is important to remember that a mineral has a precise chemical structure, chemical formula and crystal habit whereas a rock or ore does not (Dees et al., *An Introduction to the Rock-Forming Minerals*, 1992): hence, with the assumption that analytical data are available, which evidently sometimes they were not, a potter could therefore estimate reasonably precisely the chemical elemental oxide content of a mineralogical raw material but could only approximately deduce the composition of a rock which would also contain that elemental oxide as the latter will also probably contain several impurities, some of which may be even detrimental to the quality of the finished product. A good example of this particular point is the statement made by Lewis Weston Dillwyn, proprietor of the Swansea China Works between 1810 and 1820, who noted in his work books (L.W. Dillwyn, see Edwards, *Nantgarw and Swansea Porcelains: An Analytical Perspective*, 2018; Eccles & Rackham, *Analysed Specimens of English Porcelain in the Victoria and Albert Museum Collection*, 1922) that the inadvertent presence of calcium carbonate in the raw materials of his porcelain paste before firing could result in the formation of gaseous voids in the fired product because of the decomposition of the carbonate into gaseous carbon dioxide occurring around 650 °C in the kiln, which then formed bubbles in the viscous porcelain body paste, and lime which increased the concentration of the calcareous alkaline flux, compounding chemically with the aluminium silicates at higher temperatures. Chemically, this gaseous decarboxylation process is represented by the equation:

$$CaCO_3 = CaO + CO_2$$

Dillwyn then goes on to discuss the care that must be taken to monitor the alabaster content of the raw materials as this would also affect the quality of the fired porcelain: this is a rather mysterious comment to be made in the context of carbonate decomposition as alabaster is technically a translucent variety of gypsum, $CaSO_4.2H_2O$, and therefore contains no carbonate ion. However, it would still dehydrate at elevated temperatures to form anhydrite, $CaSO_4$, which then remains stable. It can be suggested that perhaps Dillwyn was here referring to the production of water vapour in this dehydration process, which in a viscous paste mixture at elevated temperatures would still create bubbles or voids similar to those encountered in the decomposition of calcium carbonate. An alternative explanation could be that earlier authors seemed to confuse the term "alabaster" with the translucent varieties of calcite occurring naturally in the absence of analytical chemical information and this would of course then generate carbon dioxide through the decomposition process already illustrated above. It is noted that Dillwyn should have appreciated the chemical distinction between gypsum and calcite and would not have been expected to have confused these two minerals chemically as he was a respected scientist and a Fellow of the Royal Society of London.

Before considering the detailed descriptions of the major individual raw materials used as components in porcelain manufacture in the eighteenth and early nineteenth centuries, it is relevant to recapitulate on the background terrestrial geological processes from which these materials have been formed in the Earth's crust. These processes not only create the primary mineral but also form others which are often found in association with them, the presence of which can be deleterious to the successful production of a ceramic body and this therefore prompted early porcelain factory proprietors to source the highest quality materials, often acquired from some considerable distances with their attendant increased transportation costs, when apparently more locally sourced and cheaper variants were also available (Anderson, *Derby Porcelain and the Early English Fine Ceramic Industry*, 2000). Minerals were used prehistorically as pigments and their modification and working up were later used to create metal implements and ceramics such as bricks and pottery (Edwards, *Porcelain to Silica Bricks; The Extreme Ceramics of William Weston Young, 1776–1848*, 2019). The oldest historical record of minerals and their commercial usage can be found in the writings of Aristotle (384–322 BCE) and Pliny the Elder (23–79 CE) which fed into the later mineral classifications of Avicenna (980–1037 CE), Albertus Magnus (1193–1280 CE) and the works of the "father of mineralogy", Georg Bauer, also known as Georgius Agricola (1494–1555 CE) (Duda and Rejl, *Rocks and Minerals of the World: An Illustrated Encyclopedia*,

1990; Nesse, *An Introduction to Mineralogy*, 2000). Mineralogy (the study of minerals) and petrography (the study of rocks) both involve the chemical composition, the physical and chemical properties and the origins of geological materials; here, the relevance to our discussion and the following survey of raw materials of importance to the porcelain and ceramics industries will focus upon their origins in the Earth's crust and variations in their chemical composition which arise from their weathering under different climatic conditions. Minerals can be defined as magmatic, sedimentary or metamorphic, depending upon the geological processes which have been operating from the deposition of the original magma which emerged from the Earth's mantle into the crustal surface as a result of tectonic and volcanic activity in the Earth's crust. Interaction of this magma with the atmospheric and hydrothermal environments caused magmatic differentiation to occur in which minerals rich in some metal oxides crystallised, leaving others in solution, with acidic and basic properties, such as quartz, feldspars, pyroxenes and micas. Further post-magmatic processes operating at the Earth's surface geology, usually termed weathering, produced the decomposition of secondary minerals and the formation of sedimentary deposits such as calcite, gypsum and dolomite from aqueous and glacial action. The further action of hydrothermal events and of lacustrine or glacial flows created localised deposits of sedimentary minerals, for example sand, halites and feldspars, separated by grain size, and occurring some distance from their place of origin. These localised deposits proved to be the source of many of the raw materials utilised by ceramics manufacturers seeking their ideal composition for porcelain paste manufacture and firing: the major problem facing their selection of these suitable raw materials was establishing firstly, what was the nature of the material, and secondly, what was the purity of the material under consideration? The foundations of chemical qualitative and quantitative analysis for the determination of the elemental presence and the percentage composition of a mineral or ore deposit was still a long way off being standard practice in the early eighteenth century as the novel science of chemistry emerged from its alchemical beginnings: chemical science was therefore then empirical both experimentally and theoretically but it could draw upon a vast base of experimental observations on the behaviour of natural materials from early alchemy and its recorded laboratory transformations. Although the alchemical texts did recognise the existence of elements such as sulfur, mercury, tin, copper, phosphorus,gold, lead, iron, arsenic and antimony along with acids and alkalis such as "spirit of hartshorn" for ammonia and "oil of vitriol" for sulfuric acid their reactions were steeped in mystique and secrecy. It was not until the arrival of Antoine Lavoisier (1743–1794) and Joseph Priestley (1733–1804) in the mid to late eighteenth century that the formal basis for the recognition and importance of elemental chemical analysis was established: sadly, Lavoisier fell into conflict with the French Revolutionary Government of Maximilien Robespierre and he was executed on the 8th May, 1794, at the height of his intellectual powers – the judge at his trial saying that the world had no need for scientists and chemists!

As a result, the modern chemical analyst and scientific historians should be aware that written accounts of recipes and experiments carried out in early ceramics firing processes in the 1700s could be subject to difficulties arising from an injudicious or perhaps even an incorrect nomenclature used to describe the raw materials used therein and certainly regarding any inferences that were made then about their interactive chemistry. For example, to enlarge upon the analogy cited above, *alabaster* is strictly a translucent mineral form of calcium sulfate dihydrate (generically called gypsum, $CaSO_4.2H_2O$; *selenite* is another translucent form of gypsum of the same chemical composition) but there was much confusion in the late eighteenth century between *alabaster* and the minerally unrelated *marble* (generically called calcium carbonate or calcite, $CaCO_3$). As Lewis Dillwyn commented in his empirical formulation recipe experiments undertaken during the period 1815–1817 in Swansea and now preserved in notebook form (L.W. Dillwyn Notebooks, see Edwards, *Nantgarw and Swansea Porcelains: An Analytical Perspective*, 2018; Eccles & Rackham, *Analysed Specimens of English Porcelain in the Victoria and Albert Museum Collection*, 1922), the use of calcite in place of alabaster had a disastrous outcome for the integrity of his porcelains because of the gaseous thermal emission of carbon dioxide in the kiln firing around 650–700 °C occurring from calcite decomposition. Around the same temperature, gypsum also dehydrates to anhydrite, $CaSO_4$. The retrieval of the historical situation can be even more complex today when it is realised that most mineral identifications and classifications carried out in the eighteenth century were based upon their visual appearance, crystal form and habit, streak colours, flame colours and hardness, so visually different mineral crystals could have an identical chemical composition (but not necessarily the same chemical solid state structures). Classic examples are the minerals kaolinite, dickite, halloysite and nacrite, all of which have the identical chemical formulation of $Al_2Si_2O_5(OH)_4$ but each one having different crystal habits. Other minerals form solid solutions with a variable but precisely defined composition between two extremes such as the $(Mg,Fe)_2SiO_4$ isomorphous series generically called "*olivine*" between the two mineral extremes of forsterite (Mg_2SiO_4) and fayalite (Fe_2SiO_4). *Olivine* therefore strictly can have an indefinite composition anywhere between these two extreme members which hence creates a natural subdivision into iron-rich and magnesium-rich olivines. Such a grouping of minerals is reflected in the terminology of feldspars as orthoclase (potassium alumino silicates of formulation $KAlSi_3O_8$) and plagioclase (sodium and calcium alumino silicates of formulation $(Na,Ca)Al_{1\text{-}2}Si_{3\text{-}2}O_8$): the former includes sanidine, microcline and amazonite whereas the latter includes albite, oligoclase, andesine, bytownite and anorthite.

Another problem which arises from the geological formation of minerals in the Earth's crust as described above is that we can expect to find several minerals in association with each other in varying quantities and this can result in sedimentary deposits which can vary extensively in purity. For example, kaolinite can be associated with montmorillonite, halloysite, bentonite and quartz, and dolomite can be associated with calcite, ankerite and quartz. Hence, it was really important that proprietors of porcelain manufactories, who as advised above, recognised the vital importance of using only the purest materials for their ceramic syntheses and who sourced their raw materials from known mine deposits which could be some distance away from their factory base. It is recorded that the Bow factory proprietors,

Heylyn and Frye in the 1740s obtained an especially pure form of kaolinite, known as "*unaker*", from the Cherokee lands in North America and that William Duesbury proprietor of the Derby factory at one stage sourced his kaolinite from Saxony. In this particular regard, it should also be noted that most British factories in the late eighteenth and early nineteenth centuries obtained their kaolin from Cornwall, where pure localised deposits were available near to St Austell, which also had adequate local supplies of the secondary ball clay from the Nordern mine which was favoured by many formulators as a mixing agent for kaolin to increase the plasticity of the clay, assist their working of the paste and to provide a better kiln stability for the prototype porcelain artefacts under firing. Another potential problem resulting from the association of minerals in clays is the possibility of finding organic matter, which may have been degraded to amorphous carbon or graphite over geological time, which could have arisen originally from the biological colonisation of the sediments in early lakes and seas, encapsulated in the clay components. A recent research paper (Moroz et al. 2019) illustrates this nicely with some analytical work on the iron-rich phyllosilicate, nontronite, formulated as $Na_{0.3}Fe_2Si_3AlO_{10}(OH)_2.4H_2O$, found in smectic clays and in which small conical particles of graphite were identified in association with the silicon oxide layered tetrahedral clay structure. A carbonaceous inclusion of this sort found as in "impurity" in kaolinite clays would therefore have serious implications for its deleterious effect upon any porcelain which utilised it as a raw material: although invisible to the naked eye, the graphite particles found in association with the nontronite being as small as 200–400 nm (1 nm, 10^{-9} of a metre, is only one millionth of a mm), the conversion of these graphitic carbon particles into carbon dioxide gas at a kiln temperature approaching 600 °C would produce a trail of bubbles in the porcelain paste at elevated temperatures which would clearly affect the translucency of the finished, fired porcelain artefact. Simple calculation reveals that only 1 microgram of carbon which became volatilised into carbon dioxide at STP would yield a bubble of volume 0.005 cm^3, of approximate diameter 1 mm and easily visible to the naked eye, entrapped in a porcelain matrix. The detrimental effect of the incorporation of carbon and other carbonaceous matter into the porcelain paste during its manufacture was appreciated by the early proprietors of the porcelain manufactories and it is fair to say that their major concern in this respect addressed the potential origin of carbonaceous matter in their bone ash raw materials component additive. Bone ash was made from the thermal calcination of animal or fish bones which could have proteinaceous impurities present, such as hair or scales, in addition to their incipient keratotic component which is associated naturally with the inorganic calcium hydroxyapatite bone matrix. The calcination process is designed to remove this organic component of bone, and any added organic impurities such as contaminant residual flesh and hair, by volatilisation at high temperature, whilst converting the calcium hydroxyapatite into tricalcium phosphate, the mineral whitlockite, $Ca_3(PO_4)_2$. It is for this reason that William Duesbury, the proprietor of the Derby China Works in the 1780–90s, sourced his calcined bone ash from the most reputable supplier of quality material in London who used ox bones solely for this purpose (Anderson, *Derby Porcelain and the Early English Fine Ceramic Industry*, 2000) and William Billingsley at the

Nantgarw China Works in 1817–1819 used ox bones calcined by his local miller, David Jones, under his personal supervision and direction at the relatively high temperature of 1100 °C. Billingsley specifically instructed Jones in the importance of this part of his operation, especially the care that needed to be taken in this calcination process and particularly the meticulously fine grinding and milling that was required afterwards to produce a satisfactory bone ash raw material for the incorporation into his porcelain manufacturing purposes. We shall see later that Josiah Spode, in contrast, for the production of his famed and eponymous bone china in 1795 that soon became the English ceramics industry standard, did not use a two-stage process as adopted for the manufacture of phosphatic soft paste porcelains such as those made at Derby and Nantgarw but rather used a one-stage process in which all the raw materials were compounded together and heated in the kiln in a single-stage process more akin to the production of the Chinese hard paste or true porcelains. We can but conclude that Spode must have had a supreme faith in the quality of his raw materials as any detrimental organic impurities which may have been removed by the initial calcination procedure could have otherwise contaminated the whole batch of his bone china during its kiln firing.

There now follows a description of the major raw materials used in English and Welsh porcelain manufacture in the eighteenth and nineteenth centuries, with the inclusion of the Chinese analogue equivalents where appropriate.

1.2.1 Petuntse or China Stone

These are actually two distinct materials geologically, but both are micaceous feldspathic materials with the presence of small, indefinite amounts of kaolinite, quartz and fluorspar, arising from the hydrothermal environmental degradation of the parent orthoclase and plagioclase rocks (Edwards & Atkinson, *Ore Deposit Geology*, 1986). Both rocks are specifically defined as being iron-deficient to account for their use as pure white components for porcelain manufacture in admixture with china clay. Deposits of *petuntse (bai-dunzi),* which in Chinese means "little white bricks", were located originally in the Jiangxi province in South East China; these were mixed with kaolin (china clay) in a 1:1 ratio for the highest quality porcelain product and fired at the high temperature of 1450 °C in the neighbouring dragon kilns at Jingdezhen. Analytical data reveal that the porcelains from Jingdezhen in south-east China have the characteristic elemental oxide composition that is high in silica and potash but low in alumina compared with porcelains produced from the kilns and factories situated in the northern China region. The output from the Jingdezhen china manufactory kilns alone must have been very large since a record of an order from the Chinese Imperial palace in 1433 (probably from the Emperor Zheng He) was placed for delivery to the Emperor in Peking of over 400,000 porcelain items of the highest quality for the Emperor's personal use there!

Chemically, *china stone* and *petuntse* are both represented formulaically as $KAl_2(AlSi_3O_{10})(OH)_2$, but this formulation can be rather misleading as they are both rocks of necessarily variable composition and cannot therefore strictly be assigned a chemical formula as would a pure mineral, but this chemical formulation does indicate that they are both hydrated potassium aluminosilicates, unlike kaolin which is a potassium deficient aluminosilicate of formulation $Al_2Si_2O_5(OH)_4$. The term *china stone* was first coined by William Cookworthy, a Quaker clergyman (1705–1780), who discovered small localised deposits of the rock in Cornwall near St Austell in 1745, which he alternatively termed "*Cornish stone*" in the mistaken belief that it was the equivalent of the *china stone* then being used for the manufacture of Chinese porcelain: in fact, *petuntse* and *Cornish stone* are similar in appearance visually but are of slightly different composition mineralogically. Dr Richard Pococke in 1750 reported (W.B. Honey, *Old English Porcelain: A Handbook for Collectors*, 1977) that the deposit of china stone found near Lizard Point in Cornwall was "*most valued for making porcelane..... and they get £5 per ton for the manufacture of porcelane now carrying on at Bristol*". There is still some dispute as to whether or not this deposit referred to soaprock/steatite, the incorporation of which into a porcelain paste recipe was soon valued for its ability to confer a robustness and a resistance to mechanical shock and minimising the tendency to shatter upon the resultant porcelain artefacts. *Cornish stone* has a high silica content, typically 75%, with about 16% alumina, 7% alkaline earth oxides of sodium and potassium, and 2% lime; it has extremely low percentages (small fractions of 1% only or less) of iron, titanium and magnesium oxides. It contains a mixture of igneous rocks in varying states of decomposition – in its early stages, that is geologically young, *Cornish stone* has a blue cast which is attributed to the fluorspar content, calcium fluoride CaF_2, but this is leached away in naturally alkaline environments to leave a white, fine-grained material. The nearest mineral equivalents to *Cornish stone* are orthoclase, potash feldspar $K_2O.Al_2O_3.6SiO_2$, and albite, a soda feldspar, $Na_2O.Al_2O_3.6SiO_2$. A synthetic version of *Cornish stone* is now being supplied to modern porcelain manufacturers in the form of a nepheline syenite, which contains silica, kaolin, wollastonite, feldspar and dolomite in its composition: note that this synthetic variant actually contains dolomite, $CaMg(CO_3)_2$, which therefore would render a positive signal for the analytical determination of magnesia in the resultant porcelain, which certainly would not have been present to any significant amount in the old porcelains made from natural *Cornish stone*.

1.2.2 Soapstone/Soaprock/Talc/Steatite

The terms, *china stone, petuntse, soapstone, soaprock, talc* and *steatite* were often used generically and rather loosely by porcelain manufacturers in the eighteenth and nineteenth centuries in their recipe descriptions and care must therefore be taken by modern analysts to interpret properly their historic usage in formulations, especially since soapstone, soaprock, steatite and talc are all high magnesia-content minerals and rocks, which confers the title *magnesian* upon the category of the resultant porcelain after firing. In contrast, *china stone*, *petuntse* and *Cornish stone*, although not identical rocks chemically are all low in magnesia content, which now can afford a discriminatory capability upon the interpretation of the chemical analytical data from the study of porcelains which claim them as constituents in recipes. Of these minerals and rocks, perhaps the most closely represented by its accepted chemical formulation is talc, which has the chemical formula $Mg_3Si_4O_{10}(OH)_2$, although traces of chlorite, $(Fe,Mg,Al)_6(Si,Al)_4O_{10}(OH)_8$, and magnetite, Fe_3O_4, can occasionally be found in admixture. This has given rise to the statement that talc is the purest natural form of steatite. Soapstone, soaprock, talc and steatite are all termed metamorphic schists, which are found over wide geological locations where they have been formed from thermal and pressured geological metamorphism at subducted tectonic plates. Typically, they are all magnesium-rich and contain approximately 64% silica, 32% magnesia and 5% water, with minor amounts of alumina and iron oxide. Evidently, this group of minerals and rocks is not chemically identical with those feldspathic silicate materials in the *petuntse* and *china stone* category. In passing, the name of William Borlase (1696–1772) deserves mention here: a noted polymath, being Cornish antiquary, geologist and naturalist whilst holding the position of Rector at Ludgvan in Cornwall He investigated the stretch of coastline between Mullion Cove and Pentreath on the Lizard peninsula which had outcrops of serpentine, a soaprock silicate rich in magnesium typically the magnesium phyllosilicate serpentinite of formulation $(Mg,Fe)_3Si_2O_5(OH)_4$ formed from the aqueous alteration of the olivines fayalite and forsterite, samples of which he sent to Dr John Woodward at the Royal Society in the 1720s, deeming that the soapy clay found here at Kynance Cove and Gew Graze (Soapy Cove) might be suitable for the manufacture of porcelain (Borlase, *Observations of the Antiquities, Historical and Monumental, of Cornwall*, 1754). The experimental results were obviously promising as the first commercial extraction and quarrying of soaprock at Kynance and Gew Graze were taken up by Benjamin Lund in 1748. In 1752 the Worcester porcelain manufactory then acquired the licence to quarry soaprock at the same site by the acquisition of the Bristol works. By 1760 Worcester had become the largest consumers of soaprock mined at the site, which also later supplied raw materials to Vauxhall, Liverpool, Caughley and Swansea.

1.2.3 Kaolinite/Kaolin/China Clay

is a layered phyllosilicate of chemical formulation $Al_2Si_2O_5(OH)_4$, or $Al_2O_3.2SiO_2.2H_2O$, wherein the tetrahedral arrangements of SiO_4 units are linked structurally through shared oxygen bonds to octahedral AlO_6 units. A microcrystalline form of kaolin found in Anglesey in the early 1800s gave rise to the alternative terminology of kaolinite. The generic term kaolin is derived from the Chinese "*gaoling*", a ridge feature near Jingdezhen in Jiangxi province, SE China, where the clay was first discovered: the term "*kaolin*" was first used in Western writing by Francois Xavier d'Entrecolles in 1727. It is formed geologically from the weathering of feldspars. We should also here include some comment upon the term *ball clay*, which was used as a component of porcelain pastes by several manufacturers in the eighteenth and nineteenth centuries: occasionally, the recipes cited do not specify the alternate use of ball clay and kaolin, but ancillary historical information about the mine location of the raw material enables us to differentiate between them – for example, the Gewgraze and Norden mines in Cornwall gave a high quality kaolin clay, whereas the neighbouring St Austell mine gave a rich ball clay. Whereas kaolin is geologically derived from the primary weathering of feldspathic rocks, ball clay is a secondary clay component, and as a result it is of a different siliceous structure, rendering the paste more fusible at a lower temperature in the kiln and this prompted china works proprietors to adopt mixtures of ball clay and china clay in their formulation recipes for greater plasticity of the paste. Kaolin transforms thermally into other silicaceous and aluminosilicaceous materials on increasing the firing temperature in the kiln (Dees et al., *An Introduction to the Rock-Forming Minerals*, 1992): firstly, the loss of bound water occurs up to 520 °C from the kaolin to form metakaolin, of chemical formula $Al_2Si_2O_7$, which at 950 °C then transforms into an aluminium-silicon spinel, $Si_3Al_4O_{12}$, with the loss of silica. At the higher temperature of 1050 °C, the transformation into two new components occurs, namely, mullite platelets and a polymorph of silica called *beta*- cristobalite, according to the equation:

$$3Si_3Al_4O_{12} = \underset{\text{Mullite platelets}}{2\left(3Al_2O_3 + SiO_2\right)} + \underset{\textit{Beta}-\text{cristobalite}}{5SiO_2}$$

Finally, at 1400 °C, the mullite component further transforms into needle-like crystals of the same chemical formulation. In the presence of other porcelain paste components such as calcined bone ash, calcium hydroxyapatite, the thermally transformed silica and silicates from the china clay and ball clay react to give chemical species such as wollastonite, bytownite and whitlockite, $Ca_3(PO_4)_2$.

1.2.4 Glass Frit

This material is described as such generically and conceals the presence of several key ingredients and minor components in the porcelain paste recipe. At first, it would be quite acceptable for readers to consider that *glass frit,* or *cullet,* which are both finely powdered or crushed glass residues from molten glass manufacture that have been synthesised and allowed to cool, would be a precise definition, but unfortunately this is not so. Firstly, the glass itself could comprise a *soda glass* made from soda ash and silica, effectively a sodium silicate (*crown glass*), or perhaps a potash glass and silica, effectively a potassium silicate, or possibly even a *flint glass*, which could contain soda or potash, or both alkalis, plus an indefinite quantity of lead oxide additive: *flint glass* was thus heavier and had a higher refractive index than *crown glass*. Several recipes for porcelains in the eighteenth and nineteenth centuries mention the incorporation of glass frit into the paste to increase the translucency of the vitreous material – for *crown glass* this can be virtually undetectable in the final analytical determination as the soda, calcium, potassium and silica also occur in other components of the paste, but the presence of a lead glass additive in *flint glass* is immediately detectable analytically from the signature of lead (II) oxide in the fired porcelain. The difficulty does not end there, however, since china factory proprietors often refer to a "frit" or a "fritting" of their paste after a preliminary calcination procedure has taken place. This has nothing whatsoever to do with the addition or treatment of a glass additive but rather the subsidiary but very necessary process of grinding finely the calcined aggregates and composites from the first kiln firing, such as quartz sand, bone ash and china clay, in preparation to receive the secondary addition of other components, such as borax, glass frit, and a further addition of china clay and lime. So, the term "frit" here has two unrelated meanings, namely, the fine pulverisation of an aggregate obtained from the preliminary thermal treatment of a porcelain body, or alternatively, the addition of a new component of glass frit, called *cullet* to an existing composite of fired raw materials.

Another source of glass frit which could be added to a porcelain paste formulation in the secondary stage was a "blue glass" frit which has been described as *smalt* in several recipes. Lewis Weston Dillwyn at the Swansea China Works between about 1815 and 1817 undertook some experiments to improve his porcelain quality (L.W. Dillwyn *Notebooks*, see Edwards, *Nantgarw and Swansea Porcelains; An Analytical Perspective*, 2018; and Eccles and Rackham, *Analysed Specimens of English Porcelain in the Victoria and Albert Museum Collection*, 1922), during which he discovered his esteemed and highly successful Swansea "duck-egg" porcelain body. He had learned that the addition of *smalt*, a deep blue glass, could remove any vestige of yellowness from the finished product, which he correctly assumed arose from the presence of iron (III) oxide impurities in the quartz sand selected for his recipe. Dillwyn is quite specific in his notes when he refers to the use of several percent of *smalt* as an additive to his experimental porcelain pastes. A further source of confusion arises because of the imprecise terminology and usage for *smalt, cobalt blue* and *Bristol blue* in contemporary writings which has given rise to a misconception about which one of these was actually used in the recipe formulation; upon closer inspection, in addition to an apparently loose usage in terminology, they are evidently all very different chemically also and this too can create an analytical problem.

Smalt is a synthetic blue, glassy material, formally a potassium cobalt aluminosilicate, which is of an indefinite chemical composition of several elemental oxides approximating to 65% SiO_2, 15% K_2O, 5% Al_2O_3, and 10% CoO. It has often been confused in early ceramic accounts with "*Bristol blue*", which is also a glassy synthetic material, comprising mainly cobalt and silica with lead oxide, and *cobalt blue*, which is a cobalt aluminate. The discovery of *smalt* has been attributed to Christophe Schurer (cited by W. Ganzenmuller, *Beitrage zur Geschichte der Technologie und der Alchemie*, 1956), a Bohemian glassmaker known for his addition of cobalt oxide or cobalt carbonate to molten glass to produce a vivid blue colouration, which he carried out around the period 1540–1560, but it has also been identified as a pigment in earlier fresco paintings of Ghirlandaio executed around 1480 (Muhlethaler and Thissen, in *Artists' Pigments: A Handbook of their History and Characteristics*, 1997; Gettens and Stout, *Painting Materials: A Short Encyclopaedia*, 1942; Harley, *Artists' Pigments ca. 1600–1835*, 1982). It is interesting that despite this long history of recognition that the addition of certain metal ores to glass produced a blue colour the official discovery that cobalt is the blue colourant in glass is credited to the Swedish chemist Georg Brandt in 1735. It was quickly adopted into the pigment palette of later Renaissance artists, such as Hans Holbein, who used it from 1540. Strictly, *smalt* is really a cobalt aluminosilicate whereas *Bristol blue* is a cobalt silicate and analytically they are, therefore, quite distinctly different materials chemically. Many authors have interpreted the use of blue *smalt* in porcelain body manufacture with the purchase of *Bristol blue* by china works proprietors, which was an attractive blue pigment used in the decoration of Chinese and early English porcelains. The name *Bristol blue* first appears actually in 1780 when Lazarus and Isaac Jacobs first made the glassy pigment in their glassworks in Bristol (Elphinstone and Hall 1963): typical analysis figures for the elemental oxide composition of *Bristol blue* are cobalt oxide 10%, lead oxide 24% and silica 66%. Much of this pigment was shipped through the port of Bristol and could be purchased there directly by ceramic artists or through agents in London. William Weston Young, the proprietor of the Nantgarw China Works in 1820–1823, noted in his diaries that he made several trips personally to Bristol to purchase *Bristol blue* pigment (Edwards, *Porcelain to Silica Bricks: The Extreme Ceramics of William Weston Young, 1776–1847*, 2019): it has been assumed that Young required the *Bristol blue* supplies for a pigment, however, we cannot predict what this was actually used for – it is true that he had also at that time engaged the services of Thomas Pardoe to decorate the residual stock of Nantgarw porcelain left in the white after William Billingsley and Samuel Walker had left to seek employment with John Rose at the Coalport China Works in 1819–1820. Was this blue glassy pigment then required for Young's experimental forays into recreating the Nantgarw china body, or for an additive encapsulation for Young and Pardoe's Nantgarw glaze (the so-called *Nantgarw No.2 glaze*), or was it simply for use as a pigment required for the remnant white porcelain decoration? The confusion arising between the terminology for *Bristol Blue, smalt* and *cobalt blue* still persists to this day and in a recent paper (Jonynaite et al. 2010) have described the quantitative Fourier-transform infrared spectroscopic analytical determination of *smalt* based upon the molar ratios of silica and cobalt oxide in the range 9–50% which perhaps now would be better described as a lead-free *Bristol blue* analysis, with a formulation given as Co_2SiO_4. They also refer to the preparation of a *cobalt olivine blue* of the same formulation, which also may contain traces of alumina and zinc oxide, the formula being of relevance to olivine $(Fe,Mg)_2SiO_4$ whose extremes are the minerals fayalite, Fe_2SiO_4,and forsterite, Mg_2SiO_4. A cobalt-doped willemite, of formulation $Zn_{2-x}Co_xSiO_4$, is used in modern ceramic technology to replace *cobalt blue* and *Bristol blue* (de Waal 2009) *Cobalt blue* pigment dates from its discovery by Thenard in

1803/4 and it analyses as a spinel, $CoAl_2O_4$, effectively represented by $CoO.Al_2O_3$; it quickly replaced *lapis lazuli* as the blue pigment of choice for many artists. J.M.W. Turner was the first artist to use *cobalt blue* as an oil-based pigment in 1806. It was simply made by dissolving cobalt arsenate or phosphate in molten alumina: the use of the phosphate deepened the blue colour to a desirable mauve or purple colour. The preference for *cobalt blue* as a decorative pigment and intense depth of colour for porcelain enamelling could be related to its very high refractive index: the refractive indices for *smalt, Egyptian blue* and *cobalt blue* glasses progressively increase from 1.5 to 1.6 to 1.75, respectively. *Smalt* is the least stable of these three glassy pigments, although an increased content of potash alkali used in its preparation does confer some resistance towards decomposition; the substitution of potash for soda in its synthesis also gives a deeper blue colour to the glass.

Another confusion arises with the existence in antiquity of *Egyptian blue*, which equates with the natural calcium copper silicate mineral *cuprorivaite,* $CaCuSi_4O_{10,}$ or $CaO.4SiO_2.CuO$. The blue colours are frequently found to be very similar to the *cobalt blue* and *smalt* discussed above: *Egyptian blue* was synthesised in antiquity, it is believed around 3000 BCE, by heating powdered limestone or natron with malachite and quartz sand and is first encountered in 4th Dynasty Egyptian tomb wall paintings around ~2500 BCE (Laurie et al. 1914; Riederer, in *Artists' Pigments: A Handbook of their History and Characteristics*, 1997). The Chinese had synthesised an analogous synthetic material called *Han blue* in which the calcium was replaced by a barium salt, probably barytes, $BaSO_{4,}$ or witherite, $BaCO_3$, thus it is formulated as $BaCuSi_4O_{10}$, and there is also a purple analogue known as *Han purple* of formulation $BaCuSi_2O_{6,}$ which is deficient in two moles of silica per mole of pigment in comparison with the *Han blue* . *Han purple* has been located as residual pigment traces on the "Terracotta Army" artefacts of Emperor Qin Shi Huang, dating from 500 BCE, in Xian which have been fired at the same temperature used to prepare the *Han purple* pigment. It is believed that *Han purple* is formed from the increased thermal decomposition of *Han blue* by taking the temperature up to 1000 °C from 900 °C and maintaining it there for about 48 h. The elemental oxide compositions of *Han blue* and *Han purple* are, respectively, 43 and 27% silica, 37 and 48% barium oxide, and 19 and 25% copper oxide (Fitzhugh and Zycherman 1983, 1992; Wiedemann and Bayer, in *Conservation of Ancient Sites on the Silk Road: Proceedings of the International Conference on the Conservation of Grotto Sites*, 1997, 2001). It has recently been ascertained that *Han blue* exists as a natural mineral of precisely identical formulation, called *effenbergerite.* Despite its high temperature stability relative to *Han blue, Han purple* is more chemically reactive at ambient temperatures because it contains a copper-copper bond which makes it structurally very different (Wiedemann and Berke 2001) from *Han blue*. The relevant elemental oxide analytical data and comparative compositional details for the blue glasses, *cobalt blue, smalt, Egyptian blue, Bristol blue* and *Han blue* are presented in Table 1.2.

Table 1.2 Comparison of blue glass frit used in porcelain recipes and decoration

Glass frit	Alternative name	Year	Chemical name	Composition	References
Cobalt blue	Bristol blue	1803	Cobalt aluminate	81% Al_2O_3	Thenard (1803)
	Thenard's blue			15% CoO	Roy (2007)
	Vienna blue			1% (Na,K)O	Easthaugh (2004)
				3% H_2O	
Smalt	Kaiserblau	1540	Cobalt silicate	70% SiO_2	Muhlethaler and Thissen (1997)
	Bleu d'email			15% K_2O	Harley (1982)
	Azzuro di smalto			5% Al_2O_3	
				10% CoO	
Egyptian blue	Blue frit	3000 BCE	Calcium copper silicate	70% SiO_2	Riederer (1997)
	Caeruleum			8% CaO	Easthaugh (2004)
	Cuprorivaite			16% CuO	
				3% Na_2O	
				2% Al_2O_3	
				2% Fe_2O_3	
Han blue	Chinese blue	1045 BCE	Barium copper silicate	43% SiO2	Fitzhugh and Zycherman (1983, 1992)
	Effenbergerite			37% BaO	Wiedemann and Bayer (1997; Wiedemann and Berke 2001)
				19% CuO	Easthaugh (2004)
Bristol blue	Cobalt blue	1780	Cobalt lead silicate	15% CoO	Elphinstone (1963)
				24% PbO	
				61% SiO_2	

1.2.4.1 The Presence of Arsenic in Blue Glass Colourants for Porcelain Decoration

Recent molecular spectroscopic studies by Professor Philippe Colomban and his associates on early porcelain and pottery ceramic silicate glazes using pieces from the Kangxi (1661–1722), Yongzheng (1723–1735) and Qianlong (1735–1796) periods has revealed that the Chinese -European transfer of technology regarding porcelain production and its decoration was not perhaps as unidirectional from China to Europe as might have been originally assumed. As Professor Colomban has indicated (Colomban 2013; Colomban et al. 2004) cobalt ores can be sourced from terrestrial and marine locations, *viz.*, deep sea nodules which are a mixture of manganese, iron, cobalt and nickel oxyhydroxides displaced in metamorphic layers by tectonic plate action and intrusive veins in granitic rocks such as skullerite ($CoAs_3$), smaltite ($Co,Ni,FeAs_2$), cobaltite (CoAsS) and erythrite ($Co_3As_2O_8.8H_2O$). The essential difference between these two sources of cobalt ores which can be used to differentiate them analytically is the presence of a significant arsenic component in the terrestrial sources which is not present in the marine sources. Different methods were available in the eighteenth century to work up the terrestrial unrefined ores, a famous location for these being Erzgebirge in Saxony, into additives for the glass and porcelain industries which involved heating the raw ores with coal, potash and glass to produce *zaffre* or "*smalt*" or chemical digestion with sulphuric or nitric acids and alkalis to produce a refined cobalt oxide; the former method retained an arsenic component but the latter resulted in a loss of arsenic. Colomban (2013) has suggested the following processes by which the arsenic was incorporated into the glazes from the thermal treatment of the ores:

$$CoAs_3 = Pb_3O_4 + nSiO_2 \rightarrow 3PbAsO_4 + \left(Co,nSiO_2\right) \text{ glass}$$

and that the formation of solid solutions of variable composition containing lead and arsenic were formed in the glazes such as the range from the relatively simple formulation $(Ca,Pb)_3(AsO_2)_2$ to the complex $Na_{1\text{-}x\text{-}y}K_xCa_{y/2}Pb_8(AsO_4)_6$ as found in early majolicas.

Analytical studies using molecular spectroscopic and SEM/EDAXS techniques carried out on a range of Chinese porcelains and glazed wares from the Kangxi, Yongzhen and Qianlong periods (1680–1796) by Colomban et al. (2017a, b, 2018) reveal that there are distinctive signatures to be observed which confirm the presence of arsenic either as a metallic element or through the detection of Raman spectral bands characteristic of arsenic -oxygen bond stretching at 885 cm^{-1}. They found evidence of arsenic in the blue coloured decoration of early European porcelains from Rouen, Paris and St. Cloud (1680–1700) and also in the even earlier products of the Medici porcelains from Florence, dating from 1575–1585. However, the most revealing discovery was the observation of an arsenic component in eth blue colours of the Chines *falangcai* opaque enamelled porcelains from the Kangxi period made for the Beijing Court, where a lead -arsenate/apatite as they found in European cobalt ores from Erzgebirge,where the arsenic had reacted thermally with a lead glaze to form the lead arsenate/apatite complex, with a characteristic Raman spectral feature at 823 cm^{-1}. In contrast, porcelain from the Imperial Chinese workshop at Yongzhen (1723–1735) used specifically Asian cobalt sources as used previously in the Ming Dynasty (1348–1645) porcelain production and no trace of lead arsenate could be seen. Two glazes were identified on the Yongzhen porcelains studied, namely a lead-rich glaze and a mixed lead-sodium/potassium glaze, the latter perhaps signifying a mixture of flint and soda glass in the frit additive used. Both glazes evidenced a Raman feature at 970 cm^{-1} and the mixed glaze an additional band at 1020 cm^{-1} and in addition SEM/EDAXS showed the presence of boron oxides in both glazes. Colomban et al. conclude that in China technological improvements and research was extant at this time, unlike some authors who maintained that the successful Chinese porcelain glaze and body formulations remained static. Finally, they found that the later Qianlong (1735–1796) porcelains used a heterogeneous mixture of cobalt ores for their blue decoration, one containing an arsenic rich (European) component and the other a manganese, iron-rich, copper, arsenic free (Asian) component. This information was revealed analytically and had not been suspected to have occurred historically.

1.2.5 Soda Ash/Potash

The use of alkaline oxides as a flux to aid the melting of the components of the porcelain paste resulted in the achievement of a lower temperature in the biscuit porcelain kilns and this was especially appreciated for the manufacture of soft paste porcelains: hence, kiln temperatures of around 1200–1300 °C could now be effected instead of the 1400 °C required for firing the Chinese hard paste or true porcelain. Generally, soda ash from calcined plants, or potash from calcined seaweed, comprising sodium and potassium carbonates, respectively, were used alone or in admixture for this purpose: both would be converted upon heating in the kiln into the corresponding oxides, Na_2O and K_2O, in which form they would contribute to the relevant analytical data. Potash was sometimes alternatively referred to as *pearl ash* in the formulation recipes (Edwards, *Nantgarw and Swansea Porcelains: An Analytical Perspective*, 2018a). A key procedure was to thermally heat the plant material sources of the soda and potash to a sufficient temperature to volatilise the associated carbonaceous organic material, which otherwise might char in the kilns and form carbon, which could then result in dark blemishes being left in the porcelain paste. If a blemish was noted and the item was otherwise perfect then a random decoration could be applied to mask the defect in the artefact before sale. Such is the case for the porcelain plate shown in Fig. 1.2, a large porcelain platter from a Nantgarw porcelain dinner service supplied to William Ramsden Hawksworth Fawkes of Farnley Hall in 1817–1819, where a firing blemish on the underside of the verge has been hidden by the application of three mosquito-like insects in brown

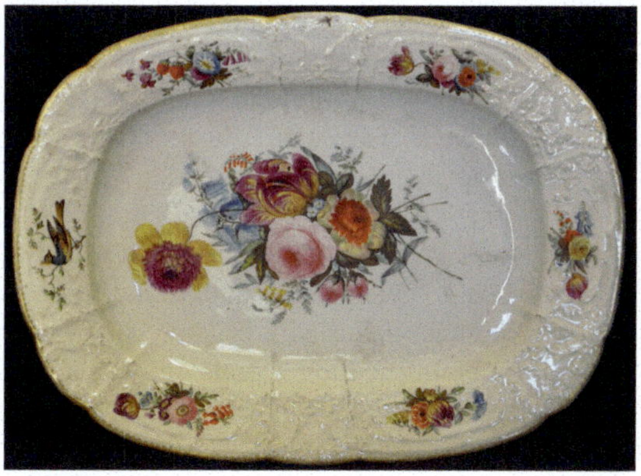

Fig. 1.2 Large rectangular meat platter from the *Farnley Hall* service of Nantgarw porcelain, commissioned by William Ramsden Hawksworth Fawkes MP between 1817 and 1819, unmarked except for an impressed 4 and 7, and superbly decorated in London with dentil edge gilding. Central bouquet of garden flowers and six vignettes in the moulded verge of flowers, berries and an exotic bird. Note the mosquito at the uppermost edge placed to cover a blemish in the porcelain. The plate shows signs of use and is a remarkable survivor – one of only 37 pieces now extant from the large original dinner service of over 120 pieces. (Reproduced with permission of Guy Fawkes Esq., Farnley Hall, Otley, North Yorkshire)

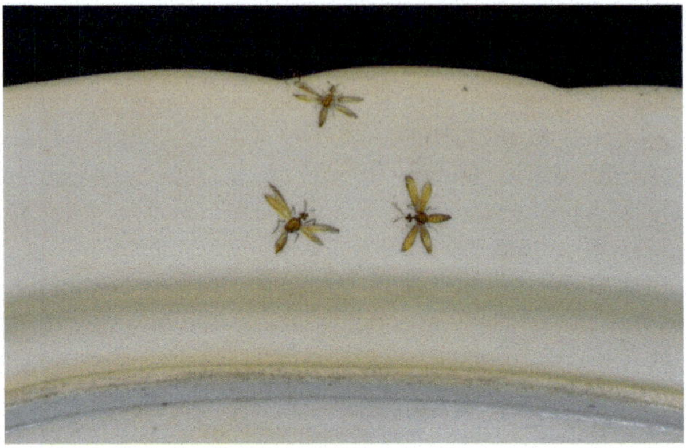

Fig. 1.3 Reverse of the large meat platter shown in Fig. 1.2 from the *Farnley Hall* service of Nantgarw porcelain, unmarked except for an impressed 4 and 7, superbly decorated in London with dentil edge gilding, showing three enamelled mosquitoes strategically placed to cover surface blemishes. (Reproduced with the permission of Guy Fawkes Esq., Farnley Hall, Otley, North Yorkshire)

and yellow enamels (Fig. 1.3). This platter was the largest size of any item produced in the Nantgarw factory and the rejection for sale of an otherwise perfectly shaped and fired item would have not been a commercially viable proposition (Edwards, *Swansea and Nantgarw Porcelain: a Scientific Reappraisal*, 2017) which justified the masking of the small surface blemish.

Several alternatives to soda ash and potash were used by porcelain manufactories, such as lime (CaO) prepared from the calcination of limestone, dolomitised lime prepared from dolomite and containing an equimolar mixture of magnesia and lime (CaO and MgO), and borax (sodium tetraborate decahydrate, $Na_2B_4O_7.10H_2O$). The latter could be sourced from the naturally occurring mineral which was used alternatively as washing soda crystals. Although a natural sodium carbonate exists geologically as *natron,* $Na_2CO_3.10H_2O$, which is found in salt lake sedimentary deposits such as those in Wadi Natrun, Egypt, these deposits do not yield pure sodium carbonate upon extraction as they contain components of *trona,* a mixed sodium carbonate/bicarbonate formulated as $Na_3H(CO_3)_2.2H_2O$: these ancient lacustrine deposits are also host to cyanobacterial extremophilic biological colonies and are therefore seriously contaminated with organic carbonaceous material. The ancient Egyptian funerary mummification rituals involved a desiccation of the human cadaver to prevent putrefaction using about 250 kg of natron from Wadi Natrun over a period of some 40 days and it has been recently shown that geological specimens contain spectroscopic biosignatures of existing biological colonies of *Natromonas pharaoensis*, which undoubtedly contributed greatly to the subsequent biological degradation noted for many Middle Kingdom Dynastic Egyptian mummified remains in the tomb or in museum storage (Edwards et al. 2007).

1.2.6 Bone Ash

The idea of incorporating calcined bones into the raw materials for the production of porcelain has been attributed to the founders of the Bow porcelain factory in the late 1740s, Messrs. Frye and Heylyn, although Josiah Spode certainly adopted the concept fully from the 1770s for the manufacture of his porcelain, which he thereafter called "Staffordshire bone china". This appendage then became synonymous with the fine English porcelain for which Spode became famous. It would be wrong to think, therefore, that Josiah Spode had the monopoly of the first discovery of this component in porcelain recipes and in fact there were many other factories which also advocated the use of calcined bones in their production of quality porcelain pastes around this time. William Billingsley, arguably one of the greatest producers of the finest quality porcelain in the late eighteenth and early nineteenth centuries, adopted finely ground bone ash in all of his manufacturing ventures from Derby through Pinxton and on to his early trials at Mansfield, Brampton-in-Torksey, Worcester and then Swansea: in the production of his highly esteemed Nantgarw porcelain between 1817 and 1819, Billingsley regarded bone ash as the vital ingredient, so much so that he kept its incorporation into his paste as a secret known only to himself and his co-partner and kiln manager, Samuel Walker. His other partner, William Weston Young, who did so much to encourage interest and the financial support,and then acquire the initial investment for the setting up of the Nantgarw China Works, who eventually took on the failing factory operation after the departure of Billingsley and Walker for the Coalport China Works of John Rose in 1820, had no idea at all of its incorporation into the formulation recipe at Nantgarw. Apparently, Billingsley was so secretive about his raw materials usage that he mixed these personally in the cellar of his home at Tyla Gwyn, Nantgarw, and his local miller, David Jones, was sworn to secrecy about his role on Billingsley's workforce in the preparation of this vital calcined bone ash component (Edwards, *Nantgarw and Swansea Porcelains: An Analytical Perspective,* 2018a; Edwards, *Porcelain to Silica Bricks; The Extreme Ceramics of William Weston Young, 1776–1848*, 2019) . The selection of bones for calcination to produce the hydroxyapatite used for porcelain production was a specialised operation undertaken by factory proprietors who would source their raw materials carefully because they realised that contamination of this component with organic residues would invariably result in defects in their fired porcelain artefacts. William Billingsley at the Nantgarw China Works specified the preferential use of ox bones for this purpose, whereas horse bones were not favoured as highly, and William Duesbury II at the Derby China Works sourced his bones from London (Edwards, *Nantgarw and Swansea Porcelains: An Analytical Perspective*, 2018a; Anderson, *Derby Porcelain and the Early English Fine Ceramic Industry*, 2000). Although defined as hydroxyapatite, which has the chemical formulation of $Ca_5(OH)(PO_4)_3$, calcined bone ash has had several alternative formulations advanced in the literature, including the formula $Ca_3(PO_4)_2.Ca(OH)_2$; this creates a problem analytically for the determination of the bone ash content of the derived porcelain paste recipe by back-calculation from the phosphorus oxide determination because the conversion factors are sensibly different according to the chemical formula used. In the first formulation, the CaO, P_2O_5 and H_2O percentages are 55.9, 42.4 and 1.7%, respectively, whereas in the second formulation the CaO, P_2O_5 and H_2O percentages are 58.3, 36.9 and 4.7%, respectively. This confers a significant potential error in determination of the raw material bone ash ingredient from the phosphorus pentoxide percentage in the fired porcelain paste as high as ~13%. The bone ash problem has been discussed in greater detail elsewhere, along with the correlation between the different published representations of the phosphorus content in the fired paste by different analysts as phosphoric acid, phosphate and phosphorus pentoxide, which itself creates some potential areas of hidden difficulty in the comparison of older literature data in this respect (Edwards, *Nantgarw and Swansea Porcelains: An Analytical Perspective*, 2018a). In Table 1.3 is presented the elemental oxide deconstruction of the major raw materials which comprise the porcelain pastes of most factories in the eighteenth and nineteenth centuries. It should still be noted that the chemical formulae relate to pure materials or minerals and that the quality of the raw materials selected by the china works proprietors is an unknown parameter which was very dependent upon their geographical sourcing: in particular, the different states of hydration of many raw materials will obviously affect their component masses when being tared for admixture into the pre-fired paste and this will also depend critically upon their storage conditions at the factory. In this case, calcined bone ash, lime, magnesia, pearl ash and soda ash will be components that are particularly prone to the absorption of water from a damp atmosphere and this will naturally affect their individual component weights, and china clay will also be hygroscopic in this respect. It is now the concensus of considered opinion that fully calcined bone ash is perhaps best represented as tricalcium phosphate, $Ca_3(PO_4)_2$, wherein the complete elimination of water and hydroxylated material, as well as the organic keratotic moieties, has been removed by the high temperature calcination process if that had been effected around 1000 °C. If calcined at lower temperatures, however, then the preservation of the hydroxyl groups attached to the calcium ions could still be maintained and the formulation would be expected to be different.

1.2.7 Ball Clay

The use of ball clay in the manufacture of porcelain can be traced back to the middle of the eighteenth century, when Josiah Wedgwood of Etruria in Staffordshire realised that it conferred a greater plasticity upon his fired ceramics: his records show that his factory purchased some 1400 tons of this raw material per annum in the 1770s, a rather large quantity compared with other ceramic factories at that time. Ball clay, a geologically rare sedimentary deposit is found only in three sites in the SW of England and very few other places globally, is of a highly variable composition, typically represented by 20–80% kaolinite, 10–25% mica and 6–65% quartz. It is fine-grained and also includes a variety of accessory minerals such as titania

Table 1.3 Elemental oxides/% and the composition of raw materials and related compounds for porcelains

Material	Alternative name	Chemical Formula	RMM[a]	MgO	SiO_2	CaO	P_2O_5	Al_2O_3	H_2O	K_2O	Na_2O	CO_2	PbO	CoO
Steatite	Soapstone, talc	$Mg_3Si_4O_{10}(OH)_2$	379	31.9	63.5				4.7					
		$3MgO.4SiO_2.H_2O$												
Kaolin	China clay	$Al_2Si_2O_5(OH)_4$	258		46.6			39.5	14.0					
		$Al_2O_3.2SiO_2.2H_2O$												
Ball Clay	Dorset clay	$SiO_2.Al_2O_3.yCaO.xH_2O$	[b]	1	57	1		33		2	2	3		
	Devon clay													
Feldspar	Potassium aluminosilicate	$KAlSi_3O_8$	557		64.7			18.3		16.9				
		$K_2O.Al_2O_3.6SiO_2$												
Hydroxyapatite	Bone ash	$Ca_5(OH)(PO_4)_3$	1004			55.9	42.4		1.7					
		$10CaO.3P_2O_5.OH$												
Calcite	Limestone	$CaCO_3$	100			56.1						44.0		
		$CaO.CO_2$												
Dolomite	Dolostone	$CaMg(CO_3)_2$	184	21.7		30.4						47.9		
		$CaO.MgO.2CO_2$												
Magnesite	Magnesium carbonate	$MgCO_3$	84	48.0								52.0		
		$MgO.CO_2$												
Petuntse	China stone	$KAl_3Si_3O_{10}(OH)_4$	862		41.8			35.5	6.2	10.9				
		$K_2O.3Al_2O_3.6SiO_2.3H_2O.(O)_3$												
Flint glass	Cullet; lead glass	$PbO. SiO_2. (Na_2O.K_2O)_x$	[c]		53.5					8.3	0.5		37.0	
Crown glass	Soda glass	$SiO_2.Na_2O. K_2O. CaO$	[c]		70	15				8	8			
Potash	Potassium carbonate	K_2CO_3	138							68.1		31.8		
	Pearl ash	$K_2O.CO_2$												
Soda ash	Sodium carbonate	Na_2CO_3	80								58.5	41.5		
	$Na_2O.CO_2$													

(continued)

Table 1.3 (continued)

Material	Alternative name	Chemical Formula	RMM[a]	MgO	SiO_2	CaO	P_2O_5	Al_2O_3	H_2O	K_2O	Na_2O	CO_2	PbO	CoO
Phosphoric acid		H_3PO_4	196				72.4		27.6					
		$P_2O_5.3H_2O$												
Phosphate		PO_4^{3-}	190				73.5		26.5					
		$P_2O_5.(O)_3$												
Gypsum	Alabaster	$CaSO_4.2H_2O$	172			32.6			20.9					
	CaO. $SO_3.2H_2O$													
Smalt	Cobalt silicate	Indefinite	[c]		70			5		15				10
	Kaiserblau													

[a]RMM: The Relative Molar Mass in amu (atomic mass units; $C^{12} = 12.00$)

[b]Ball clays are also of a highly variable composition, comprising mainly silica, mica and quartz: the Dorset ball clays have a typical composition of 62% kaolinite, 22% mica and 11% quartz whereas the Devon ball clays have a typical composition of 44% kaolinite, 25% mica and 28% quartz. Many industry handbooks give the composition of a generic ball clay as a wide ranging: 20–80% kaolin, 10–25% mica and 6–65% quartz, with additional carbonaceous material

[c]Flint glass, crown glass and smalt are synthetic and of variable chemical composition: the percentage figures cited are therefore typical average experimental values

(TiO_2, namely, anatase and rutile) and has very low levels of iron oxide and carbonaceous impurities such as lignite, which would be detrimental to its use for the manufacture of fine porcelain. In its incorporation into kaolin as a raw material in porcelain pastes it acts as a binding agent and confers greater plasticity upon the moulded items before firing, enabling a finer china piece to be thrown or formed. The three sites in SW England where it is located are in the Bovey and Petrockstowe Basins in Devon and at Wareham in Dorset. Transportation to the major porcelain manufacturing sites in London, the Midlands and the North of England was costly and probably utilised its being shipped to nearby ports, the use of navigable rivers and an extensive canal system and finally a road network using waggons; this could take several weeks to accomplish deliveries to the factory sites, during which the minerals in each consignment were exposed to rain and winds and potential contamination.

1.3 Celadons, Faience and Majolica

Inspection of the analytical and historical ceramics literature reveals that we need to distinguish between terms such as celadons, faience, majolica and maiolica which are often found to occur in generic descriptions of "porcelains", and indeed were sometimes made alongside porcelain or as precursors to genuine porcelain production in potteries. A generic descriptor of "*faience*", which is often used to encompass glazed earthenwares, is fired pottery which has been either tin-glazed or lead-glazed, using appropriately cassiterite (tin (IV) oxide, SnO_2) or cerussite (lead (II) carbonate, $PbCO_3$) as a glaze component, and can comprise utilitarian articles such as bowls, plates, mugs and decorative artefacts such as tiles and figurines. *Celadons* are rather special in that they normally comprise pottery which has been glazed with a particular jade green colour slip and then fired at a very high temperature; this celadonic glaze is sometimes found on porcelain as well as earthenware bodies. The distinction between faience and porcelain is, therefore, simply based upon the observation that porcelain is translucent but faience is not, but as Sir Arthur Church has pointed out (*English Porcelain*, 1894), faience manufacturers deliberately made their products superficially porcelain-like with their high-gloss surface glazing and, hence, these items could be mistaken for porcelains by a casual observer (Church cited Wedgwood as a typical example of this practice!).

1.3.1 Celadons

The historical origin of celadons is rather conjectural: they are generally believed to have originated in the northern Chinese Song Dynasty (960–1127) in Longquan and in some respects they offer an interface between glazed stoneware and porcelain. Their characteristic feature is a beautiful clear jade-green coloured glaze which arises from the addition of up to 2.5% iron oxide, Fe_2O_3, to the tin- or lead-glazing slip on a earthenware clay body, which is then fired in a kiln at a very high temperature approaching 1300 °C in a reducing atmosphere, whereby the Fe(III) is reduced to Fe(II), effectively therefore becoming FeO, and imparting a green colour to the glaze. The appearance of jade gave this ceramic a particular affection and desirability in China, where it was very highly prized (Vainker, *Chinese Pottery and Porcelain*, 1991; Wood, *Chinese Glazes: Their Origins, Chemistry and Recreations*, 1999). Celadons were also being manufactured from these earliest times in Korea (Goryeo Dynasty, 918–1392), Thailand and Japan; the Koreans even applied this celadonic glaze to their porcelains. The etymological origin of *celadon* is also lost in time and several roots are thought possible including a French descriptor from the seventeenth century, the Sanskrit *sila dhara (*translated as "green stone") and a corruption of Saladin (Salah ad-Din) who in 1171 sent 40 superb pieces of celadon ceramics as a gift to Sultan Nur ad-Din Zengi of Syria. The glaze was the outstanding accomplishment of these early celadons and this also had a specially applied *craquelure* effect which rendered it even more appealing to the eye of the connoisseur. Celadons reached Europe via the Islamic trade routes and also were well received by an appreciative audience here: it has been recorded that the celadonic glaze was also applied to Chinese porcelains, especially in their main centre at Jingdezhen – which means that they must then surely be considered as a potentially separate category of true Chinese porcelains in their own right, if not interesting hybrids.

1.3.2 Faience and Majolica

The discovery of faience, a tin-glazed ceramic on a fired beige earthenware body, is believed to have occurred in Persia in the ninth century and that of majolica (known as *maiolica* in England) is credited to Luca della Robbia in Italy in 1438. Both are tin oxide-glazed eathenwares to which vivid colours and pigments have been applied for decoration (Fortnum, *Maiolica*, 1876). A later successful variant called *palissy* was manufactured by the Minton China Works in 1850, comprising their robust china stone body and a highly coloured lead oxide-glaze, which caused a sensation when it first appeared in the French International Exhibition in Paris in 1855 (Atterbury and Batkin, *Dictionary of Minton*, 1990). The historical derivation of the word *majolica* or *maiolica* is now lost but it is believed that it could refer to its original importation into Europe from Moorish-held Majorca in the fourteenth century.

Egyptian faience is a rather special type of material in that it is not really faience as described above but a vitreous frit which is similar to glass and does not even contain clay: it was manufactured from about 4000 BCE and is found in ceramic beads and ornaments. It is therefore probably more akin compositionally to *Egyptian blue*, a glassy synthetic pigment comprising copper silicate and also known as *cuprorivaite,* than it is to a tin oxide-glazed earthenware.

Majolica was manufactured widely across Europe from the sixteenth century and the major factories can be listed as: Italy (Savona, Turin), England (Lambeth, Staffordshire), Spain (Manises, Talavera), France (Lyon, Quimper, Strasbourg), Germany(Hanau, Nurnberg) and the Netherlands (Delft).

References

J. Addis, Porcelain-stone and kaolin-late Yuan development at Hulian. Trans. Orient. Ceram. Soc. **45**, 54–66 (1981)

J.A. Anderson, *Derby Porcelain and the Early English Fine Ceramic Industry*, PhD thesis, University of Leicester, UK, October 2000

P. Atterbury, M. Batkin, *Dictionary of Minton* (Antiques Collectors Club, Woodbridge, 1990)

P. Bemrose, The Pomona potworks, Newcastle, staffs. Trans. Engl. Ceram. Circ. **9**, 1–18 (1975)

K. Boney, *Liverpool Porcelain of the 18th Century and its Makers* (B.T. Batsford, London, 1957)

W. Borlase, *Observations of the Antiquities, Historical and Monumental, of the County of Cornwall, and Consisting of Several Essays on Druid Superstitions. The First Inhabitants, Customs and Vocabulary of the Cornu-British Language* (Jackson, Oxford, 1754)

P. Bradshaw, *English Porcelain Figures of the 18th Century 1745–1795* (Antique Collectors Club Art Books, Suffolk, 1981)

P. Bradshaw, *Derby Porcelain Figures 1750–1848* (Faber & Faber, London, 1990)

J. Carswell, *Blue and White Chinese Porcelain Around the World* (British Museum Press, London, 2000)

Sir A.H. Church, *English Porcelain: A Handbook to the China Made in England During the 18th Century as Illustrated by Specimens Chiefly in the National Collection*, A South Kensington Museum Handbook (Chapman & Hall Ltd., London, 1885 and, 1894)

P. Colomban, Rocks as blue (green and black) pigments/dyes of glazed pottery and enamelled glass artefacts: The potential of Raman spectroscopy. Eur. Mineral. J. **25**, 863–879 (2013)

P. Colomban, G. Sagon, L.Q. Huy, N.Q. Liem, L. Mazerolles, Vietnamese 15th century blue-and-white Tam Tai and lustre porcelains and stonewares: glaze, composition and decoration techniques. Archaeometry **46**, 125–136 (2004)

P. Colomban, F. Ambrosi, A.-T. Ngo, T.-A. Lui, X.-L. Feng, S. Chen, C.-L. Choi, Comparative analysis of Wucai Chinese porcelains using mobile and fixed Raman microspectrometers. Ceram. Int. **43**, 14244–14256 (2017a)

P. Colomban, Y. Zhang, B. Zhao, Non-invasive Raman analysis of Chinese huafalang and related porcelain wares: searching for evidence of innovative pigment technologies. Ceram. Int. **43**, 12079–12088 (2017b)

P. Colomban, V. Milande, T.-A. Lu, Non-invasive on-site Raman study of blue decorated early soft paste porcelain. The use of arsenic – rich (European) ores compared with huafalang Chinese porcelains. J. Eur. Ceram. Soc. **38**, 5228–5233 (2018)

D. de Waal, Micro-Raman and portable Raman spectroscopic investigation of blue pigments in selected Delft plates. J. Raman Spectrosc. **40**, 2162–2170 (2009)

W.A. Dees, R.A. Howie, J. Zussman, *An Introduction to the Rock-Forming Minerals*, 2nd edn. (Longmans, Harlow, 1992)

L.W. Dillwyn, *Notes and Workbooks of Recipes at the Swansea China Works, 1815–1817,* reproduced in H. Eccles & B. Rackham, *Analysed Specimens of English Porcelain*, 1922, and in Edwards, Nantgarw and Swansea Porcelains; An Analytical Perspective, 2018

J.-B. du Halde, *The General History of China containing a Geographical, Historical, Chronological, Political and Physical Description of the Empire of China, Chinese Tartary, Corea and Thibet,* English Translation and Edition (John Watts Printer and Publisher, London, 1736)

R. Duda, L. Rejl, *Rocks and Minerals of the World: An Illustrated Encyclopedia* (Aventinium Nakladatelstvi Publishing/Tiger Books, Prague/Twickenham, 1990)

N. Easthaugh, V. Walsh, T. Chaplin, R. Siddall, *Pigment Compendium: A Dictionary of Historical Pigments* (Elsevier Butterworth-Heinemann, Oxford, 2004), p. 181

H. Eccles, B. Rackham, *Analysed Specimens of English Porcelain in the Victoria and Albert Museum Collection* (Victorian and Albert Museum, London, 1922)

H.G.M. Edwards, *Swansea and Nantgarw Porcelains: A Scientific Reappraisal* (Springer, Dordrecht, 2017)

H.G.M. Edwards, *Nantgarw and Swansea Porcelains: An Analytical Perspective* (Springer, Dordrecht, 2018a)

H.G.M. Edwards, in *Derby Porcelain: The Golden Years 1780–1830*, ed. by M. Denyer, (Penrose Antiques, Thornton, 2018b)

H.G.M. Edwards, *Porcelain to Silica Bricks: The Extreme Ceramics of William Weston Young (1776–1847)* (Springer, Dordrecht, 2019)

R. Edwards, K. Atkinson, *Ore Deposit Geology* (Chapman and Hall, London, 1986)

H.G.M. Edwards, K.J. Currie, H.R. Ali, S.E. Jorge Villar, A.R. David, J. Denton, Raman spectroscopy of natron: shedding light on ancient Egyptian mummification. Anal. Bioanal. Chem. **383**, 683–689 (2007)

N. Elphinstone, E.T. Hall, Bristol blue glass. Archaeometry **6**, 26–30 (1963)

E.W. Fitzhugh, L.A. Zycherman, An early man-made blue pigment from China- barium copper silicate. Stud. Conserv. **28**, 15–22 (1983)

E.W. Fitzhugh, L.A. Zycherman, A purple barium-copper silicate pigment from early China. Stud. Conserv. **37**, 145–154 (1992)

C.D.E. Fortnum, *Maiolica* (University of Michigan Press, New York, 1876)

W. Ganzenmuller, *Beitrage zur Geschichte der Technologie und der Alchemie* (Weinheim, Verlag Chemie GmbH, 1956)

R.J. Gettens, G.L. Stout, *Painting Materials: A Short Encyclopaedia* (D. Van Nostrand, New York, 1942)

G.A. Godden, *Godden's New Guide to English Porcelain* (Miller/Octopus Publishing Group Ltd, London, 2004)

R. Harley, *Artists' Pigments ca. 1600–1835*, 2nd edn. (Butterworth Scientific, London, 1982), pp. 53–56

M. Hillis, *Liverpool Porcelain, 1756–1804* (Maurice Hillis Publishing, 2011)

W.B. Honey, *Old English Porcelain: A Handbook for Collectors*, 3rd edn., revised by F.A. Barrett (Faber & Faber, London, pp. 4–5 and 211–216, 1977)

D. Jonynaite, J. Senvaitene, A. Beganstevene, A. Kareiva, Spectroscopic analysis of blue cobalt smalt pigment. Vib. Spectrosc. **52**, 158–162 (2010)

M. Keating, *The Hunt for White Gold* (Hodder & Stoughton, London, 2011)

W.D. Kingery, The development of European porcelain, in *Ceramics and Civilisation, High-Technology Ceramics, Past, Present and Future*, ed. by W. D. Kingery, vol. III, (Ohio Press, Athens, 1986), pp. 153–180

A.P. Laurie, W.F.P. McLintock, P.D. Miles, Egyptian Blue. Proc. R. Soc. Ser. A **LXXXIX**, 418–5429 (1914)

B. Muhlethaler, J. Thissen, Smalt, in *Artists' Pigments: A Handbook of their History and Characteristics*, ed. by A. Roy, vol. 2, (National Gallery of Art/Oxford University Press, Washington, DC/New York/Oxford, 1997), pp. 113–130

T.N. Moroz, H.G.M. Edwards, V.A. Ponomarchuk, A.N. Pyrayev, N.A. Palchikand, S.V. Goryainov, Raman spectra of a graphite -nontronite association in marbles from Oltrek Island (Lake Baikal, Russia). J. Raman Spectrosc. **50**, 1–9 (2019)

W.D. Nesse, *Introduction to Mineralogy* (Oxford University Press, New York/Oxford, 2000)

J.V. Owen, On the earliest products (ca. 1751–1752) of the Worcester porcelain manufactory: Evidence from sherds from the Warmstry House site, England. Hist. Archaeol. **32**, 63–75 (1998)

J.V. Owen, Antique Porcelain 101: A primer on the chemical analysis and interpretation of eighteenth century British wares, in *Ceramics in America*, ed. by R. Hunter, (Chipstone Foundation, Milwaukee, 2002), pp. 39–61

M. Polo, *Il Milione*, cap. CLVIII, dell edizione a cuva di L.F. Benedetto, translated, 1928

P. Rado, *An Introduction to the Technology of Pottery* (Pergamon Press, Oxford, 1969)

W.R.H. Ramsay, E.G. Ramsay, A case for the production of the earliest commercial hard paste porcelains in the English-speaking world by Edward Heylyn and Thomas Frye in about 1743. Proc. R. Soc. Vic. **120**, 236–256 (2008)

J. Riederer, Egyptian Blue, in *Artists' pigments: a handbook of their history and characteristics*, ed. by E. W. Fitzhugh, vol. 3, (National Gallery of Art/Oxford University Press, Washington, DC/New York/Oxford, 1997), pp. 23–46

A. Roy, Cobalt blue, in *Artists' pigments: a handbook of their history and characteristics*, ed. by B. H. Berrie, vol. 4, (National Gallery of Art/Archetypal Publications, Washington, DC/London, 2007), pp. 151–178

G. Savage, *Dictionary of Antiques* (Book Club Associates, London, 1970)

G. Savage, H. Newman, *An Illustrated Dictionary of Ceramics* (Thames & Hudson, London, 1976)

N. Sundius, W. Steger, Sung Sherds, in *The Constitution and Manufacture of Chinese Ceramics from Sung and Earlier Times*, Section 2, (Almqvist & Wiksell, Stockholm, 1963), pp. 375–506

L.J. Thenard, Sur les couleurs suives d'en procede pour preparer une couleur aussi belle que l'outremer. J. des Mines **15**, 128–136 (1803/1804)

M.S. Tite, I.C. Freestone, M. Bimson, A technological study of Chinese porcelain of the Yuan dynasty. Archaeometry **26**, 139–154 (1984)

J. Twitchett, *Derby Porcelain: 1748–1848, An Illustrated Guide* (Antique Collector's Club, Woodbridge, 2002)

S.J. Vainker, *Chinese Pottery and Porcelain* (British Museum Press, London, 1991)

B.M. Watney, *Liverpool Porcelain of the 18th Century* (R. Dennis Publishers, London, 1997)

H.G. Wiedemann, G. Bayer, Formation and stability of Chinese barium-copper-silicate pigments, in *Conservation of Ancient Sites on the Silk Road: Proceedings of the International Conference on the Conservation of Grotto Sites*, ed. by N. Agnew, (Getty Conservation Institute, Los Angeles, 1997), pp. 379–387

H.G. Wiedemann, H. Berke, Chemical and physical investigation of Egyptian and Chinese blue and purple, in *The Polychromy of Antique Sculpture and the Terracotta Army of the First Chinese Emperor: Studies of the Materials, Painting, Technology and Monuments and Sites Conservation III*, (ICORMS, Paris, 2001), pp. 154–169

N. Wood, *Chinese Glazes: Their Origins, Chemistry and Recreations* (University of Pennsylvania Press, Philadelphia, 1999)

N. Wood, Plate tectonics and Chinese ceramics – new insights into the origin and distribution of China's ceramic raw materials, in *Revue Annuelle de la Societe Francaise d'Etude de la Ceramique Orientale, Actes du Colloque "Le Bleu et Blanc" du Proche – Orient a la Chine*, ed. by M. Crick, (Findakly, Suilly-la-Tour, 2000), pp. 15–24

G. Yanyi, Raw materials for making porcelain and the characteristics of porcelain wares in north and South China in ancient times. Archaeometry **29**, 3–19 (1987)

Chapter 2
The Development of British Porcelain from the Eighteenth into the Nineteenth Centuries

Abstract The development of the early English porcelain factories from local ceramic potworks is considered and an assessment is made of the generic and specific literature published for two small early nineteenth century Welsh factory exemplars, Nantgarw and Swansea, to illustrate the paucity of analytical information that is available to the researcher in comparison with the historical accounts and description of the factory products. The occurrence of fakes and forgeries is described and the foundations laid for the holistic forensic evaluation of analytical data on the reported percentages of elemental oxides correlated with established views of the factory production recipes.

Keywords Porcelain literature · Nantgarw China works · Swansea China works · Elemental oxide determinations · Porcelain fakes and forgeries · Basis of forensic evaluation · Early analyses

The literature available currently for studies of early British porcelain manufactories in the eighteenth and nineteenth century is quite vast and comprises many general texts which set out to exemplify the products of each factory and there are relatively rather fewer specialist texts which cover the history and details of porcelain production at each site. Whereas the former books and literature tend towards a description of specific porcelain items and their decoration to enable collectors and curators to identify typical factory products, for example, *A Guide to English Porcelain* (Godden 1992), *An Illustrated Encyclopaedia of British Pottery and Porcelain* (Godden 1980) and *The Philips Guide to English Porcelain of the 18th and 18th Centuries* (Sandon 1989), in contrast, the latter works include more detailed, additional valuable historical information about the proprietors and their workforce which provides what can be termed as a more holistic appreciation of the manufacturing operation. Examples of the latter category of specialised texts include, *Derby Porcelain:1748–1848, An Illustrated Guide* (Twitchett 2002), *Swansea and Nantgarw Porcelain: A Scientific Reappraisal* (Edwards 2017), and *Rockingham Porcelain* (Cox and Cox 2005). The reader may therefore question the

H. G. M. Edwards, *18th and 19th Century Porcelain Analysis*,
https://doi.org/10.1007/978-3-030-42192-2_2

need for yet another text to add to this already large bibliographic corpus, but the purpose of the current book is rather different to those which have preceded it in that the accent of the present text is focussed primarily upon the comprehensive analytical data which have become available for the porcelain wares of what can be regarded as some of the prime factories of the eighteenth and early nineteenth century in England and Wales. Moreover, the prime objective of the current text is the comparative interpretation of the analytical results from a forensic evidential standpoint, and can be stated simply: the input of analytical science to who made what and when – and specifically can one use these analytical data and their interpretation to define the source origin of an unknown porcelain specimen or artefact? An integral part of this exercise is the inclusion where possible of documentary information about the factory productions which can be correlated holistically with the analytical data to engender a better appreciation of the results for what could then lead to a formative forensic assessment of the data interpretation and a considered attribution of the factory of production, weighing together the scientific and historical evidence – the holistic approach.

Of the English factories enumerated in Table 1.1 which originated in the eighteenth century only a few survived as ongoing manufacturing concerns into the nineteenth century, when several more new manufacturing ventures then joined the list; a more complete list, numbering in all some 63 factories, including the major European factories manufacturing porcelain during the same period, is to be found in a previous publication (Edwards, *Nantgarw and Swansea Porcelains: An Analytical Perspective,* 2018, Appendix I) for which the detailed research therein for the two small Welsh porcelain manufactories of the early nineteenth century at Swansea and Nantgarw generated the realisation for the necessity of a much wider analytical consideration of porcelains to be undertaken. Table 2.1, hence, gives a comprehensive list of English and Welsh porcelain manufactories which were newly established in the nineteenth century and also contains the survivors of those early English factories from the eighteenth Century: in summary, only ten porcelain factories of the 30 English porcelain factories established in the eighteenth century listed here in Table 1.1 (namely, Coalport, Derby, Lowestoft, Minton, New Hall, Pinxton, Worcester, Ridgway, Spode and Davenport) survived into the nineteenth century and now appear in Table 2.1, and to this list has been added four new porcelain factories, namely Nantgarw, Swansea, Rockingham and Daniell, which were founded in the early nineteenth century. In common with many other porcelain manufacturing enterprises and with their eighteenth century forebears these new factories also ceased to function only a few years after their foundation . The four sole survivors of all these English factories into the twenty-first century are Derby (as the Royal Crown Derby China Works), Worcester (the Royal China Works), the Spode Bone China Works and the Coalport China Works: as stated earlier, the earliest of these to survive to the present day is the Worcester China Works, which was established in 1751 and which subsumed the then recently failed porcelain factory, called "Lund's Bristol" at Bristol founded by William Miller and Benjamin Lund in 1749. There is much conjecture as to the origin of this first Bristol factory but it is believed that Benjamin Lund moved there from Limehouse (Cushion and Cushion,

Table 2.1 English and Welsh porcelain factories in the early nineteenth century

Factory	Date of foundation	Date of closure[a]	Founder(s)
Coalport	1795	Present	John Rose
HR Daniel	1822	1846	Henry Daniel
Davenport	1794	1887	John Davenport
Derby	1748	1848 /present	William Duesbury
Lowestoft	1757	1802	Philip Walker, Obed Alred, Robert Browne
Minton	1793	Present	Thomas Minton
Nantgarw	1817	1823	William Billingsley, Samuel Walker, William Weston Young
New Hall	1782	1814	Jacob Warburton, Samuel Hollins
Pinxton	1796	1813	John Coke, William Billingsley
Ridgway	1794	1940	Job & George Ridgway
Rockingham	1826	1842	Thomas & John Wager Brameld
Spode	1776	Present	Josiah Spode
Swansea	1814	1820	Lewis Weston Dillwyn
Worcester	1751	Present	Dr John Wall, William Davis

[a]This date is often synchronous with the death of the founder, as at that time the factory either closed or was morphed into a different enterprise, such as earthenware manufacture, or was taken over and absorbed into another factory to manufacture porcelain and other materials

A Collector's History of British Porcelain, 1992); Lund was granted a licence to quarry steatite at Gewgraze on the Lizard peninsula in Cornwall on 7th March 1749 (Hobbs 1995), but it is equally uncertain that Limehouse actually made steatitic porcelain as no evidence of this has yet been found analytically (Freestone, in *Limehouse Ware Revisited*, 1993). Nevertheless, the analysis of very early Worcester shards from the Warmstry House site (Owen 1998) clearly shows the significant presence of magnesia in several of these, whose origin can be correlated to the use of steatite, probably from a Cornish source, in the porcelain paste composition. In passing, it should be recorded that the Derby China Works lays claim to a foundation date of 1748, which predates that of the Worcester manufactory, acquiring the Chelsea factory in 1770 and thereby becomes a major player under the proprietorship of the Duesburys and then Robert Bloor until 1848 when it suffered closure with a re-opening on a new site in Derby some two years later and then carrying into the present day.

The development of porcelain manufacturing in England and Wales in the nineteenth century evolved into two main themes based upon *hard paste* and *soft paste* porcelain bodies, whose differential composition depended upon whether or not the ingredients of soapstone and bone ash had been included as paste components. Eccles and Rackham (*Analysed Specimens of English Porcelain in the Victoria and Albert Museum Collection,* 1922) carried out some of the earliest chemical analyses on "English porcelains", a misnomer in title for a study which also included Welsh and Chinese porcelain specimens as well as output from the English factories! Also included by Eccles and Rackham in their seminal study in addition to *hard paste* and *soft paste* porcelains is a separate category for *bone china* and yet another in the form of a fourth definitive category for a porcelain body composition which they termed "*hybrid*" porcelain, and into this classification they incorporated Coalport, Davenport and New Hall porcelains. It should be realised that in the early days of porcelain manufacture in the nineteenth century, chemical analytical data were conspicuously absent for the products of most if not all factories and hence the proprietors of the manufactories had little idea of what ingredients, raw materials, and components their competitor factories were using, unless the formulations or recipes were available directly at source, and in many cases these were very carefully guarded secrets and subject to an economic strategy. Chemical analyses of their rivals' ceramic products were definitely not generally accessible for comparison purposes. Professor Mark Pollard (2013 and 2015), in his authoritative accounts of early ceramics manufacture in Europe, has attributed the trigger and subsequent rapid growth of analytical chemical science to the 1790s in France, when the emergent Revolutionary Government of Maximilien Robespierre literally wished to turn church bells into bronze cannon for the furtherance of their military ambitions, necessitating the provision of chemical analyses for the identification of superior bronze metals, and this was accomplished only several decades after the foundation of many of the porcelain factories under discussion here. Professor Philippe Colomban (2013) has identified some of the earliest chemical analyses of ceramics undertaken to the work of Reamur in France, and in England Sir Humphry Davy (1815) reported the first chemical analyses of archaeological material from wall painting fragments recovered by Canova in his Pompeii excavations in the years prior to 1815. Hence, much of the porcelain manufacture and modifications being carried out in the late eighteenth and early nineteenth centuries was largely based upon conjecture, empirical experimentation and acquired inside knowledge by the proprietors and their workforce, but nevertheless, some of the very finest examples of ceramic art appeared at this time from many factories which are justifiably highly prized by museums and collectors today. An excellent example of this empiricism applied to porcelain manufacture made in pursuit of the achievement of the finest quality paste or porcelain body is provided by the diaries and work book notes of Lewis Weston Dillwyn, which he undertook in collaboration with his kiln manager Samuel Walker between 1815 and 1817, during which he made his renowned duck-egg china for which the Swansea China Works established an enviable reputation with its Regency Georgian clientele (Dillwyn, *Notes*, in Eccles and Rackham, *Analysed Specimens of English Porcelain in the Victoria and Albert Museum Collection,* 1922; and Appendix II in Edwards, *Nantgarw and Swansea Porcelains: An Analytical Perspective*, 2018). Dillwyn was a respected scientist, a Fellow of the Royal Society of London, and his work books are testament to his logical and methodical approach in his attempts to improve the robustness of his esteemed translucent duck-egg china years before analytical chemistry made an impact in ceramics manufacture: sadly, these efforts towards an improvement in his china were to no avail at the time and Dillwyn's new Swansea porcelain "trident" body, although harder and more resistant to thermal and mechanical shock than its delicate duck-egg analogue precursor, suffered from its almost complete rejection by a discerning clientele who had set such a high value upon the texture and beautiful translucency of their porcelain as exemplified by the fabled "duck-egg" china. As a result, the Swansea China Works became insolvent in 1820 and the remaining stock of Dillwyn Swansea porcelain in the white was decorated locally by a skilled workforce and finally dispersed at auction sales thereafter up to 1823.

2.1 The Literature of Nineteenth Century Porcelain

The literature concerning the 30 china factories given in Tables 1.1 and 3.3 factories in Table 2.1 is extensive and these are but a representative sample of a larger number of British factories which existed during this period: when this list is extended to contemporary mainland European, North American and Asian ceramics factories, the corpus of literature becomes overwhelmingly large. It became apparent to the author during his personal research specifically into the two Welsh porcelain manufactories of Nantgarw and Swansea, which operated over only a relatively small time frame during the second decade of the nineteenth century, probably effectively comprising at most only some three or four years apiece between 1815 and 1820, that the relevant literature which existed could be classified into several types. This comprises: dedicated books on the Nantgarw China Works and the Swansea China Works, several general texts on eighteenth and nineteenth century porcelains which often merely mentioned the output of these two factories, specific research and historical articles on the key personnel associated with these two factories such as Lewis Weston Dillwyn and William Billingsley, specific research papers on the products of the two factories and several rather diffuse publications accounting for the personal reflections of people and their family relatives who had actually worked there. In its entirety, this corpus of literature can be seen as rather daunting for potential readers who wished to gain some knowledge into the fascinating stories which lay behind the creation, operation and final cessation of production at the Swansea China Works and the Nantgarw China Works. The situation is rendered even more complex by the seemingly contradictory and confusing information relayed by some of the earlier historical accounts of the two factories, which although written over a temporally adjacent time period spanning just a few years, when personnel who had worked there were still alive and able to recount their experiences, now do not always stand the test of their veracity in comparison with later emergent and associated historical documentation. Hence, the claims and counter-claims of early documentation which have been perpetuated into more recent writings have been re-examined and have sometimes found to be wanting in terms of the establishment of evidence, which has unfortunately frequently required several accounts being necessarily subsequently downgraded to unsupportable myth, legend or perhaps even fairy tales! A full appreciation of the historical background is even more critical for a researcher who is specifically interested in the analytical science which has been undertaken on these two specified factories of Swansea and Nantgarw: the isolation, extraction and proper identification of key analytical research work on their porcelains which is often buried within other more general works can itself pose a non-trivial challenge. In an earlier publication (Edwards, *Nantgarw and Swansea Porcelains: An Analytical Perspective*, 2018) a definitive assessment of all the generally available published analytical literature for Nantgarw and Swansea porcelains was made and it was quickly realised that many other dedicated and more general texts relating to these factories used common source material which often related back some one hundred years or more. The purpose of this discussion, therefore, is to make the reader aware of the type of information released in the panoply of information now available to them and to enable them to assimilate properly the true facts relating to these two factories at Swansea and Nantgarw from which one can generate a wider extrapolation to the literary situation which one might expect to encounter for the larger factories which have been in operation for longer periods of time and which may even still be operational, such as Derby, Worcester and Coalport.

In Table 2.2 is given the comprehensive literature source of no fewer than 40 books and articles for Nantgarw and Swansea porcelains, their dates of publication and author(s), supported by their full reference citations. This must be viewed as a large collection of literary effort describing various aspects of the products of these two admittedly small porcelain manufactories and the personalities of their proprietors and also of their known named artists and workforce. Although many of these articles and texts are still readily available, several were published in only very limited supplies for local consumption and are now no longer available for consultation except for collectors' copies which now can be acquired at an appropriately elevated purchase cost, unless modern reprinted facsimile editions have been made, and others are located only in museums or in private archives which are sometimes difficult to access. It is important, therefore, that a proper assessment can be accomplished of any analytically relevant material encountered, especially in earlier sources, so that it can be evaluated against current documentary evidence in a truly forensic evidential procedure: this means, obviously, that the primary statements from eye-witnesses to events occurring nearly two centuries ago should be considered against later emergent documentation and that these are not taken at an absolute face value without supportive documentary corroboration which has been found to exist from other independent sources. More importantly for the current discussion, of the 40 items of source material available for these two factories listed here only 7 can be considered to be truly focused on the analytical chemical work which has been carried out on their factory products and artefacts: although many more of the texts cited in Table 2.2 actually mention the analytical studies accomplished on their porcelains in passing, most of these refer solely to the earliest chemical work of Herbert Eccles and Bernard Rackham (*Analysed Specimens of English Porcelain in the Victoria and Albert Museum Collection,* 1922) which was carried out in 1922, almost 100 years ago, and very few texts written since have actually re-evaluated the implications of these early analyses when taken in comparison with the later published studies.

It is anticipated, therefore, that this specific case study of two Welsh porcelain manufactories, which actually produced their finest quality porcelains for only

Table 2.2 Source literature for Nantgarw and Swansea porcelains

Type	Title	Year of publication	Author(s)
Dedicated book	Ceramics of Swansea Nantgarw	1897	W. Turner
	The Pottery Porcelain of Swansea Nantgarw	1942	E. Morton Nance
	Nantgarw Porcelain	1948	W. John
	The Swansea and Nantgarw Potteries	1949	K. S. Meager
	Swansea Porcelain	1958	W. John
	Swansea and Nantgarw Porcelain from the Clyne Castle Collection	1971	K.S. Meager
	The Nantgarw Porcelain Album	1978	W. John, G.J. Coombes, K. Coombes
	Swansea Porcelain	1970	E. Jenkins
	Nantgarw Porcelain	1993	R. Williams
	Swansea Porcelain	1988	A.E. Jones and Sir L. Joseph
	Swansea Nantgarw Porcelains: Scientific Reappraisal	2017	H.G.M. Edwards
	Nantgarw Porcelain: The Pursuit of Perfection	2017	H.G.M. Edwards
	Welsh Porcelain at Plas Glyn-y-Weddw	2018	F. Gambon
	Nantgarw & Swansea Porcelains: Analytical Perspective	2018	H.G.M. Edwards
	Gartre'n Ol: Coming Home – Nantgarw Bicentenary Exhibition	2019	Crochendy Nantgarw
	Swansea and Nantgarw Porcelain	1988	D. Jones and P.E. Jones
	Welsh Ceramics in Context, Part 1	2003	J.Gray (ed.)
	Welsh Ceramics in Context, Part 2	2005	J.Gray (ed.)
General texts	English Porcelains in the V&A Collection	1894	Sir A. Church
	Ceramics of Great Britain	1886	Ll. Jewitt
	The Old Derby China Factory	1886	J. Haslem
	The Complete Potter	1847	J. Taylor

(continued)

Table 2.2 (continued)

Type	Title	Year of publication	Author(s)
Associated personnel	William Billingsley	1968	W.John
	Lewis Weston Dillwyn: Duck-Egg Vision	2018	H.G.M. Edwards
	William Billingsley: Artist, Porcelain Manufacturer	2016	H.G.M. Edwards and M. Denyer
	Not Just a Bed of Roses: William Billingsley	1996	J. Robinson and R. Thomas
	William Weston young	1968	E. Jenkins
	Porcelain to Silica Brick: Extreme Ceramics of WWY	2019	H.G.M. Edwards
	An English Porcelain Maker in West Troy – Samuel Walker	1998	W.F. Broderick
	Billingsley, Brampton and Beyond	2010	P.T. Gardner
Research articles	Archaeological Investigation of the Nantgarw Site	1932	I. Williams
	Analysed Specimens of English Porcelain	1922	H. Eccles and B. Rackham
	Historical Pigments; Swansea and Nantgarw Porcelains	2015	H. Edwards
	Swansea and Nantgarw Analyses	1998	J. V. Owen et al.
	Nantgarw Porcelain: Kiln defects	1999	J.V. Owen, M.L. Morrison
	The Development of Welsh Porcelain Bodies	2005	M. Hillis
	Dillwyn Notebooks	1815–1817	L. W Dillwyn
	The Dismantling of Kiln II, Nantgarw	1997	K. Murphy, R. Ramsay, D.A. Higgins
	The Silica Brick and its Inventor, WW Young	1942	R. Jenkins
	Thomas Pardoe and William Weston Young	2005	A. Renton
	Nantgarw Soft Paste Porcelain	2020	P. Colomban et al.

References Cited in Table

W.F. Broderick, An English Porcelain Maker in West Troy. Hudson Valley Regions Rev. 5(2), 23 (1988, September)

Sir A.H. Church, *English Porcelain: A Handbook to the China Made in England During the 18th Century as Illustrated by Specimens Chiefly in the National Collection*, A South Kensington Museum Handbook, Chapman Hall Ltd., London, 1885 and 1894

P. Colomban, H.G.M. Edwards and C. Fountain, Raman spectroscopic and SEM/EDAXS analysis of Nantgarw soft paste porcelain, *J. Eur. Ceram. Soc.*, to be published, (2020)

L.W. Dillwyn, *Notes on the Experimental Production of Swansea Porcelain Bodies and Glazes Made by Lewis Weston Dillwyn with Samuel Walker at the Swansea China Works Between 1815 and 1817*. Presented to the Library of the Victoria Albert Museum, South Kensington, London by

(continued)

Table 2.2 (continued)

John Campbell in 1920. Reproduced in Eccles Rackham, Analysed Specimens of English Porcelain, 1922, see reference below

H. Eccles, B. Rackham, *Analysed Specimens of English Porcelain in the Victoria and Albert Museum Collection* (Victorian and Albert Museum, London, 1922)

H.G.M. Edwards, Historical pigments (Nantgarw and Swansea Porcelains), in *Encyclopaedia of Analytical Chemistry*, ed. by R. Meyers, Y. Ozaki, (Wiley, Chichester, 2015)

H.G.M. Edwards, *Nantgarw Porcelain: The Pursuit of Perfection,* Penrose Antiques Ltd. Short Guides, Series Editor: M.D. Denyer, Penrose Antiques Ltd. (Thornton, 2017). ISBN: 978-0-244-90654-2

H.G.M. Edwards, *Swansea Porcelain: The Duck Egg Translucent Vision of Lewis Weston Dillwyn,* Penrose Antiques Ltd., Short Guides (Thornton, 2017)

H.G.M. Edwards, M.C.T. Denyer, *William Billingsley The Enigmatic Porcelain Artist, Decorator and Manufacturer,* Penrose Antiques Ltd. Short Guides, Neopubli (Berlin, 2016). ISBN: 978-3-7418-6

H.G.M. Edwards, *Swansea and Nantgarw Porcelains: A Scientific Reappraisal* (Springer, Dordrecht, 2017)

H.G.M. Edwards, *Nantgarw and Swansea Porcelains: An Analytical Perspective* (Springer, Dordrecht, 2018)

H.G.M. Edwards, *Porcelain to Silica Bricks: The Extreme Ceramics of William Weston Young (1776–1847)* (Springer, Dordrecht, 2019)

F. Gambon, *Porslen Abertawe a Nantgarw (Swansea and Nantgarw Porcelain),* Catalogue of the F.E. Andrews Collection in the oriel Gallery, Plas Glyn-y-Weddw, Llanbedrog, Pwllheli, Gwynedd, North Wales, John Andrews Charitable Trust Publications/JACT Publications, Oriel Gallery, Llanbedrog, 2017

P.T. Gardner, *Billingsley, Brampton and Beyond: In Search of the Weston Connection – The Provenance of a Porcelain Service Over 200 years Old is Investigated* (Troubadour/Matador Publications, Leicester, 2010)

Gartre'n Ol: Coming Home – The Bicentenary Exhibition of Nantgarw Porcelain on the Departure of William Billinglsey and Samuel Walker from Nantgarw in 1819, Crochendy Nantgarw/Nantgarw Chinaworks, Nantgarw China Works Trust Ltd., 2019

J. Gray (ed.), *Welsh Ceramics in Context, Part 1* (Royal Institution of South Wales, Swansea, 2003)

J. Gray (ed.), *Welsh Ceramics in Context, Part 2* (Royal Institution of South Wales, Swansea, 2003)

J. Haslem, *The Old Derby China Factory* (George Bell, London, 1876)

M. Hillis, The development of Welsh Porcelain bodies, in *Welsh Ceramics in Context, Part II*, ed. by J. Gray, (Royal Institution of South Wales, Swansea, 2005), pp. 170–192

E. Jenkins, William Weston Young, in *Glamorgan Historian*, ed. by S. Williams, vol. 5, (D. Brown Sons, Cowbridge, 1968), pp. 61–101

E. Jenkins, *Swansea Porcelain* (D. Brown Sons, Cowbridge, 1970)

R. Jenkins, The Silica Brick and its inventor, William Weston Young, 1776–1847. *Trans Newcomen Soc* **XXII**, 139–147 (1942)

L. Jewitt, *The Ceramic Art of Great Britain from the Prehistoric Times Down to the Present Day*, vol I and II (Virtue Co. Ltd., London, 1878)

W.D. John, *Nantgarw Porcelain* (The Ceramic Book Company, Newport, 1948)

W.D. John, *Swansea Porcelain* (The Ceramic Book Company, Newport, 1958)

W.D. John, *William Billingsley* (Ceramic Book Co., Newport, 1968)

W.D. John., G.J. Coombes and K. Coombes, *The Nantgarw Porcelain Album,* The The Ceramic Book Company, Newport 1975

A.E. Jones, S.L. Joseph, *Swansea Porcelain: Shapes and Decoration* (D. Brown and Sons, Ltd., Cowbridge, 1988)

D. Jones, P.E. Jones, Swansea and Nantgarw Porcelain. *The Antique Collector*, 34–41 (1988)

(continued)

Table 2.2 (continued)

K.S. Meager, *The Swansea and Nantgarw Potteries: Together in a Catalogue of the Collection of Welsh Pottery and Porcelain on Exhibition at the Glynn Vivian Art Gallery, Swansea,* Swansea Corporation, 1949
K.S. Meager, *Swansea and Nantgarw Porcelain from the Clyne Castle Collection*, 2nd edn. (Glynn Vivian Art Gallery, Swansea, 1971)
E. Morton Nance, *The Pottery and Porcelain of Swansea and Nantgarw* (B.T. Batsford Ltd., London, 1942)
K. Murphy, R. Ramsey, D.A. Higgins, The Dismantling of Kiln II, Nantgarw China, Pottery and Pipe Works, Mid-Glamorgan, 1995. *Post- Mediaeval Archaeol.* **31**, 231–247 (1997)
J.V. Owen, J.O. Wilstead, R.W. Williams, T.E. Day, A tale of two cities: compositional characteristics of some Nantgarw and Swansea porcelains and their implications for kiln wastage. *J. Archaeol. Sci.* **25**, 359–375 (1998)
J.V. Owen, M.L. Morrison, Sagged phosphatic Nantgarw porcelain (ca. 1813–1820): Casualty of overfiring or a fertile paste? *Geoarchaeology* **14**, 313–332 (1999)
A. Renton, Thomas Pardoe and William Weston Young, in *Welsh Ceramics in Context, Part II*, ed. by J. Gray, (Royal Institution of South Wales, Swansea, 2005), pp. 120–146
J. Robinson, R. Thomas, *Not Just a Bed of Roses: The Life and Work of the Artist, Ceramicist and Manufacturer, William Billingsley, (1758–1828),* Usher Gallery, Lincoln, G.W. Belton Ltd., Gainsborough, 1996
J. Taylor, *The Complete Practical Potter* (Shelton, Stoke-upon-Trent, 1847)
W. Turner, *The Ceramics of Swansea and Nantgarw: A History of the Factories with Biographical Notices of the Artists and Others, Notes on the Merits of the Porcelains, the Marks Thereon etc., and an Appendix on the Mannerisms of the Artists*, Bemrose Sons Ltd., Old Bailey London, 1897
I.J. Williams, *The Nantgarw Pottery and its Products: An Examination of the Site*, The National Museum of Wales and the Press Board of the University of Wales, Cardiff, 1932
R. Williams, *Nantgarw Porcelain 1813–1822: The Inaugural Lecture to the Friends of Nantgarw China Works.* The friends of Nantgarw China Works Museum/Taff Ely Borough Council, September, 1993

approximately two or three years each during the closing years of the second decade of the nineteenth century, is illustrative as an exemplar of the wealth of historical and descriptive information which has emanated therefrom and which now faces the modern ceramics researcher who wishes to sift through the source material to filter out and extract the relevant analytical evidence that has been obtained for their porcelain products.

2.2 Forensic Analytical Foundation

The first reported chemical analyses of English porcelains are generally and universally credited to Sir Arthur Church, who presented his work orally in the Cantor Lectures series (1881), followed by his authoritative book, *English Porcelain: A Handbook to the China Made in England in the Eighteenth Century as Illustrated by Specimens Chiefly in the National Collections,* frequently abbreviated to *English Porcelain,* which was published in 1894. Prior to this, Jewitt (Jewitt, *The Ceramic Art of Great Britain from the Prehistoric Times Down to the Present Day*, 1878) published an equally authoritative account of the history of British ceramic art and its manufacture but necessarily written without making reference to any analyses undertaken on the pieces described therein. Church's book was also the first attempt at a holistic literary appreciation of the selected porcelain manufactories in that historical and personal information about the factories and their key workers were recounted by the author along with selected chemical analytical data, although the latter were admittedly rather sparse in comparison with the historical data presented alongside them. Church's book indeed opened up the way forward for the application of analytical chemical science to porcelains and in his book he cites no fewer than 23 English and Welsh factories in his coverage of a nominated temporal period from 1744 to about 1808, namely, Chelsea, Chelsea-Derby, Bow, Derby, Worcester, Plymouth, Bristol, Longton Hall, New Hall, Minton, Spode, Wedgwood, Davenport, Lowestoft, Liverpool, Brancas-Lauraguais, Caughley, Coalport, Pinxton, Church Gresley, Rockingham, Nantgarw and Swansea. It should be noted that Church's stated period of coverage is somewhat arbitrary in itself since the porcelain products of several of his specified factories were not made until well after his cutoff date of 1808 – namely, Rockingham, Davenport, Swansea and Nantgarw! Nevertheless, this is an impressive list of factories, despite the inclusion of several which would not have been expected strictly from the title of the book: two of the factories are not English, and four did not even exist in the eighteenth century and actually did not make porcelain until the first or second decades of the nineteenth century. The embarkation of Josiah Wedgwood into porcelain manufacture was itself but a short-lived event which started in 1805; the production output of this porcelain was barely minimal and lasted only a few years – Sir Arthur Church recognised this fact but still included Wedgwood in his coverage because he personally regarded the famed Wedgwood blue "jasper ware" as "*porcelain-like*". Perhaps the most disappointing aspect scientifically of Church's book for the reader seeking analytical information is the paucity of analytical data contained therein for most of these nominated factories: of the 23 porcelain factories discussed in the text, only for Chelsea, Chelsea-Derby, Bow, Bristol and Brancas-Lauraguais are any analytical results given (Table 2.3), and these are derived from single specimen analyses only. In some cases he does record tantalisingly the formulations of raw materials that several factories used in their porcelain production recipes, which presumably he had obtained from

Table 2.3 Chemical analyses/% of Sir Arthur Church (1894)

Factory[a]	Date	SiO_2	Al_2O_3	CaO	P_2O_5	MgO	K_2O	Na_2O	Fe_2O_3	Bone Ash[c]
Chelsea	1755	40.2	8.4	27.4	20.3	–	0.9	1.0	1.2	52.0
Ch-Derby	1770	40.3			13.9					35.5
Bow	1744	40.0	16.0	24.0	17.3	0.8	0.6	1.3		44.3
Bristol	1770	62.9	33.2	1.3			[b]	[b]		
B-Laura	1768	58	36	1			3	1	1	

[a]*Ch-Derby* Chelsea –Derby, *B-Laura* Brancas-Lauraguais
[b]Bone ash component calculated as 2.56 × P_2O_5 content
[c]Analyses specify total alkalis i.e. K_2O and Na_2O as 2.6%

the appropriate factory records or from discussions undertaken with surviving personnel and/or from the notes made by the factory proprietors themselves to which he had access. Hence, for only somewhat less than 22% of the factories studied in his book does Sir Arthur Church actually provide any factual analytical data. As can be seen from Table 2.3, the analytical data reported by Sir Arthur Church are also incomplete and no experimental details of the methodology or adopted chemical procedures are given, such as sample size, digestive treatment and number of replicate analyses, from which one could deduce some possible experimental error standard deviations and therefrom establish some mathematical sigma confidence intervals for the quantitative analytical data. As will be apparent, these qualifications will be very necessary when it comes to the unequivocal assignment of a factory attribution based upon compositional analyses, which would intrinsically depend upon the percentages of silica, phosphorus oxide or alumina and other key materials present and the reliability of their quantitative determinations.

The well-respected antique porcelain expert, David Battie, has made the following statement (Battie, *The Eyes Have It,* 1994), which might now be considered to be somewhat controversial but which actually has a direct bearing on the topics being addressed in this current text:

> *The one thing that English (porcelain) factories of the 18th Century were not was consistent. Many factories experimented on a regular basis to improve their ingredients (and even the Chinese were less consistent than was once thought) which would throw a machine. And, of course, many of these tests are invasive – they cannot now, and probably never will, be conducted without damaging the object.*

The conclusions of this statement fuelled Battie's further premiss at the closure of his article that "*The ultimate test is the expert eye. You, a collector with perhaps only a few years' experience, can still beat any technology available today. And you probably always will.*"

This would have had much agreement and popular support from expert opinion at that time, some 25 years ago, but in the intervening period the application of scientific analysis to artworks and especially to oil paintings and manuscripts has had some spectacular successes which have paradoxically sometimes reinforced expert opinion, and sometimes not, in that there is sometimes a divergence created between the conclusions of analytical science relating to an artwork and an avowed, expressed expert opinion. This manifestly requires the analyst to address closely the discriminatory evidence and interpretation derived from the experimental data – in a word, can the analytical data definitively and unambiguously discriminate between the potential factories of origin of a piece of porcelain – so providing the ceramics historian and investigator with a hard and consistent forensic yardstick, independent of a specialist opinion which may nevertheless still occasionally be found to be in conflict with the analytical conclusions! The respective evaluations of expert opinion and analytical interpretation to identify porcelains of unknown origin, equating with the forensic theme of "*soft*" and "*hard*" evidence respectively, will be considered later in its proper context, but several authors have opened a platform for debate in this direction in which the expert observational skills of "*connoiseurship*" are compared with the often limited and sometimes conflicting analytical data interpretation from wasters, shards and occasionally finished pieces of porcelain which relate to the same porcelain factory. The phraseology used by art experts in this context is important, and it will be seen that the meaning of terms such as "*attribution*", "*probability*" and "*consistent with*" will be need to be explored on a more rational basis in the assignment of porcelain pieces to potential source factories of origin. An important and very appropriate comment in David Battie's statement is the first one, in which he alludes to the inconsistency in manufacturing tolerances adopted by the earliest factories which, of course will have a major impact upon the interpretation of quantitative analytical data and selection of "standards" by which a particular factory will be measured and evaluated.This is a fundamental issue, which seems to have been glossed over by some later historical commentators, and will be considered later in this book.

The first comprehensive and detailed list of chemical analytical data on English porcelains actually appeared nearly thirty years after the publication of Sir Arthur Church's book of 1894, in the form of the monograph of Herbert Eccles and Bernard Rackham entitled "*Analysed Specimens of English Porcelain in the Victoria and Albert Museum Collection*" (1922). In contrast to Church's book, this monograph also truly establishes the first real attempt at recounting the analytical data for a comprehensive list of English and other porcelain factories: in summary, porcelains from 16 different factories are studied in this publication, comprising one Chinese (K'ang Hsi), two Welsh (Nantgarw and Swansea) and thirteen English porcelain manufactories, namely, Chelsea, Bristol, New Hall, Longton Hall, Worcester, Liverpool, Bow, Lowestoft, Derby, Pinxton, Caughley, Davenport and Coalport (Table 2.4). The Eccles & Rackham monograph is a fount of historical information for the analytical scientist but is admittedly not very readable for historians of china factories in that it is very chemically focussed and data-oriented. An interesting and relevant item of information for this current survey which emerges from reading these individual analyses in the monograph is that Eccles and Rackham have actually used their analytical data to identify one piece of porcelain of rather doubtful attribution submitted from the Victoria Albert Museum Collection, comprising a mug painted in underglaze blue enamel with a Chinese scene which had been questionably given an expert attribution devoid of any analytical information to the Bristol factory manufacture for the period 1750–1755. Under the heading of soapstone porcelains, the mug (with a Victoria and Albert Museum acquisition number C.253-1915) is labelled as "*of uncertain origin*" and was badly damaged (it is depicted as such in No. 25, Plate X in Eccles and Rackham, *Analysed Specimens of English Porcelain in the Victoria and Albert Museum Collection*, 1922), and shows the presence of extensive metal riveting in its restoration. In their text (Eccles and Rackham, *Analysed Specimens of English Porcelain in the Victoria and Albert Museum Collection,* 1922, page 17), Herbert Eccles and Bernard Rackham conclude that their elemental oxide data analysis of this mug numerically approaches that of the preceding specimen in their list which had been presented for analysis in its composition, namely a Bristol sauceboat (with a Victoria and Albert Museum acquisition number C.857.1920), and they surmise that the high percentage of

Table 2.4 Chemical analyses/% of Eccles and Rackham (1922)

Factory	Date	SiO_2	Al_2O_3	CaO	H_3PO_4	MgO	K_2O	Na_2O	PbO	Bone Ash	S'Stone
Hard paste porcelains											
K'ang Hsi	1622–1722	71.8	23.0	0.6	0.2		1.9	2.1	0.6[a]		
Bristol	1750	70.0	24.4	1.5	0.2		1.4	1.9	1.5		
New Hall	1790–1800	70.1	24.3	1.9	0.4		0.6	2.9			
	1800	73.6	19.3	4.0	0.2		0.9	2.1	0.7		
Worcester	1800	75.4	18.9	2.8	0.2	0.2	1.3	2.0			
Soft paste porcelains											
Chelsea	1750–1755	64.8	6.0	25.0	0.2		2.6	1.8	0.6		
	1755–1760	69.1	5.9	20.5		0.7	3.3	0.7			
Longton H	1750–1760	76.2	4.3	9.3					6.5[b]		
Liverpool	1760	62.8	4.5	20.1	2.1	0.2	1.2	1.1	8.6[b]	5	
Bone China											
Bow	1750–1760	43.6	8.4	24.5	19.0	0.6	0.9	1.2	1.8	48	
	1750–1760	42.8	8.8	28.3	18.1		0.7	1.1	0.5	46	
	1750–1760	55.1	16.5	15.1	11.5	0.4	0.7	0.3		30	
	1760–1770	50.4	7.8	24.5	13.7		0.5	0.7	1.5	35	
Chelsea	1760–1765	45.5	12.1	26.0	14.3		0.9	0.7		37	
Lowestoft	1760	41.4	9.6	25.4	18.8					48	
Derby	1780–1790	41.9	16.0	24.3	15.0	0.2	0.9	1.1	0.4	38	
	1780–1790	43.2	12.4	25.3	12.4					32	
Pinxton	1800	41.9	13.8	24.8	14.1	0.6	0.6	0.8	4.0[b]	36	
Nantgarw	1811–1819	46.0	17.0	19.7	13.9		2.7	0.4		36	
	1811–1819	38.9	18.1	22.5	17.1		2.6	0.1		44	
Swansea	1814–1820	47.8	26.5	13.3	9.9		3.1	0.2		25	

Soapstone porcelains											
Bristol	1750	67.6	4.6	2.6	2.0	13.3	1.2	1.6	8.0[b]	5	40
Bristol?	1750–1755	65.5		4.0	0.7	10.0				2	30[c]
Worcester	1760	72.8	6.9	4.0		11.9					36
Worcester	1760	74.3	5.9	2.4		12.4					37
	1760	70.1	7.2			15.2					46
	1760–1770	75.1	5.6	1.2		13.6					41
	1760–1770	70.0				13.0					39[d]
Worc' BFB	1807–1830	76.1	7.6	1.3		9.7	4.1	1.4			29
Worc' FBB	1823	76.4	7.7	1.8	0.4	6.1			5.4[b]		18
Caughley	1780	74.2	8.5	2.8	0.2	7.6	1.3	2.3	3.7[b]		23
Swansea	1810–1817	84.0	8.7	1.0		2.5	2.9	0.6			8
	1810–1817	81.6	8.9	0.7	0.3	4.3	3.3	0.6			13
Hybrid porcelains											
Davenport	1830	38.6	22.0	22.4	13.8		0.5	1.5	1.0	35	
New Hall	1820–1825	38.5	18.2	24.9	15.5	0.2	0.7	1.6	0.9	40	
Coalport	1820–1830	42.9	15.1	23.2	16.3					42	

[a]Attributed to analytical contamination from the lead glaze
[b]Indicative of the incorporation of flint glass cullet into the porcelain body
[c]This specimen was of doubtful or questionable origin and Eccles Rackham on the basis of their analyses assigned it to the Bristol factory but additionally, because of its similarity to Worcester, they suggested that it could have been of Worcester sourcing but decorated at Bristol
[d]Average of five different specimens analysed

soapstone (derived by multiplying the magnesia percentage content by 3X) in these samples of between 30% and 40% is very close to those values derived from the Worcester factory specimens dating slightly later in period to 1760! They, therefore, proposed that this particular Bristol factory output of Benjamin Lund represents a time when its takeover by the Worcester china manufactory had just been completed – and that the porcelain specimen could therefore have even been Worcester in origin, or perhaps even purchased from Worcester and then decorated locally at Bristol at the time of the takeover of the factory! The present author maintains that this account represents the first recorded and published attempt at using chemical elemental oxide analysis of porcelain body pastes to differentiate between the porcelain outputs of English porcelain factories from which a potential analytical attribution could then be made, and the conclusion of Eccles and Rackham was apparently acceptable to the artistic ceramic experts at that time. However, when we inspect the analytical data from which Eccles Rackham derived their conclusion relating to the mug's Bristol provenance, some need for caution is perhaps now required. The data for the Bristol (?) mug of doubtful provenance and its standard Bristol sauceboat comparator are both given here in Table 2.4 as specimens numbered 23 and 22, respectively, in the Table: reference to these data reveals that although the silica percentages are reasonably close at 65.5 and 67.6% for the mug and the sauceboat, respectively, little else matches for the elemental oxide analytical data derived from the two specimens. In essence, the questionable Bristol (?) mug contains no alumina, potash, soda and lead analytical components, unlike the sauceboat which was selected as a definitive example of the Bristol genre, and there are also significant discrepancies found to occur in the lime, magnesia and phosphoric acid contents, too. On this basis, it would perhaps be considered rather hasty and certainly questionable today to conclude that the mug and sauceboat were both originating from the Bristol factory. Then we read further that Eccles Rackham have concluded that perhaps the "Bristol" mug (?) was really of Worcester manufacture anyway, even though analytically the figures provided in their other data analyses of the Worcester specimens in their list were even more remotely different numerically from those analogous values cited for the mug and for the standard Bristol sauceboat! So, it should be remarked that although this analytical test case was clearly seen to be ground-breaking at the time, in retrospect it seems rather more questionable in its authenticity of application now, and the conclusions derived from the data one hundred years ago are patently much more indefinite and inreliable now. It would be a very brave analyst indeed who would conclude from such comparative data given in Table 2.4 that the mug under consideration was firstly of a Bristol origin like the "standard" sauce boat and even more outrageous to then imply, secondly, that the differences in composition could be assigned to its Worcester origins, which are seen to be even more analytically remote.

It is relevant to bear in mind that the basis of the quantitative chemical analysis procedures at that time, *viz.* the early 1900s, all involved a "wet chemical digestion" procedure for which a significantly large sample was taken from each specimen which was then reacted with concentrated acids or alkalis, such as hydrofluoric acid (for the dissolution of silica as water-soluble hexafluorosilcates) and *aqua regia* (a one to three mixture of concentrated nitric and hydrochloric acids, in which the nitrosyl cation, NO^+, was presumed to be the activator for the dissolution of gold, giving rise to its name) or caustic soda, to dissolve other silicaceous components but not silica itself. The key elements that were determined during this analytical process were silicon (as silica), aluminium (as alumina), magnesium (as magnesia), calcium (as lime), phosphorus (as phosphate, phosphoric acid or phosphorus pentoxide), sodium (as soda), potassium (as potash), and occasionally sulfur (as sulfate), and lead (as lead oxide). Several problems immediately become apparent in the sampling process, firstly, that the quantity of sample required for the analyses is quite considerable (and for this, often repetitive or replicate measurements would have been needed), and, secondly, that this inevitably involved the partial destruction of the porcelain items (although generally, it must be acknowledged that these were sometimes already badly damaged or faulty pieces anyway) or of shards recovered from derelict factory sites. The samples cited by Sir Arthur Church in his book (*English Porcelain*, 1894) were taken from completed, finished and decorated china pieces in the Museum ceramics collection, including those from the Lady Charlotte Schreiber Collection at the Victoria & Albert Museum, South Kensington, London. In the preface to his book, Church makes some interesting statements such as "*... a single specimen of Minton early porcelain is sufficient to represent a whole group of factories, including those carried on by Davenport, Spode and Wedgwood – so far at least as their productions in porcelain are concerned*", and "*I have added new facts, and helped to settle doubtful matters by careful study of collections of English china, and by chemical and microscopical analysis of individual specimens*". It does surely seem a *non-sequitur* presumption nowadays to assume that a single piece of Minton porcelain is analytically representative of all Davenport, Spode and Wedgwood factory porcelains produced – especially the latter manufactory, whose porcelain production was very small indeed in comparison with their earthenware and jasper ware factory output. The statement of even greater interest here for the current text, however, is actually the second one, in which Church records that he is able to "*settle doubtful matters*" by chemical analysis – which must surely therefore be the first published reference, dated January 1885 in his Preface, to the potential discriminatory and forensic use of chemical analysis for the correct source attribution of suspected or "doubtful" porcelains. It is very unfortunate that Sir Arthur Church does not actually elaborate on this idea or even provide an example of his attribution of a porcelain piece to a particular factory from his analytical evidence alone, as this would indeed have been a ground-breaking definition historically and a notable landmark for analytical science! In fact, although he mentions this exciting prospect in the Preface to his book, he does not then proceed to enlarge upon it at all in the following text and the author cannot find any reference to this most interesting and crucial development in a careful reading through of the 90+ pages of Church's text (*English Porcelain*, 1894): in fairness, therefore, the honour of first using analytical science to assist in the attribution of a "doubtful" porcelain item to a particular porcelain manufactory, where details have actually been provided for the close examination and evaluation by subsequent readers, must surely therefore rest with Herbert Eccles and Bernard Rackham (*Analysed Specimens* *of English Porcelain in the Victoria and Albert Museum Collection*, 1922), who at least have registered for their suspect Bow mug specimen the analytical data for their factory attribution and conclusions for inspection in the published literature, even if now perhaps this is potentially open to further debate and possible question. It has already been mentioned that many of the analyses carried out by Church and by Eccles and Rackham were reported for single specimens and comprised solitary measurements only: on this basis it is rather difficult to assess the experimental errors of each factory determination, which can then be processed into an average percentage with a standard sigma mathematical deviation. However, in Table 2.5 the data have been collected for the analyses reported by Eccles and Rackham which involve a multiplicity of their analytical determinations carried out on replicate samples of the same type from several factories; these actually comprise only 7 factories from the 16 in total studied in their monograph – namely, New Hall, Chelsea, Bow, Derby, Nantgarw, Worcester and Swansea, and cover their four nominated categories of porcelain, *viz.*, hard paste, soft paste, bone china and soapstone. Nevertheless, only two specimens are reported for each of these factories, except for the Worcester factory which has been subdivided into three periods of production: namely, 1750–1760, 1760–1770, and the Barr, Flight & Barr/Flight, Barr & Barr periods from 1800–1830, for which 7 specimens in total were taken. Also, the multiple Swansea data include one specimen of the "trident" body, which was created by Lewis Dillwyn as a more robust variant on his esteemed duck-egg bone china but this was apparently unacceptable to the clientele, whose lack of appreciation of its quality in comparison with the normal duck-egg china caused the factory to close down in 1820. It has not, therefore, been possible to compile these last data as an average with that of the single duck-egg china version of Swansea porcelain in the same Table because only one specimen of duck-egg porcelain was analysed in Eccles and Rackham's study. Hence, the two Swansea results stated by Eccles and Rackham were derived from different porcelain body compositions and cannot therefore be included as true replicates for this purpose of data averaging.

This foundation then leads us directly into the major themes of this book: namely, how does chemical analysis establish the source factory for a porcelain item, can it be used thereby to differentiate between factory output, how rigorous is the discriminatory analytical process involved and, finally, can it distinguish between a genuine item and a *fake* or a *forgery*? The relevant statement to this effect made by Sir Arthur Church in the preface to his book of 1894 (*English Porcelain*) discussed above, and actually made and noted as early as 1885, is ground-breaking in its concept and was uttered almost 50 years before the mantra of forensic science, the *Locard Exchange Principle*, was compiled in 1928 – "*Every contact leaves a trace*"- a maxim that applies universally to our twenty-first century forensic analytical science principles. The growth of modern forensic science can formally be traced back to this date, even though isolated incidents relating to the application of chemical detection of hazardous materials did occur earlier in Victorian times. The author believes that this statement by Sir Arthur Church has largely been ignored by subsequent chroniclers and even the classic much-quoted and often cited, cutting-edge analytical studies of Herbert Eccles and Bernard Rackham on porcelains published

Table 2.5 Analyses of Eccles and Rackham (1922) for multiple samples – averages and standard deviations/%

Type	Factory	Specimens	SiO_2	Al_2O_3	CaO	H_3PO_4	MgO	K_2O
Hard paste	New Hall	2	72+/−2	22+/−3	3+/−1	0.3+/−0.1	–	0.8+/−0.2
Soft paste	Chelsea	2	67+/−2	6.0+/−0.1	23+/−3	–	–	3.0+/−0.4
Bone China	Bow	2	43.5+/−0.4	8.6+/−0.2	26+/−2	18.5+/−0.5	–	–
	Derby	2	42.6+/−0.7	14+/−2	24.8+/−0.5	14+/−2	–	–
	Nantgarw	2	43+/−4	17.6+/−0.5	21+/−2	15+/−2	2.7+/−0.1	0.3+/−0.2
Soapstone	Worcester[a]	3	72+/−2	6.7+/−0.8	2.2+/−0.2	–	13+/−2	–
	Worcester[b]	2	72+/−3	23+/−3	–	–	13.5+/−0.3	–
	Worcester[c]	2	76.3+/−0.2	7.7+/−0.1	1.5+/−0.3	0.2+/−0.2	8+/−2	2+/−2
	Swansea[d]	2	83+/−1	8.8+/−0.1	0.8+/−0.2	0.2+/−0.2	3.4+/−0.9	3.1+/−0.2

All standard deviations are given to the first significant figure

[a]1750–1760

[b]1760–1770

[c]Barr, Flight Barr, Flight, Barr Barr, 1800–1830

[d]This is the Swansea "trident" body; the soft paste duck-egg porcelain body comprised only one sample and therefore does not feature in this table

several decades later (*Analysed Specimens of English Porcelain in the Victoria and Albert Museum Collection,* 1922) does not attempt to use their detailed analytical information for this purpose, but for the one notable exception of the "doubtful" and suspected Bristol (?) porcelain mug discussed earlier, which has also escaped the attention of many researchers hitherto for its historical and ground-breaking chronological importance.

In passing, it is appropriate that the technical distinction between the terms *fake* and *forgery* is explored here: an artefact which has been modified by the addition of or removal of parts to make it appear to be something else, which is generally presumed to be of an increased resultant monetary value, is a *fake*—such as a genuine Renaissance oil painting which has in modern times acquired the signature of an esteemed, well-known artist, such as El Greco or Raphael. A *forgery*, on the other hand, is where an artwork has been deliberately created maybe even using chronologically acceptable and contemporary materials and pigments in simulation of an original artwork by an esteemed artist, again created for the purpose of a deceitful transaction which will realise a significant financial gain for the forger by duping the purchaser into believing that it is an original artwork from someone else. A third category in the genre of "fakes" has been proposed by Craddock in his seminal text on fakes and forgeries (*The Scientific Investigation of Copies, Fakes and Forgeries,* 2009) in which he identifies a *pastiche*, a synthetic mock-up of one or more antique or archaeological artefacts which comprises two or more genuine pieces of antiques which have been adapted or manufactured into a completely new and potentially unique artefact: two famous examples of this type of *pastiche* forgery include the fabulous "mermaid" and the "archaeopteryx" specimens which for many years were both accepted as genuine archaeological relics until X-ray analytical studies showed that they were actually composed of pieces from different animals joined together to create novel species of fauna! In the realm of antique furniture, such a *pastiche* might involve the cutting down and refurbishment of an old chest of drawers to make it more amenable spatially for a modern home, the carving of designs on a plain wooden settle backrest or kist, or the replacement of an original table top with a novel one of unusual size, veneer or timber construction: items bearing such distortions and amendments are frequently referred to as being "*associated*" in the antiques trade. A potential ceramics case study analogue would be that of a plain spill vase mounted upon a florally encrusted plinth in an attempt to make a unique and obviously very rare example of a porcelain product (with a thereby consequently increased premium value at resale) attributed to the output of a particular factory. Both fakes and forgeries generally have to pass verification tests by panels of experts, who should also take into account the scientific evidence presented to them as well as considering any historical documentation relating to provenancing issues in their assessment.

Despite this codicil, the mantra first proposed by Professor Virginia Orna in 1996 that:

> *Science cannot authenticate an artwork but can only expose a fake*

is still maintained to apply generically nowadays, and perhaps with good reason, as will be discussed later – the clarity of the depth of meaning of "authentication" is the real issue here. In the field of ceramics the *Orna principle* can be translated as: analytical science will never be able to prove that a piece of porcelain is a genuine product of a particular factory but it could be used to confirm that it is a fake – however, it will be shown that this is not strictly true, and we shall evaluate later this mantra when specifically applied to porcelains, demonstrating that science can frequently identify the genuine products of a factory even when these are of an unusual shape or design. Certainly, through the adoption of the "fingerprinting" of unique and key chemical signatures in chemical databases, analysis could potentially identify the presence of interlopers which have been mis-attributed to the products of a particular factory, such as replacement pieces for broken items in an original factory tea or dinner/dessert service whose source factory is no longer in production. However, all is not straightforward in this exercise and certain rigorous scientific principles must first be established before objective conclusions can be made on the basis of an analytical science assessment, whereby the origin of a piece of porcelain can then be firmly and unambiguously established to the acceptance of all: a necessary and key theme here will be the analysis of the factors which arise during the analytical chemical determination of the components of a porcelain body and its glaze and an understanding of the changes in composition of their factory porcelain bodies that have occurred with time for inclusion in an overall *holistic analysis*.

2.3 What Information Can Be Inferred from the Chemical Analysis of Porcelains?

Having dissolved the fired porcelain specimen into solution, be it a perfect piece, a damaged piece or a shard, the analytical chemist at the end of the nineteenth century and start of the twentieth century could proceed by either or both of two routes to determine qualitatively and/or quantitatively the elemental oxide composition: namely, using gravimetric or volumetric analysis, both of which had their operational procedures developed fundamentally in the late eighteenth century with further refinement taking place in the nineteenth century. Jons Jacob Berzelius (1779–1848) is usually credited with establishing the basis for chemical gravimetric analysis (Jorpes, *Jacob Berzelius-His Life and Work*, 1966), which was then further refined and advanced by T.W. Richards in the late 1800s, for which he was awarded the first Nobel Prize to an American scientist in 1914. Francois Descroizelles, in France, is acknowledged as establishing the fundamental basis of volumetric analytical procedures (Duval 1951). Instrumental analyses, which now form the mainstay of modern analytical work with ceramic specimens, did not reach an impetus for their application until the 1950s, and an essential additional requirement demanded of analytical work currently is the ability to acquire the requisite compositional data analytically but also non-destructively from porcelain specimens, or at least to minimise the sampling required from them using modern microanalytical techniques. This is a major quantum leap forward in concept from the earliest experiments of Church and of Eccles & Rackham a century or more ago.

Inspection of the analytical data published by Sir Arthur Church in 1894 (*English Porcelains*) and by Herbert Eccles and Bernard Rackham in 1922 (*Analysed Specimens of English Porcelain in the Victoria and Albert Museum Collection*), reproduced here as average percentages in Tables 2.3 and 2.4, respectively, reveals several important features which clearly now require some further explanation and discussion:

- The quantitative analyses undertaken and reported by Church and by Eccles & Rackham produce experimental determinations of the percentages of several elemental oxides and compounds – namely, silica SiO_2, alumina Al_2O_3, magnesia MgO, soda Na_2O, potash K_2O, and lime CaO. An important additional parameter in the analytical data list is the percentage of elemental phosphorus, arising uniquely from any bone ash additive made to the porcelain paste recipe, which has been tabulated as phosphorus pentoxide, P_2O_5, by Church (*English Porcelain*, 1894) and as phosphoric acid, H_3PO_4, by Eccles and Rackham (*Analysed Specimens of English Porcelain in the Victoria and Albert Musuem Collection*, 1922). This has been complicated further by the more modern determinations of Tite and Bimson (1991) and Owen and Morrison (1999) who both expressed the phosphorus content in terms of the pentoxide of phosphorus and then by Owen et al. (1998), who determine the phosphorus content as phosphate ion, strictly the orthophosphate ion, PO_4^{3-}. Naturally, the percentage of elemental phosphorus in the analyte varies between these three compounds, namely, 31.3% for phosphoric acid, 32.3% for phosphate and 45.1% for phosphorus pentoxide, which renders the comparative back-calculation of the percentage of the bone ash component in the recipe for the raw materials difficult unless the differential comparative multiplication factors are properly applied for each method used.
- Because of the high-temperatures involved in the kiln firing process for the production of porcelain, the molecular chemical composition of the fired ceramic body has no chemical or molecular resemblance at all to the original starting materials (Edwards, *Nantgarw and Swansea Porcelains: An Analytical Perspective*, 2018). Hence, in the quantitative estimation of the percentages of the relevant elemental oxides, some degree of back-calculation is necessary to determine the raw material components that were adopted for use in the original formulation recipes. This can be rather difficult to accomplish for all analytical determinations as the chosen elemental oxides can occur in more than one component: of the raw material components of a porcelain body paste, a classic example of this is silica, SiO_2, which occurs in several of the raw materials adopted for porcelain production, such as fine quartz sand, flints or chert, soapstone, kaolin, china stone and feldspar as well as potential minor additives such as flint glass cullet and smalt (Dana, *Textbook of Mineralogy: An Extended Treatise on Crystallography and Physical Mineralogy,* 1955). The global analytical determination of a percentage value for silica, therefore, does not permit

the analyst to interpret any specific quantitative changes relating to these individual components in the raw materials, especially minor ones. Fortunately, it is a very different case for some other elemental oxides, as can be described below:

Phosphorus pentoxide (or its chemical derivatives such as phosphate ion and phosphoric acid): this uniquely defines the presence of the bone ash component as a raw material in the recipe formulation as bone ash comprises calcium hydroxyapatite, $Ca_5(OH)(PO_4)_3$, which at higher temperatures of calcination dehydroxylates and is then better represented as the mineral whitlockite, $Ca_3(PO_4)_2$.

Magnesia: this uniquely is taken to define the presence of soapstone, or steatite, $Mg_3Si_4O_{10}(OH)_2$, in the formulation recipe.

Lead oxide: this uniquely defines the presence of flint glass cullet added to the raw material components in which the lead is added in the form of lead (II) oxide, as litharge or massicot. Care should be taken by the analyst to exclude glaze components from the body sampling of finished specimens, since ceramic glazes can contain significant amounts of lead oxide which can contaminate the chemical digestion procedure for the porcelain body and thereby skew the analytical results.

Soda: which technically should only arise from an alkaline sodium carbonate component in the recipe. Generally, soda ash, essentially sodium carbonate Na_2CO_3, was derived from the burning of kelp or seaweed and potash from plants.

In contrast, the other elemental oxides occur in more than one component of most porcelain bodies, such as silica, which has been detailed above, lime CaO, which can arise from hydroxyapatite and calcite, potassium oxide, which can arise from potash, pearl ash, china stone and feldspar, and alumina, which is found in kaolin, feldspar and china stone. Details of the analytical prevalence of these elemental oxides and their occurrence in raw materials used in porcelain recipes are given in Table 2.6.

- A serious consideration that must be addressed in the quantitative analytical determination of these elemental oxides and which directly affects the interpretation of their raw material origins relates to the initial purity of the raw materials that were acquired for the recipe formulations in porcelain manufactories. Historical documentation recounts many tales of the extraordinary lengths that china works proprietors went to for the acquisition of the purest, high quality raw materials for their paste components, as it was appreciated empirically that the highest quality raw materials afforded the best finished articles after firing due to the problems associated with impurities affecting their high temperature chemical conversions to ceramic bodies in the kiln. For example, William Duesbury of the Derby China Works in the late eighteenth century insisted on using high quality bone ash from a particular source in London as it was appreciated that the presence of residual organic material in improperly calcined bone ash was a major factor in the creation of defects, blemishes and bubbles in porcelain bodies

Table 2.6 Compositional analyses/Average % for Nantgarw and Swansea porcelain bodies

Analysts[a]	Specimens	SiO_2	Al_2O_3	CaO	P_2O_5	K_2O	Na_2O	MgO	Fe_2O_3	PbO
Nantgarw soft paste										
E and R (1922)	3	42+/−3	19+/−2	21+/−2	15+/−2	2.3+/−0.6	0.5+/−0.4	0.3+/−0.3	–	–
T and B (1991)	3	44.5+/−0.7	13.0+/−0.5	21.9+/−0.7	16.7+/−0.7	2.2+/−0.3	0.8+/−0.2	0.6+/−0.1	0.2+/−0.2	–
O et al. (1998)	8	44+/−5	12.7+/−0.4	23+/−3	17.4+/−0.8	3+/−5	0.4+/−0.0	0.6+/−1.4	0.2+/−0.2	–
O and M (1999)	7	43+/−4	13+/−1	23+/−2	17+/−2	2+/−1	0.5+/−0.1	0.4+/−0.2	0.2+/−0.0	–
C et al (2020)	4	48+/−5	16+/−1	13+/−4	18+/−2	1.3+/−0.1	1.0 +/−0.4	0.5+/−0.3	0.3+/−0.1	–
Weighted average	25	44+/−2	15+/−1	21+/2	17+/−1	2.1+/− 0.8	0.6+/−0.2	0.5+/−0.3	0.2+/−0.1	–
Nantgarw medium/high silica										
C et al (2020)	3	64+/−5	19+/−4	7+/−3	6+/−2	2+/−1	0.9+/−0.7	0.1+/−0.1	O.2+/−0.2	–
Nantgarw high silica										
O and M (1999)	4	80+/−1	9+/−2	0.5+/−0.1	0.4+/−0.4	7+/−5	1.5+/−0.1	1.8+/−0.8	0.1+/−0.1	–
C et al. (2020)	1	80	13	0.3	0	1.2	2.6	3.1	0.1	–
Weighted average	5	80.0+/−0.5	11+/−1	0.4+/−0.1	0.2+/−0.2	4.1+/−3	2.0+/− 0.1	2.5+/−0.4	0.1+/−0.1	–

Swansea soft paste										
E and R (1922)	1	47.8	26.5	13.3	9.9	3.1	0.2	–	–	–
O et al. (1998)	1	45.2	20.0	16.7	13.0	2.8	1.4	0.4	0.3	–
Weighted average	2	46+/−2	23+/−3	15+/−2	12+/−2	3.0+/−0.2	0.8+/− 0.6	0.2+/−0.2	0.2+/−0.2	–
Swansea soapstone (trident)										
E and R (1922)	2	83+/−1	8.8+/−0.1	0.9+/−0.2	0.2+/−0.2	3.1+/−0.2	0.6+/−0.0	3.4+/−0.9	–	–
Swansea high silica										
O et al. (1998)	7	72+/−2	24+/−2	0.3+/−0.1	0.2+/−0.0	1.7+/−0.1	0.6+/−0.1	0.2+/−0.1	0.7+/−0/1	–

Weighted Average: the weighted average mathematically favours the individual determinations which are closest to the arithmetic mean average numerically and is equal to the sum of the products of each determination in the data set divided by the sum of the weighted sets. This focuses in on the numerically precise value that lies closest to the weighted mean and disfavours outliers: the standard deviation in the precision also thereby is decreased correspondingly to favour the weighted mean value and the individual determinations that are closest to this value

[a]*ER* Eccles & Rackham, *TB* Tite & Bimson, *O et al.* Owen et al., *OM* Owen & Morrison, *C et al.* Colomban et al.

being fired at high temperatures in his biscuit kiln (Anderson, *Derby Porcelain and the English Ceramics Industry,* 2000). William Billingsley, at the Nantgarw China Works, regarded the preferential use of ox bones as imperative for calcining to produce bone ash and required his local miller, David Jones, to grind the calcined frit very finely, in addition to his prioritisation of his china clay acquired specifically from the Gewgraze Mine in St Austell and supplies of the finest white quartz East Anglian Lynn river sand on his list of sources for raw materials for his finest quality porcelain manufacture (Edwards, *Nantgarw and Swansea Porcelains; An Analytical Perspective,* 2018). He also utilised the superior thermal properties of Welsh anthracite steam coal sourced locally in the Rhondda Valleys to achieve the high temperatures required in his kiln for the firing of his biscuit porcelain at a temperature around 1420 °C. The presence of impurities in the raw materials was not to be tolerated if the finest quality ceramic output was desired: Lewis Dillwyn at the Swansea China Works was insistent that his porcelain paste body prepared for the final stage biscuit porcelain firing in his kilns was free from calcite because its decomposition to carbon dioxide in the kiln at a lower temperature of around 650–700 °C would create bubbles (which he termed "voids") within the drying paste which were detrimental to the translucency of the finished porcelain articles. It was, hence, a great challenge for china works proprietors to source the highest quality raw materials alongside their ability to transport them to the manufacturing sites which clearly involved some significant hidden costs. As Ramsay and Ramsay (2007) point out in their scholarly article, in the earliest days of the Bow factory the proprietors sourced china clay (which they called *unaker*) from the Cherokee lands on the eastern seaboard of North America, and Anderson (*Derby Porcelain and the Early English Fine Ceramic Industry*, 2000) has noted that the Derby factory imported their china clay initially from a mine in Saxony! This was also referred to by Sir Arthur Church (*English Porcelain,* 1894). Another rather peculiar additive to the porcelain recipes at several factories was "grog", which was a finely ground frit composed of failed porcelain items or "wasters" from other factories elsewhere, which was often bought in as a bulking supplement to the porcelain paste being prepared for firing, for example, at the Derby China Works by William Duesbury (Anderson, *Derby Porcelain and the English Ceramics Industry,* 2000). Clearly, the incorporation of an unknown quantity of chemically indeterminate "grog" to an otherwise precisely well-defined recipe for a porcelain paste will introduce an analytical imbalance in the compositional data derived therefrom as well as offering the potential for the introduction of outliers and interlopers, such as magnesia, into the otherwise seemingly precisely defined factory body materials composition. Few porcelain factory proprietors admitted to their incorporation of grog as a component in their body paste as part of their published recipes but historians nevertheless have identified this as a quite prevalent practice in the industry. A further point to make here relates to the storage on site and transportation of the raw materials prior to their use in the workshop. In particular, finely ground components such as silica sand, aluminosilicates, feldspars, lime and alkaline soda and potash would be prone to the absorption of significant quantities of water, which would of course affect the weighing procedures if they were not dried consistently and thoroughly before their use in the body paste recipe. No factory records seem to apply for an additional procedure of drying to a constant weight being adopted prior to the weighing of the raw materials components being undertaken, so it must be realised that the use of perhaps damp materials which had already been transported perhaps in open waggons or containers by land or sea for many days prior to their arrival at the factory site was a real possibility. It is of course impossible to estimate the actual percentage of water absorbed by the raw materials at this stage but it would be especially important for the case of the lime being carried any distance from the limekilns, where reaction with atmospheric water vapour and carbon dioxide would occur chemically and exothermically to reverse the lime formation process back into calcite and also by the formation of a "slaked lime", calcium hydroxide, $Ca(OH)_2$, effectively fixing the water chemically into the lime component and thereby introducing errors in the calcium oxide compositional formulation for recipe comparison with the later derived analytical data.

- For the more singular source elemental oxides highlighted above it should therefore be possible to correlate their analytical determination with the raw materials with which they are associated in the recipe formulation. The conversion factors to enable this to be achieved are presented in Table 2.7, from which we can then assess the percentages of bone ash, feldspar and soapstone in the formulations of most china factories. This, of course, has already been accomplished to a large extent by previous authors in relevant research publications but in some cases the correlations made between the elemental oxides and the raw material components are not obvious, especially where these have not been precisely stated: an important point relates to the way in which the elemental oxides have been presented hitherto, which has differed for several analytical determinations and it has not been consistently applied – the case of elemental phosphorus has been expressed as one of three possibilities for Church (phosphorus pentoxide, P_2O_5), Eccles and Rackham (phosphoric acid, H_3PO_4), Owen *et al* (phosphorus pentoxide, P_2O_5) and Owen et al. (phosphate, PO_4^{3-}) clearly illustrates this analytical dilemma. The difficulty is further compounded when it is appreciated that the earliest chemical formulations for bone ash were proved to be incorrect by

Table 2.7 Conversion factors for analytically determined elemental oxides and raw materials used in porcelain paste recipes

Elemental oxide	Chemical formula	Raw material	Conversion factor[a]
Phosphorus pentoxide	P_2O_5	Bone ash	2.36
Magnesium oxide	MgO	Steatite/soapstone	3.13
Aluminium oxide	Al_2O_3	Kaolin	2.53
Calcium oxide	CaO	Calcite	1.78
Magnesium oxide	MgO	Dolomite	4.61
Lead oxide	PbO	Flint glass cullet	1.6–5.0

[a]The elemental oxide percentage determined analytically needs to be multiplied by this conversion factor to estimate the raw material content in the original porcelain paste recipe

Morgulis & Janacek only in 1931, which must therefore compromise the results of Church and of Eccles & Rackham somewhat, whose analytical work was undertaken and published several decades before this seminal paper appeared.

- Of rather more indeterminate quantification are the experimental or operational errors incurred in the factories in the early 1800s in the compounding of the paste compositions using the basic raw materials for the recipe formulations. An examination of some typical recipes of this time indicate that it was not uncommon to prepare the porcelain paste as a batch using between 50 and 100 lbs of material, perhaps consisting of three or four major components such as china clay, soapstone, flints and bone ash and the question arises as to the precision of weighing of these materials. For example, in John Taylor's book of 1847, *The Complete Practical Potter,* he gives the recipe for Nantgarw porcelain as revealed to him earlier by Samuel Walker (Edwards, *Nantgarw and Swansea Porcelains: An Analytical Perspective,* 2018) in which 26 lbs of bone ash are mixed with appropriate quantities of china clay, river sand and other composites to make about 60 lbs of a mix for fritting and biscuit firing for the first stage process in the kiln. We can only now guess at the precision of measurement of the component weights concerned – for example, an error of up to 1 lb in weight of each component, perhaps, would give a 4% error on each major component such as china clay and sand and even more for the less important components. Simple industrial or agricultural scales would normally have been used for this purpose and the precision of weighing would be determined by the skill of the individual operator. An estimate of these inherent gravimetric errors which could have been operational for the Nantgarw and Swansea factories calculated by the present author and their resultant percentage errors potentially carried over into the porcelain paste recipes is given in Table 2.8. In Appendix I, the actual weights of the combined raw materials used in a kiln firing process are considered and it is concluded that under normal production these would have been scaled up several times.
- Despite all the riders which have been proposed hitherto, perhaps the biggest imponderable which faces the analytical scientist in the interpretation of the compositional data comparisons between different ceramics factories is the accuracy and the veracity of the original recipes themselves. Early porcelain manufacture was an empirical venture, wherein the china works proprietors would necessarily be faced with the need for making changes to their recipes to create an improvement in their ceramic product, driven by the need to have a better translucency, a greater robustness in the finished product and most importantly to minimise the loss of material during the kiln firing procedures which beset all porcelain manufactories. The latter especially was found to be greatest at the Nantgarw China Works, where up to 90% of the biscuit porcelain was lost in each batch through warping or "sagging" during the firing process; the Swansea China Works fared little better with reported losses of 70–75% in this process, and William Billingsley automatically allowed for 15–25% kiln losses at the Pinxton China Works during the close of the eighteenth century in his estimation of production costings provided to the china works proprietor, John

Table 2.8 Estimated potential gravimetric errors in raw material components for porcelain manufacture at Nantgarw and Swansea

Recipe	Raw material	Elemental oxide(s)	Weight /lb	Error/lb	%Error	References
Nantgarw	Sand	SiO_2	14	0.5	4	Taylor (1847)
	Bone ash	CaO; P_2O_5	26	0.5	2	
	Potash	K_2O	2	0.1	5	
	China clay	K_2O; Na_2O:Al_2O_3	20	0.5	3	
	Cumulative error				14	
Swansea (No. 10)	Bone ash	CaO;P_2O_5	24	0.5	4	Dillwyn (1815)
	China clay	K_2O;Na_2O;Al_2O_3	24	0.5	4	
	China stone	MgO;SiO_2;Al_2O_3	24	0.5	4	
	Lime	CaO	3	0.1	3	
	Cumulative error				15	
Swansea (No. 3)	Sand	SiO_2	22	0.5	2	Dillwyn (1815)
	China stone	MgO;SiO_2;Al_2O_3	18	0.5	3	
	Pearl ash	K_2O	12	0.2	2	
	Borax	Na_2O;B_2O_3	6	0.1	2	
	Lead glass	PbO;SiO_2;Na_2O	24	0.5	2	
	Soaprock	MgO:SiO_2	2	0.1	5	
	Cumulative error				16	

Coke. Proprietors often would not divulge this figure for obvious reasons as it could alarm shareholders and investors, but it is suspected that other operators such as Derby, Worcester and Coalport also suffered significant but undeclared losses in the kiln firing of their biscuit wares. It is generally believed that the major reason for these losses was the ineffective control of temperature in the biscuit kilns and inappropriate temperature gradients generated thereby in the kiln chamber vault containing the unfired porcelain. Experiments in high temperature studies of ceramic materials recently undertaken by Professor Victor Owen (2002) have exposed the criticality of the fineness of this maintenance of the kiln temperature – which in many cases, such as that of the Nantgarw China Works, required the maintenance of a temperature in the range between 1380 and 1420 °C for up to 4 days and an equally strict ramping of the temperature (in °C per minute) over some significant period during the heating and cooling procedures – and all of this had to be undertaken by kiln managers without the benefit of high temperature thermometry or pyrometry to control the kiln firing! The use of ceramic saggars of defined composition which melted over a desired range of temperatures as temperature indicators in kilns would have assisted the kiln manager in his visual control of the thermal processing. It is no wonder, therefore, that china works proprietors sought to continually reduce their wastage through empirical experimentation in recipe formulations (Edwards, *Porcelain to Silica*

Bricks: The Extreme Ceramics of William Weston Young, 1776–1847, 2019) through an adjustment of component compositions and the addition of other refractory waste materials, such as "grog". Lewis Weston Dillwyn of the Swansea China Works, during 1815 to 1817, did undertake several modifications to his porcelain paste, assisted by Samuel Walker, who was perhaps the most capable kiln engineer of his time, and he noted his compositional changes carefully in his work books (*L.W. Dillwyn Diaries and Notebooks*, reproduced in Eccles and Rackham, *Analysed Specimens of English Porcelain in the Victoria and Albert Museum Collection*, 1922 and in Edwards, *Nantgarw and Swansea Porcelains: An Analytical Perspective,* 2018). A close inspection of Dillwyn's experimental modifications gives an interesting insight into his logical and scientific approach to the improvement of his porcelain body. Swansea porcelain started as a glassy type body, relying heavily upon the use of ground flint glass or cullet as a component as had that made at Chelsea hitherto, then came the very fine and commercially successful duck-egg translucent porcelain and finally the more robust trident soapstone ware: it is a pity that Dillwyn's achievements in creating a more robust porcelain were operationally successful but only achieved at the expense of the texture and translucency of the resultant body which rendered it undesirable for purchase by its clientele, which eventually and rather rapidly caused the Swansea China Works factory business to fail in 1820. In summary, therefore, the emergent analytical data for any particular china factory must be considered in the light of potentially unrecorded modifications which may have been made during an otherwise stable period in its recipe formulations. In this respect, statements that have been promulgated in the historical literature that Factory X only made one type of porcelain and there is no evidence of any variants being found need to be accepted with some reservation, *cum grano salis!* In some cases statements such as this have been made without the benefit of associated and relevant analytical data and only upon the visual appearance of the artefacts themselves established by connoisseurs, which reinforces the need for a more holistic and scientific approach.

It is in this context that the next chapter will review the early analytical studies of Church (*English Porcelain,* 1894) and of Eccles and Rackham (*Analysed Specimens of English Porcelain in the Victoria and Albert Museum Collection,* 1922) on a range of porcelains and to evaluate comparatively the analytical data obtained in the scenario of the points made above: these studies will then progress to more modern analytical data sets and attempts will be made to place the results in the proper context of factory characteristics and their assignments by comparison with the published and established porcelain recipes and material formulations. Before we leave the subject of analytical determination of the elemental oxides in fired porcelains and the back-calculation of the percentage of raw materials from whence these were derived it is instructive to dissect an example of how this can be achieved: for this purpose a calculation undertaken by Professor Victor Owen (Owen 2002) is illustrative of the procedures adopted and highlights the difficulties encountered in using analytical data for silica percentages, for example, as silica is found to occur in several of the key raw materials used for porcelain production. The conversion of formulaic elemental oxide percentages to wt % of a raw material in a recipe is the crux of the matter being debated and as an illustration of the procedure adopted the estimation of the wt % values of silica and alumina in kaolin can be considered. The ratio of SiO_2 to Al_2O_3 is seen to be 2:1 from the mineralogical chemical formula for kaolin ($2SiO_2.Al_2O_3.2H_2O$); the relative chemical molar masses (molecular weights) of SiO_2 and Al_2O_3 are 60.1 and 101.9, respectively, so each part of alumina in kaolin is matched by $2 \times 60.1/101.9 = 1.18$ parts silica. There are two moles of water (relative molar mass = 18) for every mole of alumina in kaolin, so by the same calculation one part of alumina is matched by $2 \times 18/101.9 = 0.35$ parts of water. Hence, if one now multiplies the analytically determined percentages of silica and water and add these to that of alumina, the amount of kaolin in the original recipe is given. An example of this type of calculation was given by Professor Owen in the same publication (Owen 2002) for a Worcester blue and white enamelled porcelain dish he analysed, namely a specimen labelled W44C from the First Period of manufacture, dating between 1752–1753. The determined alumina percentage was 36%, equivalent to 3.6% in kaolin, so the silica and water percentages were therefore 4.3% and 1.2%, respectively, giving a total for kaolin in the recipe of 9.1% calculated, which compares quite favourably with an average recommended in the First Period Worcester recipe (a soapstone porcelain) of 8.3%. Other components are evaluated as follows for calculated and recorded percentages; talc, 51.8 versus 48.3%, flint glass, 14.2 versus 13.2%, potash, 1.4 versus 1.3%. soda ash, 3.9 versus 3.7%, bone ash, 10.1 versus 9.4%, calcite, 1.4 versus 1.3% and quartz, 15.2 versus 14.2%. When one considers the arguments that have been advanced above regarding the potential sources of error that can arise in the translation of the written recipe raw material percentages into the elemental oxide percentages determined in the fired paste body, the calculated compositional results match closely with those accepted in the literature, and we should not therefore expect to derive precise and exact comparators for the analytical data in this respect.

References

J.A. Anderson, Derby Porcelain and the early English fine ceramic industry, PhD thesis, University of Leicester, UK. (October 2000)

D. Battie, The eyes have it, *Antique Collecting*, 28/9, Foreword page. (March 1994)

A.H. Church, Cantor Lectures on Some Points of Contact Between the Scientific and Artistic Aspects of Pottery and Porcelain, Lecture IV. J Soc Arts, January 14, 126–129, (1880). (Extended in monograph by Trounce Publishers, London, 1881)

Sir A.H. Church, *English Porcelain: A Handbook to the China Made in England During the 18th Century as Illustrated by Specimens Chiefly in the National Collection*, A South Kensington Museum Handbook. (Chapman Hall Ltd., London, 1885 and 1894)

P. Colomban, The destructive/non-destructive identification of enameled pottery, glass artifacts and associated pigments – A brief overview. Arts **2**, 77–110 (2013)

A. Cox, A. Cox, *Rockingham porcelain* (Antique Collectors Club Publishing, Woodbridge, 2005)

P. Craddock, *Scientific Investigation of Copies, Fakes and Forgeries* (Butterworth-Heinemann, Oxford, 2009), pp. 201–210

J. Cushion, M. Cushion, *A Collector's History of British Porcelain* (Antique Collector's Club, Woodbridge, 1992)

E.S. Dana, *Textbook of Mineralogy: An Extended Treatise on Crystallography and Physical Mineralogy,* 4th edn, revised by W.E. Ford (Wiley/Chapman Hall Ltd, New York/London, 1955)

S.H. Davy, Some experiments and observations on the colours used in paintings by the ancients. Philos. Trans. R. Soc. **105**, 97–124 (1815)

L.W. Dillwyn, *Notes on the Experimental Production of Swansea Porcelain Bodies and Glazes.* Made by Lewis Weston Dillwyn with Samuel Walker at the Swansea China Works Between 1815 and 1817. Presented to the Library of the Victoria Albert Museum, South Kensington, London by John Campbell in 1920. Reproduced in Eccles Rackham, *Analysed Specimens of English Porcelain*, 1922, see reference

C. Duval, Francois Descroizelles: The inventor of volumetric analysis. J. Chem. Educ. **28**, 508 (1951)

H. Eccles, B. Rackham, *Analysed Specimens of English Porcelain in the Victoria and Albert Museum Collection* (Victorian and Albert Museum, London, 1922)

H.G.M. Edwards, *Swansea and Nantgarw Porcelains: A Scientific Reappraisal* (Springer, Dordrecht, 2017)

H.G.M. Edwards, *Nantgarw and Swansea Porcelains: An Analytical Perspective* (Springer, Dordrecht, 2018)

H.G.M. Edwards, *Porcelain to Silica Bricks: The Extreme Ceramics of William Weston Young (1776–1847)* (Springer, Dordrecht, 2019)

I.C. Freestone, A technical study of Limehouse ware, in *Limehouse Ware Revisited*, ed. by D. Drakard, (English Ceramic Circle, Milton Keynes, 1993), pp. 68–78

G.A. Godden, *An Illustrated Guide to British Pottery and Porcelain* (Barrie and Jenkins, London, 1980)

G.A. Godden, *Godden's Guide to English Porcelain* (Wallace-Homestead Book Co., Greensboro, 1992)

B.K. Hobbs, New perspectives on soapstone. Trans. English Ceram. Circle **15**, 368–392 (1995)

L. Jewitt, *The Ceramic Art of Great Britain from the Prehistoric Times Down to the Present Day,* vols. I and II. (Virtue Co. Ltd., London, 1878)

J.E. Jorpes, *Jacob Berzelius-His Life and Work,* translated from the Swedish by B. Steele, Almqvist & Wiksell. (Stockholm, 1966/University of California Press, Berkeley, 1970)

S. Morgulis, E. Janacek, Studies on the chemical composition of bone ash. J. Biol. Chem. **93**, 455–466 (1931)

M.V. Orna, Chemistry at the interface of archaeology and art. Chem. Aust., 470 (1996)

J.V. Owen, On the earliest products (ca. 1751–1752) of the Worcester porcelain manufactory: evidence from sherds from the Warmstry house site, England. Hist. Archaeol. **32**, 63–75 (1998)

J.V. Owen, Antique porcelain101: a primer on the chemical analysis and interpretation of eighteenth century British wares, in *Ceramics in America*, ed. by R. Hunter, (Chipstone Foundation, Milwaukee, 2002), pp. 39–61

J.V. Owen, M.L. Morrison, Sagged Phosphatic Nantgarw Porcelain (ca. 1813–1820): casualty of overfiring or a fertile paste? Geoarchaeology **14**, 313–332 (1999)

J.V. Owen, J.O. Wilstead, R.W. Williams, T.E. Day, A tale of two cities: compositional characteristics of some Nantgarw and Swansea porcelains and their implications for Kiln wastage. J. Archaeol. Sci. **25**, 359–375 (1998)

A.M. Pollard, From bells to Cannon – the beginnings of archaeological chemistry in the eighteenth century. Oxford J. Archaeol. **32**, 335–341 (2013)

A.M. Pollard, Letters from China: a history of the origins of the chemical analysis of ceramics. AMBIX **62**, 50–71 (2015)

W.R.H. Ramsay, E.G. Ramsay, A classification of Bow Porcelain from first patent to closure: c.1743–1774. Proc. R. Soc. Vic. **119**(1), 1–68 (2007). ISSN 0035–9211-1-168

J. Sandon, *The Phillips Guide to English Porcelain of the 18th and 19th Centuries* (Merehurst, London, 1989)
J. Taylor, *The Complete Practical Potter* (Shelton, Stoke-upon-Trent, 1847)
M.S. Tite, M. Bimson, A technological study of English porcelains. Archaeometry **33**, 3–27 (1991)
J. Twitchett, Derby Porcelain: 1748–1848, in *An Illustrated Guide*, (Antique Collectors Club, Woodbridge, 2002)

Chapter 3
Appraisal of the Earliest Chemical Analyses of Sir Arthur Church (1894) and of Herbert Eccles & Bernard Rackham (1922)

Abstract Elemental oxide compositional data of selected porcelains from the earliest publications using destructive wet chemical digestion methods of 100 years ago to the latest instrumental analyses and microanalyses are compared and several case studies are considered for specified factories which are correlated with known factory recipes and formulations.

Keywords Wet chemical digestion · Elemental oxide percentages · Early versus modern analyses · Caughley · Coalport · Nantgarw and Swansea factories · Recipes · Replicates · Destructive sampling

It is reasonable to start an evaluation of the analytical data obtained from eighteenth and early nineteenth century English and Welsh porcelains by considering in detail the results published by Sir Arthur Church in 1894 and by Herbert Eccles and Bernard Rackham in 1922 as these were undoubtedly the earliest forerunners of the more recent investigations carried out on the products of a number of ceramic factories. The major problem analytically centres on the rather small numbers of specimens studied for albeit a significant representation of relevant porcelain manufactories: this presumably is directly correlated with the requirement that significant quantities of sample were required for the wet chemical analysis of each porcelain specimen – which resulted in permanent and visible damage being caused to the specimens, only some of which could be described as already damaged or in a bad state of repair. In many cases, only one specimen of a particular type of porcelain was analysed from each factory, or perhaps two specimens providing analytical replicates: in the latter case it is at least possible to estimate an average percentage elemental oxide composition with a minimal attempt being made to provide a mathematically significant sigma standard deviation or error bar on the experimental results. As will be seen later, some more recent analytical work on factory porcelains has utilised multiple specimens of shards from waste tips excavated archaeologically at the factory sites and the data reported are then better suited to the estimation of experimental errors and their associated standard deviations as several

H. G. M. Edwards, *18th and 19th Century Porcelain Analysis*,
https://doi.org/10.1007/978-3-030-42192-2_3

specimens were employed in each of the studies. Several authors have commented (for example, Owen 2002) that a relatively large number of shards have found their way into archives from the casual foraging carried out by amateur collectors at old, abandoned factory sites and these have thereby now lost their archaeological stratigraphical context, which is so essential for the correct placement of their discovery and the establishment of an archaeological timeline for each artefact. In contrast, the acquisition of properly excavated and relevant archaeological shards from sites of porcelain manufactories gives a much more reliable credibility to the conclusions formed from the analytical studies, which may give novel information about the products and this could even then still be in contradiction with the established historical appreciation of porcelain manufactured during the site production.

In contrast to the wider range of the results quoted in Tables 2.3 and 2.4 for the analytical data published by Church (*English Porcelain*, 1894) and by Eccles & Rackham (*Analysed Specimens of English Porcelain in the Victoria and Albert Museum Collection*, 1922), respectively, Table 2.5 gives the analytical data from the publication in 1922 by Eccles & Rackham of their experimental determinations of elemental oxide compositions from "multiple" specimens of porcelain in the Lady Charlotte Schreiber Collection at the Victoria and Albert Museum in South Kensington, London. As stated previously, although 36 different specimens of porcelains from 16 different factories are encompassed in their study, Eccles & Rackham made analyses of multiple specimens from only 7 factories, namely, Chelsea, Bow, Derby, New Hall, Worcester, Nantgarw and Swansea: this effectively consisted of only two specimens from each factory, except for Worcester, where multiple specimens from three different periods of manufacture were identified and analysed in all amounting to seven samples, one period therefore offering three specimens for analysis. In total, therefore, Eccles & Rackham cite a total of 19 different analytical determinations incorporating replicate analyses from the seven identified factories. These are described in detail in Table 2.5.

This highlights the major problem encountered in the historical appreciation of the analytical science of porcelains, namely, that very few multiple analyses are available for the provision of effective standard deviations of experimental errors in the compositional percentages of the elemental oxides. Thus, the determination of the accuracy of the percentage composition for each type of porcelain is rendered difficult unless larger numbers of specimens have been interrogated – and the difficulty in achieving this, especially in the early days of chemical analysis, is that each analysis could be undertaken only at the expense of the destruction of a significant part of each porcelain specimen, which is unacceptable for rare *objets d'art*. An idea of the number of specimens collected in this way for destructive analytical sampling can be appreciated from Tables 1.3 and 2.1: Church quotes data for 5 specimens from 5 factories, and Eccles & Rackham quote data from 36 specimens originating from 16 factories – in all, therefore, some 41 individual finished, decorated and ostensibly perfect pieces of porcelain from museum collections were sacrificed, with possibly some exceptions which were already damaged but still perhaps were visually acceptable for display. Clearly, further recourse to multiple specimen acquisition beyond that achieved in these early analyses would have been judged impractical and extremely wasteful on the specimens in the museum collections.

Table 2.5 gives the cumulative data from the multiple analyses of Eccles & Rackham for the 7 identified porcelain factories from which it is now possible to evaluate the possibility of usage of such data to discriminate forensically and definitively between the identified factories' output. Also, in Table 2.6, the combined published analytical data for two porcelain factories which were only in production for a relatively short time in the second decade of the nineteenth century, namely Swansea and Nantgarw, are revealed, encompassing five distinct and separate analytical studies carried out over almost a century between 1922 and 2018; this effectively introduces a combination of early and modern analytical data which will form a good basis for a comparative assessment. The Swansea China Works operated from 1814 until 1819 and that of Nantgarw from 1817 until 1820, during which time it has been maintained by ceramics historians that each factory was limited in its production to only one, two or three different types of porcelain: it is widely believed currently that the Nantgarw China Works produced a phosphatic porcelain of a single composition with no variation made solely during its lifetime, and that the Swansea China Works produced first a glassy porcelain in early trials, which was superseded in 1817 by its famous duck-egg phosphatic china, which was then followed finally by its rather unsuccessful trident soapstone magnesian porcelain as a replacement introduced commercially between about 1818 and the final closure of the factory in 1820 (Edwards, *Nantgarw and Swansea Porcelains; An Analytical Perspective*, 2018). It should be appreciated, therefore, that the analytical data from the compositional studies on porcelains and shards from these two factories would be favoured by their apparently simplistic and well-conceived formulations which are known and broadly documented historically. Table 2.6 gives a summary of the analytical determinations undertaken between 1922 and 2018 for Nantgarw and Swansea china, comprising some 44 individual specimens in total. The analysts are given in the Table along with the averaged mean results of their studies and additionally an arithmetically weighted mean with the appropriate sigma error limits for each type of porcelain studied. The analytical investigations were accomplished chronologically by Eccles and Rackham (1922), Tite and Bimson (1991), Owen et al. (1998), Owen and Morrison (1999) and Colomban et al. (2020). Of the 44 specimens studied from these two factories, only 7 analyses were undertaken on finished, completed porcelain items, comprising 3 Nantgarw and 4 Swansea pieces; of the former, one specimen, a saucer, was from the named *Duncombe* service (residing in the Victoria and Albert Museum South Kensington, London) and from the latter, one large platter was from the *Biddulph* service of duck-egg porcelain (residing in the Royal Institution of South Wales, Swansea), both having an ascribed date of production of between 1817 and 1819.

3.1 Summary of the Analytical Data for the Nantgarw and Swansea Factories

All the analytical chemical data for Nantgarw and Swansea porcelains which has been achieved over the last 100 years is summarised in Table 2.6: chronologically, this includes the work of Eccles & Rackham, Tite & Bimson, Owen et al., Owen & Morrison and the latest results of the author (Colomban et al. 2020). The total number of specimens analysed is 44, of which 7 are represented by finished and decorated porcelain pieces, the remainder being broken shards recovered from waste pits or dumps at the two factory sites. The Nantgarw China Works site fell into disuse after the departure of William Weston Young following the death of Thomas Pardoe in 1823 and re-opened for earthenware production in 1833 under the proprietorship of Pardoe's son, William Henry, after which it continued until it eventually closed in the 1920s under the stewardship of a family descendant, Percival Pardoe (Edwards, *Porcelain to Silica Bricks: The Extreme Ceramics of William Weston Young, 1776–1847*, 2019; Edwards, *Nantgarw and Swansea Porcelains: An Analytical Perspective,* 2018). It was very fortunate that in 1931 Isaac Williams, a professional archaeologist working at the National Museum of Wales, Cardiff,, was commissioned to undertake the first excavation of the now derelict site, where he discovered the china waste pit with shards that could be dated from the first production of porcelain commercially at Nantgarw by William Billingsley and Samuel Walker in 1817 (Williams, *The Nantgarw Pottery and its Products: An Examination of the Site,* 1932). The latest archaeological excavation of the site was undertaken by the Glamorgan Archaeological Society in 1995 and further supplies of shards were sent to the archive at Tyla Gwyn, the original residence of William Billingsley and now the home of the Nantgarw China Works Trust Museum. These latest shards were analysed by Colomban, Edwards and Fountain (Colomban et al. 2020) and unusually, several shard specimens from the excavation of the waste pit were found to be from glazed and decorated porcelain: normally, most if not all of the shards found at the factory site were of biscuit porcelain which originated from damage incurred during the initial high-temperature firing in the primary kiln (when most wastage through "sagging" was incurred) and it would be expected that normally significantly fewer shards would have arisen from the glost kiln used for the lower temperature glazing process following decoration. At Swansea, however, the site of the China Works in the Hafod district following the factory closure in 1820 was taken over for urban redevelopment and much of the site information has now been lost, including the location of the factory waste tip, so shards are not locatable in an archaeological context as they are at Nantgarw, where Isaac Williams could identify a cylindrical pit some thirty feet south-east of the original potting shed which was still standing at the time of the excavation (Williams, *The Nantgarw Pottery and its Products: An Examination of the Site*, 1932). Williams discovered that the Nantgarw waste pit (the stratigraphy of which will be discussed in more detail later) contained six discrete levels over about a metre vertical transect, the bottom level containing wasters which could be assigned to the first porcelain production at Nantgarw of

William Billingsley and Samuel Walker (Edwards, *Porcelain to Silica Bricks: The Extreme Ceramics of William Weston Young, 1776–1847*, 2019). In contrast to the Nantgarw waste pit stratigraphy, at the Swansea site, shards have been found in scattered locations.

In Table 2.6, the identification of several types of porcelain by the recent analytical studies means that the first conclusion to be drawn from the analytical data is rather controversial because historically the limited periods of production by the Nantgarw China Works and the Swansea China Works has given rise to the supposition that:

- At Nantgarw, William Billingsley, William Weston Young and Samuel Walker between 1817 and 1819 made just one porcelain body; with the departure of Billingsley and Walker for the Coalport China Works in 1820, William Weston Young engaged Thomas Pardoe to decorate the remaining stock stored "in the white" for local sale to recoup finances in an unsuccessful attempt to save the business, which eventually closed in 1823. It has always been believed that when Billingsley and Walker found the best formulation for their exquisite porcelain they would have had neither the time nor the capability to try further improvements when they were working full time to satisfy the demands of clientele through John Mortlock's agency in Oxford Street, London, especially given the appallingly high kiln wastage they were experiencing at Nantgarw (Edwards, *Nantgarw Porcelain: The Pursuit of Perfection,* 2017; Edwards and Denyer, *William Billingsley: The Enigmatic Porcelain Artist, Decorator and Manufacturer*, 2016). It has also been maintained, or rather assumed, that Young and Pardoe were decorating the remnant china and did not try to initiate a renewal of the manufacturing process – especially since Billingsley and Walker had preserved an absolute secrecy in their recipe formulation for Nantgarw porcelain and had kept their business partner, Young, out of the picture. Young, therefore, would have been judged incapable of manufacturing Nantgarw porcelain after the departure of Billingsley and Walker as he obviously did not have the essential recipe and formulation, so, Nantgarw porcelain would be anticipated reasonably to have had just one porcelain body formulation and composition. This is certainly an issue that could be verified analytically in support of these hypotheses.
- At the Swansea China Works, Lewis Weston Dillwyn tried to manufacture porcelain unsuccessfully between 1811 and 1813, but in 1814 he engaged William Billingsley and Samuel Walker upon their departure from the Royal Worcester China Works followed by their brief unsuccessful operational sojourn in Nantgarw, hereafter called Nantgarw Phase 1. The first experimental porcelain made at Swansea in 1815 and 1816 was a rather heavy bodied glassy porcelain, which probably contained a significant proportion of flint glass frit or ground cullet, which then progressed in 1817 to the renowned and highly translucent duck-egg porcelain body after trial experiments in porcelain body formulation undertaken by Dillwyn and Walker (Edwards, *Swansea Porcelain: The Duck-Egg Translucent Vision of Lewis Dillwyn,* 2017). This new porcelain was well-received by a discerning clientele but persistent high kiln wastage forced Dillwyn

to try to improve upon its robustness: he reduced the china clay and bone ash component proportions in the body composition and added soapstone, giving rise to a "trident ware" which although technically was much more robust was texturally inferior to the duck-egg porcelain it replaced (Dillwyn, *Work Books and Notes, 1815–1817*, reproduced in Eccles and Rackham, *Analysed Specimens of English Porcelain in the Victoria and Albert Museum Collection*, 1922 and in Edwards, *Nantgarw and Swansea Porcelains; An Analytical Perspective,* 2018, Appendix II). The London-based clientele did not appreciate this bodily change and by 1820, Dillwyn was forced to sell out to Timothy and John Bevington, who arranged for the decoration and sale of unsold stocks of remaining duck-egg and trident porcelains before the final closure of the Swansea China Works took place in 1823.

- The analytical data presented in Table 2.6 indicate that actually no fewer than six different porcelain bodies are evident from the compositional percentages and these can be summarised as follows.
- Nantgarw has three distinct body compositions: soft paste porcelain (25 specimens), medium-high silica porcelain (3 specimens) and high-silica porcelain (5 specimens).
- Swansea porcelain has three distinct body compositions: soft paste porcelain (2 specimens), high-silica porcelain (7 specimens) and a soapstone trident (magnesian) porcelain (2 specimens).
- Technically, therefore, the first important conclusion is that, contrary to the perceived wisdom and belief which has been held historically over many years, it appears that there was more than one porcelain body conceived at the Nantgarw China Works – namely, the standard soft paste phosphatic body, one of a medium – high silica content and another of high silica content, the former medium silica composition having a significantly lower phosphate(bone ash) component and the latter high silica composition realistically containing no phosphate at all (analytically indeterminate with a measured phosphorus oxide percentage of 0.2+/−0.2%). From the shards analyses it also appears that a Swansea body existed which was intermediate between the phosphatic duck-egg body and the soapstone (trident) body with a high silica component and effectively containing no phosphate component. It could be conjectured that this intermediate composition body would have been the result of Dillwyn's series of experimental trials to improve the robustness of his porcelain.
- The silica content of each type of porcelain defines its category: the soft paste porcelains have an average of 45+/−3% silica (based on 27 specimens), medium-high silicaceous porcelains have 64 +/−5% silica, highly silicaceous porcelains have 76+/−3% silica and the highest silica content is determined for the soap-rock/soapstone porcelains at 83+/−1% silica. These high silica bodies are very reminiscent of the silica component composition of true hard paste (Chinese) porcelains.
- Aside from the silica percentages, significant differences in porcelain body composition can be observed for the alumina, lime and phosphorus pentoxide determinations for the various porcelain body types. For example, the alumina

composition varies between 15+/−1% for soft paste Nantgarw to 23+/−3% for soft paste Swansea and 19+/−4% and 11+/−1% for medium and high-silica Nantgarw to 24+/−2% for the high-silica Swansea bodies. The lime content varies from 21+/−2% for soft paste Nantgarw compared with 15+/−2% for soft paste Swansea and 7+/−3% and 0.4+/−0.1% for the medium and high silica bodied Nantgarw specimens against 0.3+/−0.1% for the high-silica Swansea variant. The phosphatic component percentages are revealing in that the Nantgarw soft paste porcelain specimens have an average of 17+/−1% against the value of 12+/−2% for the Swansea analogue, representing a bone ash raw material component addition of approximately 44 and 30%, respectively, for the Nantgarw and Swansea porcelains. A most interesting value for the elemental oxide composition is derived for the medium silica Nantgarw variant which still has a significant phosphatic component at 6+/−2%, correlating with a bone ash recipe component of approximately 15%, whereas the high silica variant is essentially phosphate-free for both the Nantgarw and Swansea bodies. As expected, the soft paste porcelains of both Swansea and Nantgarw and their medium to high silica variants all show an analytical deficiency of magnesia, which is frequently indeterminate experimentally within the error bars of the experimentally weighted accuracies. In contrast, the magnesia values of the soapstone Swansea trident body paste are significantly larger and equate to about a 10% formulation of soapstone in the original recipe. The presence of lead oxide is not detected in all the specimens studied: this is as expected for Nantgarw, where the available information about the original body recipe excludes the addition of flint glass cullet (which can contain up to 60% lead oxide) but is more surprising perhaps for the Swansea pastes as Dillwyn's workbooks specifically mention his use of flint glass cullet in some of his recipes. The total Group I elemental alkali content is reasonably consistent between the Nantgarw and Swansea variants, summing at approximately 3% or less.

- The exclusion of a Swansea glassy porcelain variety from the collected data arises because there was an absence of appropriate shards belonging to this category. It is known that the earliest production of such a porcelain did occur at Swansea but as yet no fragments or shards have been presented from the specimen archives for analysis: it would be expected that a significantly large percentage of lead oxide would be indicated analytically for such shards, reflecting the presence of flint glass cullet as a component therein, as indicated in Dillwyn's notes on his recipes.

The summarised analytical data therefore would indicate that Nantgarw had three distinct body formulations or types and Swansea had four, including a glassy version.

3.2 The Analytical Study of Nantgarw and Swansea Porcelains: Implications for a Wider Interpretation in Ceramics Analysis

The detailed discussion which has been carried out above on the analytical interpretations of the chemical compositional data from the Nantgarw and Swansea manufactories reveals that even for the strictly limited output from these two china works, which was accomplished during the narrow time frame between 1817 and 1819/20, there seemed to have been several hitherto unrecorded attempts made to modify the known porcelain paste recipes outside of the already known and documented changes made historically and that these actually resulted in a tangible production from the shards found at the two sites. Several factors now need to be considered to establish the correlation between the analytical results and recipe formulation: a critical parameter in this must be the precision with which the known paste recipes were made up in the china manufactory. Unfortunately, the effect of this upon the analytical reliability is something that has not been fully considered hitherto, yet its influence upon the compositional analyses could be extremely important as we have discussed in the preceding Chapter. In a previous publication, Edwards (*Nantgarw and Swansea Porcelains: An Analytical Perspective*, 2018) has drawn attention to the quantitative effect of the experimental errors incurred in making up the formulation from a given recipe and these will now be explored further here. The fundamental basis of the analytical method is the weighing of a sample which will then be analysed qualitatively and quantitatively for its chemical components: this affords the analyst the capability of then estimating quantitatively the percentage composition of the original specimen from the determined data – without the initial weight determination the elemental oxide estimations are necessarily obtained relative to each other and therefore become only semi-quantitative in nature, even with the desired instrumental calibrations being undertaken against standards. Whereas the weighing of a specimen on an analytical balance in a laboratory can be achieved to an accuracy of routinely one tenth of a milligram (representing 10 ppm on a sample of 10 g, or 0.001%), this can represent an inconsequently small error for analytical samples of only a few milligrams to one gram, representing perhaps an experimental error of only 0.01% or significantly smaller. However, in a nineteenth century china works, it is presumed that at best industrial or agricultural weighing scales would have been employed of a much lower precision than laboratory analytical balances and then perhaps a component kiln component charge in a recipe of 22 lbs. (10 kilos) would only have been weighed out to perhaps the nearest 0.5 lb. (0.25 kilo), which represents already an error in that particular component of some +/−2–3%! For smaller quantities, of only a lb. or so, approximately 0.5 kilo, then weighing to an accuracy of just one ounce (~ 30 g) will produce a compositional error of about +/− 6–7%. With these errors in mind, it can be appreciated that the sigma error bars already reproduced in the average elemental oxide compositional percentages in Table 2.6, for example, are not at all excessive! In this context, some potential errors that could be incurred from weighing practices on component compositions in several well-recorded recipe formulations for Swansea and Nantgarw paste mixtures are given in Table 2.8. It can be seen that even with the estimated errors in weighing for the individual raw materials components outlined above, a conservative calculation of the cumulative error from this source could be approximately +/−15%. Three recipes are cited in Table 2.8 for illustrative purposes only and these particular examples have been selected on account of their published formulations: the Nantgarw recipe was kept a close secret and was never revealed in William Billingsley's lifetime, but it was published by John Taylor in 1847 (*The Complete Practical Potter,* 1847), who maintained that it came directly from Billingsley's associate and partner at the Nantgarw China Works, Samuel Walker. In the case of the two Swansea recipes, both were obtained from the workbooks of Lewis Weston Dillwyn (detailing his experimental changes in trials carried out between 1815 and 1817 as reproduced in Eccles & Rackham, *Analysed Specimens of English Porcelain in the Victoria and Albert Museum Collection,* 1922) and represent two of the three well-known and identified bodies produced at the Swansea China Works between 1815 and 1819, namely, the esteemed phosphatic duck-egg porcelain and the less well-appreciated but significantly more robust "trident" soaprock ware (Edwards, *Nantgarw and Swansea Porcelain: A Scientific Reappraisal*, 2017).

However, the story does not end there because there are several hidden sources of further error that can occur in paste formulations which cumulatively add to the total error and which could increase the potential discrepancy between the analytically determined data and those provided by the china works documentation in existence. The major components of a porcelain paste could include sand, china clay (kaolin, ball clay), flints, lime, potash, soda ash, calcined bone ash and soapstone, with perhaps the addition of lead glass cullet. The purity of the raw materials establishes further possibilities for error in that different states of hydration, volatile organic content, the presence of impurities and differential levels of metal oxides such as iron (III) oxide (haematite) can also contribute to the overall picture, although it is accepted that china manufactory proprietors did appreciate the essential need for use of the highest purity materials which they would acquire from great distances even if poorer quality locally sourced analogues were available. Perhaps another and potentially significantly large source of error arising from the raw materials in making up the composition of the porcelain body is their variable hydration: bulk storage and unprotected transportation will also incur dramatic changes in this parameter, as will the initial procedures whereby the raw materials are calcined or thermally and mechanically treated prior to making up the frit. Variable amounts of water in a raw material component will inevitably cause an error which is effectively hidden from the manufacturer – and this could easily account for several percentage points additionally in the cumulative error calculation, aside from other impurities such as iron oxide in the sand, for example, and the small, and perhaps even variable, amounts of additive minerals or compounds which are then necessarily added to decolourise the mixture, such as smalt.

3.3 Analytical Data and the Definition of a Factory of Porcelain Production from the Elemental Oxide Percentage Data: A Case Study for Caughley and Coalport Porcelains

In the light of the potential problems encountered in the estimation of the experimental errors attached to the determination of the elemental oxide percentages in fired porcelain specimens, the question should be addressed as the reliability of the analytical data for the determination of the source factory. There are several aspects to be considered in this, but before the detailed interpretation of the analytical data is undertaken an example of the potential power of the application of analytical techniques to porcelain analysis can be cited. In 2003, Victor Owen and John Sandon published a paper (*Post-Mediaeval Archaeology,* 2003) in which they analysed shards from two porcelain factory sites, Caughley and Coalport, and determined the elemental oxide percentages: the Caughley specimens comprised two types of porcelain, namely a magnesian porcelain (12 shards) and a true hard paste porcelain (8 shards), whereas the Coalport specimens again comprised two types of porcelain, but now consisting of a phosphatic porcelain (17 shards) and a true hard paste porcelain (18 shards, of which one, specimen COAL19, appeared actually to be an earthenware and not a porcelain). Of particular relevance to our preceding discussion concerning Nantgarw and Swansea porcelain analyses, Owen and Sandon then went on to compare the results for the Coalport phosphatic shards, which were produced during the decade when William Billingsley and Samuel Walker were employed by John Rose at Coalport, with the porcelain that they produced for Dillwyn firstly at Swansea and then for themselves and William Weston Young at Nantgarw. In the author's opinion this provides a very appropriate and linking theme for our initial consideration here, especially if the differential percentages in the elemental oxide compositions can eventually be reliably related to factory sourcing, i.e. do the analyses of elemental oxide composition support a procedure and protocol for the discrimination between Nantgarw, Swansea and Coalport porcelains which could be used in future to assign pieces of dubious provenance or of a doubtful source origin? This is particularly important for these three factories as the assignment of pieces in museum collections and those which occasionally surface in ceramics auctions today often promote an ongoing debate as to their origin – which can happen where much of the output of a factory was unmarked: at Swansea, for example, only a few pieces of a service were marked and some "SWANSEA" marks have nowadays been questioned for their authenticity, whereas at Nantgarw only the plates and flatwares were impressed with the mark "NANTGARW C.W." and at Coalport many pieces bore only an enamelled script pattern identification mark such as "2/831". The application of false identity marks by unscrupulous fakers to convert an ordinary piece of porcelain into a valuable and perhaps "unique" example is prevalent and the potential purchaser and curator should be aware of this, *caveat emptor*!

3.3.1 Similarity Between Factory Products

To illustrate this particular similarity between the Nantgarw, Swansea, Coalport and Derby factories for style of decoration, it is instructive to compare the four selected plate specimens originating from each factory, which all have a moulded edge border of foliage, flowers stars and ribbons which is often referred to generically as "*Nantgarw-like*". Figures 3.1, 3.2, 3.3 and 3.4 show three specimens of porcelain which all have this "Nantgarw-like" moulded border along with an analogous example of the genuine Nantgarw porcelain. Figure 3.1 is a Nantgarw plate from the *Lady Seaton* service, with an impressed mark NANT-GARW C.W., *ca.* 1817–1819, formerly in the possession of Lady Seaton of Bosahan, Cornwall, which was hand decorated in London with cobalt blue enamelled floral sprays and dentil-edged gilding (Edwards, *Swansea and Nantgarw Porcelains: A Scientific Reappraisal*, 2017). Figure 3.2 shows a Swansea plate from a service locally decorated by William Pollard in the finest duck-egg porcelain body, *ca.* 1815–1817, and is referred to as having a border similar to that found on Nantgarw china with five vignettes containing enamelled exotic birds, marked SWANSEA in red stencil capitals on the reverse. Figure 3.3 shows a Coalport plate with a similar moulded border, decorated with a central landscape and flowers in five reserves in the verge, marked with John Rose's stamp, *ca.* 1820, celebrating his award for the institution of a non-toxic porcelain glaze in the early 1820s. A plate from the Derby China Works factory of Robert Bloor also dating from 1820 which is part of the service delivered to *Lord Ongley* and marked with the Bloor Derby crown and circle mark in red enamel is shown in Fig. 3.4. It is recorded that Robert Bloor had paid a visit to John Sims' atelier in

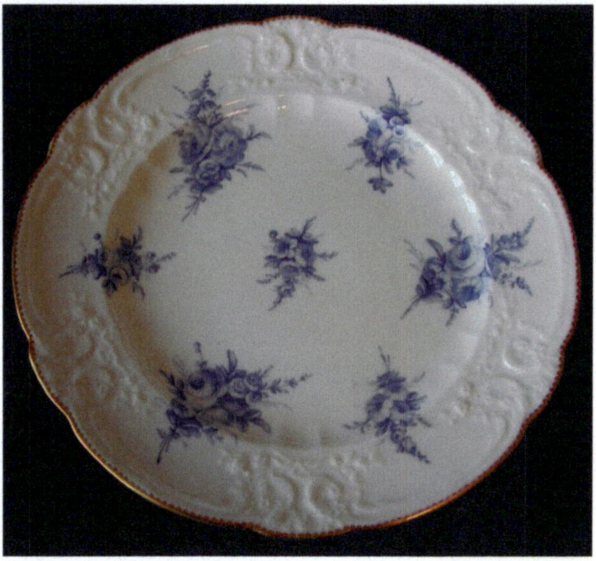

Fig. 3.1 Nantgarw porcelain, dinner plate, *Lady Seaton* service, *ca.* 1817–1819, whose nomenclature is derived from the previous ownership of Lady Seaton of Bosahan, Cornwall, marked impressed NANT-GARW C.W., underglaze blue pattern, London decorated, dentil gold edging. *Private Collection*

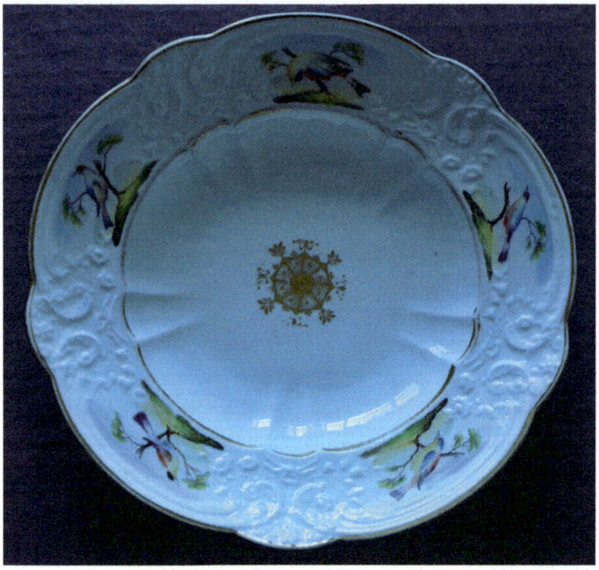

Fig. 3.2 Swansea deep dish with Nantgarw-type floral embossed moulding, decorated by William Pollard with five vignettes containing exotic birds in foliage and showing the red stencilled SWANSEA mark on its base, *ca.* 1817–1820. When viewed by transmitted light the characteristic duck-egg colouration of the highest quality Swansea porcelain is also clearly seen. *Private Collection*

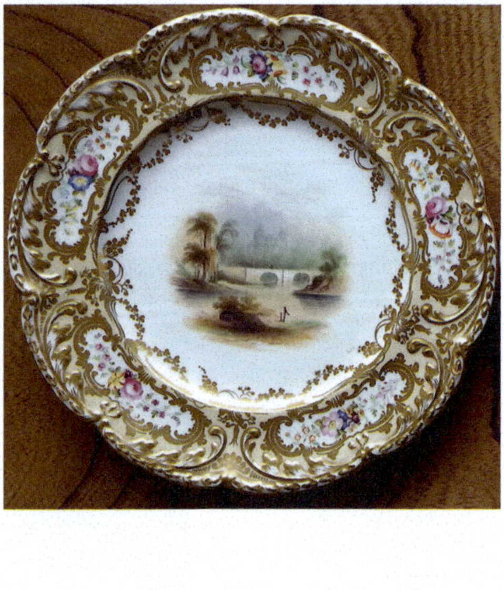

Fig. 3.3 Coalport China Works dessert plate with Nantgarw type moulded border in rococo style beautifully decorated with five floral bouquets in reserves and central landscape; probably manufactured in the 1820s having a Swansea type phosphatic paste and translucency but bearing an early 1840s pattern mark, 4/782. *Collection of Dr and Mrs Morgan Denyer*

London and saw the beautiful enamelling and decorative work being undertaken there by James Plant on Nantgarw porcelain, *ca.* 1819. He purchased several finely decorated plates of Nantgarw porcelain from the Sims workshop as examples and instructed his production team at Derby to simulate this design and the type of

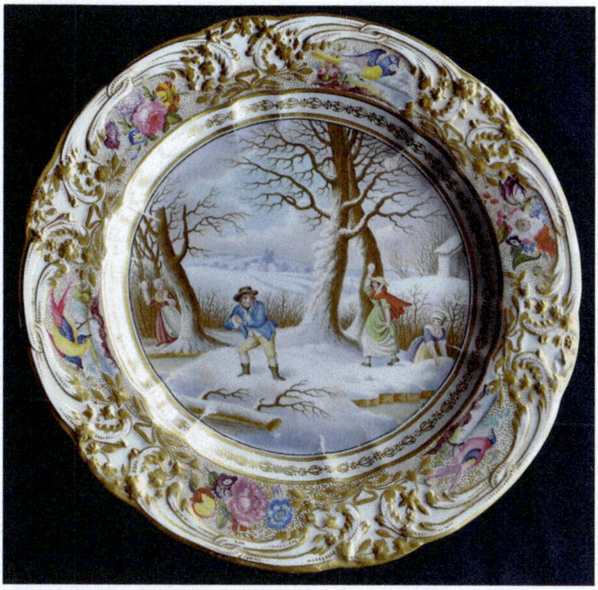

Fig. 3.4 Dessert plate from the sumptuous *Lord Ongley* service, Derby porcelain, Robert Bloor period, *ca.* 1820, with Nantgarw- type moulded border and inspired by James Plant's Nantgarw porcelain decoration at John Sims' atelier, London, *ca.* 1817–1819, showing children playing at snowballing and vignettes of birds, fruit, flowers and butterflies. Marked with the Bloor Derby crown and double circle mark in red enamel. *From the Private Collection at Muncaster Castle, Cumbria. Reproduced with kind permission of Peter Frost-Pennington Esq*

decoration for the *Lord Ongley* service, which was one of the most expensive services ever commissioned at Derby (Edwards, *Derby Porcelain: The Golden Years, 1780–1830*, 2017; Edwards, *Swansea and Nantgarw Porcelain: A Scientific Reappraisal,* 2017). The Nantgarw China Works original and the forerunner of this particular *Lord Ongley* type Derby plate is now in the collection of the National Museum of Wales, Cardiff, whereas this particular plate and the remaining surviving porcelain items from the Derby *Lord Ongley* service now resides in the collection of Peter Frost-Pennington Esq. at Muncaster Castle in Cumbria (Edwards, *Swansea and Nantgarw Porcelain: A Scientific Reappraisal*, 2017). Another example of the fine moulding and decoration on the *Lord Ongley* Derby service is provided in Fig. 3.5.

The remaining survivors of the sumptuous *Lord Ongley* service at Muncaster Castle, comprising only some eight plates and two sauce boats with their stands, are shown in their magnificent period mahogany display cabinet in Fig. 3.6, from which an idea of their beauty and original impact as a complete dessert/dinner service in the 1820s can be ascertained. In their seminal text on Swansea porcelain Jimmy Jones and Sir Leslie Joseph (Jones & Joseph, *Swansea Porcelain,* 1988) have studied in detail the intricate designs of these particular mouldings utilised by

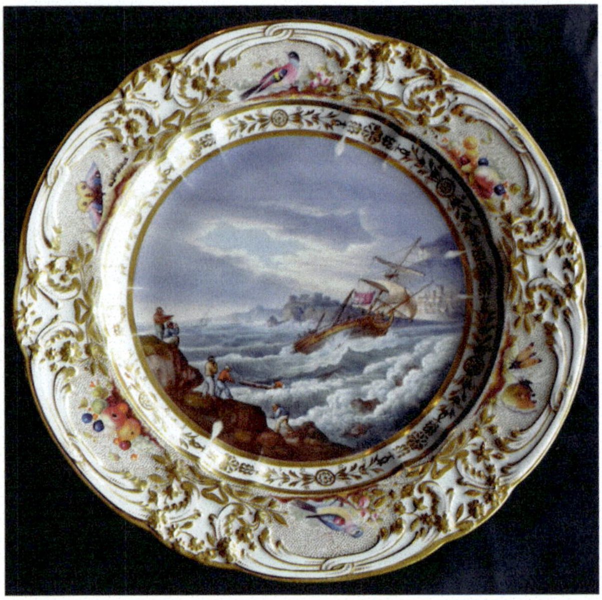

Fig. 3.5 Dessert plate from the sumptuous *Lord Ongley* service, Derby porcelain, Robert Bloor period, *ca.* 1820, with Nantgarw- type moulded border and inspired by James Plant's Nantgarw decoration at John Sims' atelier, London, *ca.* 1817–1819, showing a naval scene with a man-o'-war in heavy seas and vignettes of birds, fruit, flowers and butterflies. *From the Private Collection at Muncaster Castle, Cumbria. Reproduced with kind permission of Peter Frost-Pennington Esq*

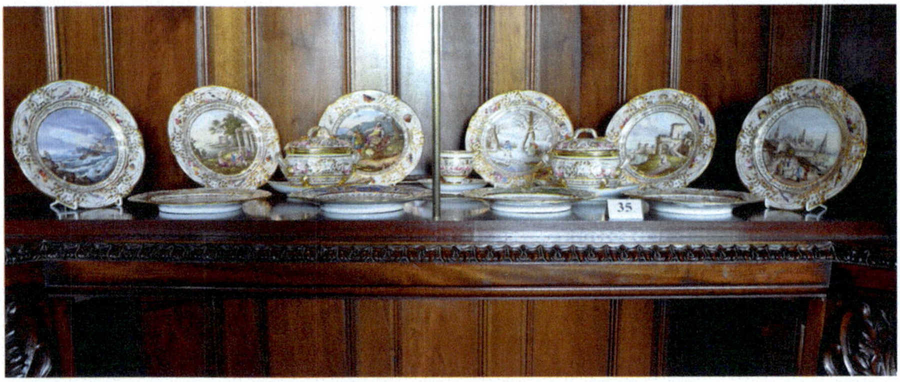

Fig. 3.6 Mahogany cabinet at Muncaster Castle, Cumbria, containing all surviving remnants of the *Lord Ongley* service: comprising eight dessert plates and two sauce boats with lids and stands. Also included with the *Lord Ongley* porcelain in this cabinet are two associated Bloor Derby plates decorated in the Sevres style and a Swansea porcelain cabinet cup and saucer. *From the Private Collection at Muncaster Castle, Cumbria. Reproduced with kind permission of Peter Frost-Pennington Esq*

Table 3.1 Nantgarw -type edge moulding – distinctive characteristics for China factories

	Factory			
Moulding Detail	Nantgarw	Swansea	Coalport	Derby
1. RH C-scroll florets	2	1	3	1
2. Ribbon bow RH upper	Twist	–	Thick	–
3. Ribbon bow RH lower	Thicker	–	Thick	Thin
4. Ribbon bow RH third ribbon	–	–	1	–
5. Floral strand LH lower; closest to ribbon	Floret	Bud	Bud +2 leaves	–
6. End of LH lower strand	Bud	Floret		Floret
7. Floral strand end RH lower floret	Floret	Floret	Floret + leaves	Floret
8. LH acanthus scroll end	–	Floret	2 leaves	–
9. Void between acanthus scrolls	2 Florets	1 Floret	3 Florets	2 Florets
10. LH acanthus scroll	Floret	–	2 Florets	–
11. LH acanthus scroll L end	3 leaves	Bud +2 leaves	2 leaves	–
12. RH acanthus scroll	5 Florets	4 Florets	3	2 Florets + Bud

specifically the Nantgarw China Works and the Swansea China Works and have cited several observable differences which characterise the two factories and which can assist in their differentiation: in other words, although these Nantgarw and Swansea china mouldings are very similar visually, to close inspection and observation, they are not identical. Careful inspection of the similar C-scroll mouldings comprising acanthus leaves with floral foliage, florets and buds setting off a coiled ribbon on each of the dessert and dinner plates illustrated in Figs. 3.1, 3.2, 3.3 and 3.4 from the Nantgarw, Swansea, Coalport and Derby factories, all dating from *ca.* 1820, reveals that there are several subtle differences which can be used to discriminate between them, as summarised in Table 3.1. Hitherto, several authors have suggested that the existence of these moulded "*Nantgarw-like*" plate mouldings has confirmed that the Swansea China Works must have acquired the Nantgarw moulds from Billingsley and Walker as a consequence of their employment there in 1814 following the failure of the first phase Nantgarw China Works enterprise. Secondly, following the eventual closure of the Nantgarw China Works in 1820 and the departure of Billingsley and Walker for the Coalport China Works, it has been assumed from a statement made by Llewellyn Jewitt (Jewitt, *The Ceramic Art of Great Britain from the Prehistoric Times Down to the Present Day*, 1878) that John Rose, the proprietor of Coalport had bought the Nantgarw moulds and had taken them to Coalport for his use there in making Coalport moulded china of the same design. Finally, it is interesting to speculate what happened to the specimen Nantgarw plates Robert Bloor had acquired from John Sims' workshop after he had received the commission for manufacture of the *Lord Ongley* dessert/dinner service at his Derby China Works, where he then had them copied for this purpose? It appears that after they had been used as exemplars at Derby for the *Lord Ongley* service they were set aside and they surfaced again in the Jermyn Street collection of the Practical Geology Museum set up by Sir Henry de la Beche, whence they again became dispersed following its closure in the 1880s (de la Beche et al., *Catalogue of Specimens in the Museum of Practical Geology, Jermyn Street, London, Illustrative of the Composition and Manufacture of British Pottery and Porcelain from the Occupation of Britain by the Romans to the Present Time,* 1876). In the light of the studies recounted here and the differentiation criteria listed in Table 3.1 it is clear that none of these historical statements and assumptions made by previous authors are strictly correct as the moulded -edge "*Nantgarw-like*" plates are of distinctly different designs at each factory which would not have occurred if they had been either cast from the same moulds or had been copied faithfully from the genuine exemplar Nantgarw specimens.

A survey of the prior relevant literature on the Nantgarw factory has turned up some interesting comments which now need to be re-evaluated in this light. Firstly, Isaac Williams in his authoritative account of his archaeological site investigation of the Nantgarw China Works in 1931 (*The Nantgarw Pottery and Its Products: An Examination of the Site*, 1932) refers on page 20 to the discovery of a biscuit shard with the typical C-scroll and ribbon moulding that has been described above in his archaeological excavation of the waste pit on site. The location and importance of the discovery of this waste pit has been considered in detail by Edwards (*Porcelain and Silica Bricks: The Extreme Ceramics of William Weston Young,* 1776–1847, 2019) and the archaeological context of the shards excavated therefrom by Isaac Williams. The porcelain shards were found in the lowest stratum of the waste tip some one metre below the surface and contained many pieces and fragments of biscuit porcelain with characteristic Nantgarw shapes, especially coffee cups and tea cups, tazzas and plates. Williams summarises his findings definitively regarding the questionable attempts made by William Weston Young to re-vitalise the manufacture of Nantgarw porcelain at the site with Thomas Pardoe after the departure of William Billingsley and Samuel Walker for Coalport in 1819/1820. He states:

> *Nothing original emerges in connection with the porcelain produced from 1819–1822 by William Weston Young, who appears to have used a composition of paste similar to that invented by Billingsley. Furthermore, Young utilized the Billingsley moulds for pressing and copied the Billingsley models for throwing; but when he (Young) did attempt anything new, as a potter, it was generally thick and heavy in construction. Beauty of form and a classical quality in design are distinguishing features of the porcelain productions of the Billingsley period at Nantgarw.*

This statement is loaded with conjecture and controversy relating to the manufacture of porcelain at Nantgarw after Billingsley and Walker had left for Coalport, and most authors have assumed that Young did not attempt to make china there (and least of all using Billingsley's formulation recipe, which we now are aware was completely unknown to him!) and that he and Pardoe concentrated upon decorating the locally remnant stock remaining in the white, applying their own glaze (hereafter called the *Nantgarw No.2 glaze*) where necessary. However, in his statement Isaac Williams is most certainly of the opinion that Young and Pardoe did in fact

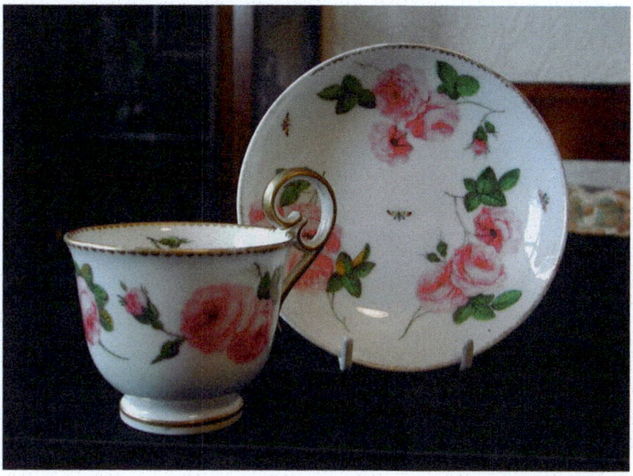

Fig. 3.7 Nantgarw coffee cup and saucer with characteristic heart-shaped handle, *ca.* 1817–1819, decorated with garden roses and butterflies by Moses Webster in the enamelling workshops of Robins and Randall for John Mortlock's agency, Oxford Street, London. *Private Collection*

manufacture porcelain at Nantgarw following the departure of Billingsley and Walker and he even comments that Young's paste is of a similar composition to that used by Billingsley, adding that Young also experimented with a new paste variety which was not successful in comparison with that produced by Billingsley! Another interesting observation, which does not agree with modern analyses of Nantgarw china, but which has been perpetuated without any tangible evidence to the contrary is that the Nantgarw recipe remained unchanged for the lifetime of the factory between 1817 and 1819 (the Billingsley/Walker period) and thereafter between 1820 and 1822 (the Young/Pardoe period) under the proprietorship of William Weston Young. Williams marks out for comment the "*singular beauty*" of the heart-shaped design handled Nantgarw coffee and tea cups, an example of which is shown in Fig. 3.7, decorated in London by Moses Webster from china supplied to John Mortlock, who was the sole appointed agent for Nantgarw china in the capital with a shop and showrooms in Oxford Street. In Fig. 4.2 of his site report, Williams shows a shard Number 8, described as "ornamental porcelain", which depicts clearly the characteristic C-scroll moulding of a Nantgarw china plate exactly like that from the *Lady Seaton* Nantgarw dinner/dessert service shown in Fig. 3.1 here. Jones and Joseph (*Swansea Porcelain*, 1988) have afforded a very useful comparison sketch reproducing exactly the Nantgarw and Swansea mouldings which have been commented on above for differentiation purposes: the Nantgarw moulding drawn there matches that of the plate shown in Fig. 3.1 and the shard discussed by Isaac Williams. Another statement in Williams' report (page 23) relates to the substance of the *Memorial* of 5th September 1814, in which Billingsley, Walker and Young attempted unsuccessfully to acquire funding from the British government for their initial start-up operation at the Nantgarw China Works (Phase I) and this too deserves some comment:

> *... the moulds from which the manufactory at Sevres have worked their forms have been faithfully transcribed in their Nantgarw ware.*

Samples of finished porcelain for inspection were offered to the governmental funding committee for comparison between Nantgarw porcelain and the celebrated Sevres French porcelain (Turner, *The Ceramics of Swansea and Nantgarw*, 1897, pp. 87–90). Williams then goes on to record that because the ornament on his fragment No.8 cited above was a faithful transcription of the well-known design occurring on plates of the soft paste porcelain made at Sevres then the attribution of the C-scroll moulded border on Nantgarw plates, being a stated copy of Sevres, must have a serious foundation. Jones and Joseph (*Swansea Porcelain,* 1988) also concur with this conclusion and point out the superficial similarities between the Swansea and Nantgarw moulded borders whilst also highlighting the subtle differences between them. They also believe that an exchange between the factories may have taken place at some time between 1817 and 1819 which could have given rise to a specimen found at Nantgarw with a Swansea border, but not the other way around! Whatever the real situation was at that time, it is clear that much copying and possible espionage went on between porcelain manufactories (Edwards, *Nantgarw and Swansea Porcelains; An Analytical Perspective,* 2018) in this period and modern analysts should be aware of this. The present author has attempted to investigate the allegation made independently by Isaac Williams and William Turner that Billingsley, Walker and Young copied closely the Sevres mouldings for their Nantgarw plates, a typical example being provided by the *Lady Seaton* service plate in Fig. 3.1: however, despite much diligent searching of the Sevres factory plates from this period, not one example could be located which bears even a close resemblance to the actual Nantgarw moulded design and execution. Evidently, the Sevres China Works did make moulded-edge plates but the C-scroll design that has come to be associated with the Nantgarw China Works with its ribbons, florets and foliage seems to actually have no analogous comparator in Sevres porcelain. The nearest Sevres equivalent found in the literature does indeed have moulded acanthus leaf "C-scrolls" but these are simply executed and have a proliferation of other unrelated floral mouldings in continuation along the verge. Moulded ribbons are completely absent from the Sevres verge mouldings, but one Sevres pattern included a wavy enamelled ribbon painted around the edge – which actually does appear in several other Nantgarw decorated plates – commonly referred to as *Sevres type* and these can be seen in Dr. W.D. John's book, *Nantgarw Porcelain Album*, published with G.J. Coombes and Katharine Coombes in 1975. It is quite possible that the mention of a physical comparator with Sevres plates in the 1814 *Memorial* and the known enamelling similarities of some Nantgarw and Sevres *decorated* plates has become transposed into a similarity in the moulded borders as we have seen suggested here but which has not been found to occur thus far in practice?

3.4 Replicate Analyses for Coalport and Caughley Porcelains

In their paper devoted to the chemical analysis of shards from the Coalport and Caughley China Works sites, Victor Owen and John Sandon (*Post-Mediaeval Archaeology*, 2003) have conducted an elegant and comprehensive study on these two factories and have made several conclusions which are far-reaching in their consequences for our wider discussion on the potential standing of analytical science in the determination of authenticity and attribution of ceramics to different manufactories. It is therefore worthwhile considering the analytical data from this study in more detail to assist in the evaluation of its potential for our analytical thesis relating to factory discrimination and paste identification. The analytical results from Owen and Sandon's comprehensive study (*Post Mediaeval Archaeology*, 2003) were acquired from 20 Caughley and 34 Coalport porcelain shards taken from their respective storage archives. John and Henry Sandon acquired their shards from the Caughley site in the 1970s, unfortunately without any associated archaeological stratification context because of some intensive mining activities which had been carried on at the site subsequent to the closure of the Salopian China Works in 1799. In contrast, the Coalport shards were acquired from the Ironbridge Gorge Museum where they had been preserved from the archaeological excavations of the stratified levels by Barker and Horton (*Post Mediaeval Archaeology*, 1999), which assists greatly in the temporal period labelling of the specimens. Table 3.2 gives a summary of the compositional data of the elemental oxides from the Caughley shards, which were subdivided into magnesian and true/hard paste porcelains, and the Coalport shards, which were subdivided into phosphatic and true/hard paste porcelains. The Coalport shards could also be further subdivided into four distinct periods of factory manufacture as identified from their archaeological context, namely, type I *ca.* 1795–1820, type II *ca.* 1815–1825, type III *ca.* 1820–1830 and type IV post-1830, with several shards being classified as being of an indeterminate type and period. Here, the mathematical averages of the analytical elemental oxide percentages determined for each type of porcelain, and for each manufacturing period in the case of the Coalport porcelain shards, have been presented for ease of data assimilation in the Table and are cited to the first decimal place.

In essence, therefore the analytical data can be summarised as follows:

- The Caughley China Works manufactured two types of porcelain during its 27 years of existence, firstly, a soapstone-rich or magnesian soft paste porcelain which Owen and Sandon thought resembled the Worcester body paste but contained more alkaline flux and a lead oxide content which the proprietor Thomas Turner had invoked to give a lower melting, cheaper but more vitrified product in competition with the much esteemed Worcester body. Prior to his starting porcelain production at Caughley in 1772, Thomas Turner had previously worked in a managerial capacity at the Worcester China Works where he had acquired practical knowledge of their recipes and production. Secondly, in 1796, during the latter stages of the factory operation at Caughley, Turner changed his production body paste to a hard paste porcelain but he still utilised the soft paste kiln firing sequence, which proved to be not altogether successful in practice and this procedure eventually soon left the factory then open to acquisition by John Rose of the neighbouring Coalport China Works in 1799. The Caughley magnesian porcelain shards are therefore representative of the earliest porcelain manufactured at the site which was followed later by the hard paste body version which was

Table 3.2 Analytical data /% for shards from Coalport and Caughley factories (Owen and Sandon 2003)

Factory	Shards	Type	SiO_2	Al_2O_3	Fe_2O_3	MgO	CaO	Na_2O	K_2O	P_2O_5	PbO
Caughley	12	Magnesian	76+/−2	5+/−1	0.4+/−0.1	9+/−2	1.6+/−0.3	1.6+/−0.4	2.6+/−0.5	0.4+/−0.2	3+/−2
	8	Hard paste	75+/−2	20+/−3	0.2+/−0.1	0.2+/−0.1	0.6+/−0.2	1.5+/−0.1	2.9+/−0.2	0.3+/−0.1	0.1+/−0.0
Coalport	2	Phosphatic I	47+/−4	12.7+/−0.1	0.2+/−0.0	0.5+/−0.1	21+/−2	0.9+/−0.2	1.6+/−0.1	17+/−2	–
	1	II	44.4	13.4	0.3	0.5	21.4	1.2	2.0	16.9	–
	5	III	46+/−3	12.6+/−0.8	0.2+/−0.0	0.4+/−0.1	21+/−1	1.0+/−0.2	1.6+/−0.4	17+/−1	–
	7	IV	44+/−6	14+/−2	0.2+/−0.1	0.5+/−0.1	21+/−2	0.8+/−0.3	1.9+/−0.5	17+/−2	-
	2	*	45+/−3	12.4+/−0.3	0.2+/−0.0	0.5+/−0.1	21+/−2	1.0+/−0.1	1.6+/−0.2	18+/−1	–
	4	Hard paste I	73+/−2	19+/−1	0.3+/−0.1	0.2+/−0.1	2+/−1	1.3+/−0.2	2.6+/−0.1	1.3+/−0.1	-
	2	II	75+/−1	19.0+/−0.3	0.3+/−0.0	0.2+/−0.1	1.2+/−0.0	1.2+/−0.0	2.5+/−0.2	0.7+/−0.1-	
	11	*	75+/−1	19+/−1	0.2+/−0.1	0.2+/−0.1	1.0+/−0.8	1.4+/−0.2	2.7+/−0.2	0.6+/−0.5	–

I 1795–1820; II 1815–1825; III 1820–1830; IV Post 1830; * Indeterminate period manufactured between 1796 and 1799. The major difference between the analytical determinations of these two types of Caughley porcelains can be found in the figures for alumina, magnesia, lime and lead oxide – which are, respectively, averaging 5+/−1%, 9+/−2%, 1,6+/−0.3% and 3+/−2% for the magnesian porcelain and 20+/−3%, 0.2+/−0.1%, 0.6+/−0.2% and 0% for the later hard paste porcelain bodies. The other elemental oxide percentages are not sensibly changed: so, the silica, iron oxide, soda, potash and phosphate composition percentages are found to occur within one experimental standard deviation of each other.

- The analyses of the Coalport porcelain shards reveals that only two types of porcelain were manufactured by John Rose from 1795 to post-1830 – these being a hard paste body similar to that of Caughley and a phosphatic body, rich in calcined bone ash as a component. However, here the high silica content of the hard paste porcelain body is not reproduced in the phosphatic version, which has a significantly smaller value in silica content, which is almost halved in its compositional percentage. Likewise, the alumina content is higher in the hard paste porcelain body, whereas the lime is significantly higher in the phosphatic porcelain, which also has the highest phosphate content at 17+/−1% for all periods of the factory manufacture. The magnesia and alkaline oxides content in the porcelain bodies are not too sensibly different for the periods of manufacture and for the type of porcelain produced.
- A most interesting comment has been made by Owen and Sandon in conclusion that their results seem to indicate that the influence of William Billingsley's arrival at Coalport in early 1820 seemed to have had no effect at all upon the analytical composition of the porcelain made thereafter at the Coalport China Works: again, this contrasts with the established opinion of several authors and historians that John Rose not only acquired some hardware from the financially troubled Nantgarw China Works at auction upon its closure but that he recruited Billingsley and Walker expressly to improve his own porcelain quality. This is clearly not supported by the analytical figures presented by Owen and Sandon here. Others have suggested previously that Rose utilised the services of Billingsley and Walker solely for their expertise and advice in his enamelling workshops and kiln operation at Coalport, which now seems much more likely to be the true situation as reflected in the analytical data reported by Owen and Sandon. A rather obtuse suggestion, which actually may hold forth and apply here, is that Rose employed Billingsley and Walker for the express purpose of denying his competitors the chance to secure their valuable expertise and this is perfectly understandable. Analytical figures for the type III Coalport phosphatic porcelain shards, for the period 1820–1830, when Billingsley and Walker were employed there, are certainly supportive of Owen and Sandon's conclusion that John Rose did not modify his Coalport body to approach the Nantgarw formulation, which, of course, presupposes the idea that they would have provided this information to their new employer which they had so closely guarded hitherto. They had maintained the secrecy of the Nantgarw formulation recipe very closely whilst in operational production at Nantgarw and had not even revealed it to their

Table 3.3 Comparison of Swansea, Nantgarw and Coalport phosphatic porcelains elemental oxide/% compositions, *ca.* 1820

Factory	Type	SiO_2	Al_2O_3	Fe_2O_3	MgO	CaO	Na_2O	K_2O	P_2O_5
Nantgarw	Soft paste	44	15	0.2	0.5	21	1	2	17
Swansea	Duck-egg 1[a]	53	20	0.2	0.5	13	1	3	10
	Duck-egg 2[b]	45	20	0.3	0.4	17	1	3	13
Coalport	II/III	45	14	0.2	0.5	21	1	2	17
Swansea	Soft paste	46	23	0.2	0.2	23	1	3	15

[a]Owen and Sandon (2003): Swansea gilded plate
[b]Owen et al. (1998): *Biddulph* service

former partner at the Nantgarw China Works, William Weston Young, who was trying to recover the failing business there and recoup some of the financial losses they had sustained over the 2 years, i.e. 1820–1822, following Billingsley and Walker's departure for Coalport.

- The comprehensive analyses of the Coalport phosphatic porcelains can now be directly compared with those of the Nantgarw and Swansea porcelains, which we have already considered earlier. In summary, the Coalport phosphatic porcelains have a similar bone ash to clay ratio as their Nantgarw analogues of 1.3, but also interestingly have a rather different internal alkaline composition in that the potash: soda ratio is 1.9 for Coalport porcelain compared with the significantly different Nantgarw analogous ratio of 5.2.
- Owen and Sandon also noted that no fewer than 7 of their hard paste Coalport porcelain shards had a surprisingly high phosphate percentage ranging between 0.7 and 2.7% P_2O_5, when technically they should have been realistically devoid of phosphate content at all, as found in the Caughley hard paste body version, to which the other Coalport hard paste shards conform exceptionally well analytically. It was concluded that John Rose did manufacture this unusual phosphatic hard paste body of his own design periodically during the period 1795 until the 1830s and that it was not as suggested by some historians an analogue of the bone china hybrid porcelain then being manufactured so successfully by Josiah Spode, which he patented in 1795, and which contained a much higher percentage of phosphate in the form of a bone ash component.
- Another feature of interest in Owen and Sandon's study of the Coalport shards is a comparison made with the fine Swansea duck-egg paste exemplified by a 10″ (25 cm) dinner plate they had acquired from an unknown source marked SWANSEA in red enamelled stencil on its base and decorated with gilt roses in what appears to be a characteristic Swansea basket-weave and cartouche moulded pattern (Jones and Joseph, *Swansea Porcelain*, 1988) which can be dated to between 1815 and 1817. They confirmed that compositionally this plate differs from its contemporary Nantgarw and Coalport analogues, since it contains approximately only half the amount of phosphate and 50% more alumina content than the Nantgarw body paste (Table 3.3) and in addition the glaze contains an estimated 25% lead oxide. The plate has a pronounced greenish translucency in transmitted light which immediately assigns it to the Lewis Dillwyn's duck-egg

porcelain type that was so esteemed as a product from the Swansea factory (Edwards, *The Duck Egg Translucent Vison of Lewis Weston Dillwyn*, 2017). They then compared this Swansea plate with another analysed previously (Owen et al. 1998), a specimen from the named and authenticated *Biddulph* service which was provided by the Royal Institution of South Wales, Swansea, from their collection: it is also intriguing that the post-1830s Coalport paste had changed to this Swansea-type formulation with a higher alumina and lower phosphate body composition and that the glaze also had about half the lead oxide content at 13%. This variation is consistent, maintain Owen and Sandon, with the description of the Swansea exhibits cited by Kildare Meager at the Glynn Vivian Art Gallery Exhibition (Meager, *The Swansea and Nantgarw Potteries*: *A Catalogue of the Collectrion of Welsh Pottery and Porcelain on Exhibition at the Glynn Vivian Art Gallery, Swansea,* 1949). In Table 3.3 the compositional data are cited for the average Nantgarw soft paste body from the excavated shards, the average soft paste duck-egg Swansea body from excavated shards, the average for the Coalport II/III period shards which would have been contemporary with the Swansea and Nantgarw factory output and the two individual duck-egg analyses for finished Swansea porcelain specimens, citing the *Biddulph* service (Owen et al. 1998) and another Swansea plate analysed as part of the Owen and Sandon (2003) study of Caughley and Coalport porcelains. This combination affords the opportunity for the first evaluation to be made analytically of the feasibility of using the compositional data from distinct factories to discriminate between these porcelains.

3.5 Compositional Analyses of Swansea, Nantgarw and Coalport Porcelains

These three factories which existed in the broad temporal period 1815–1828 are all associated with William Billingsley and Samuel Walker: several previous authors have suggested that there must have been an interchange of technical information concerning porcelain recipes and formulation mixtures between Billingsley and Walker, on the one hand, with Lewis Dillwyn of the Swansea China Works and John Rose of the Coalport China Works on the other. It is equally clear and very well-documented that Billingsley and Walker never revealed their Nantgarw recipe, even keeping this knowledge from William Weston Young, their partner in Phase II of the re-opening of the Nantgarw China Works in 1817, without whose financial support and business acumen they would not have achieved their primary objective there. The secrecy which existed behind the formulations and recipes known only to porcelain factory proprietors was well-known and the Nantgarw formulation was not eventually made public until John Taylor (*The Complete Practical Potter,* 1847) revealed the details of the Nantgarw recipe in his 1847 publication, information which he claimed came directly from Samuel Walker, almost 20 years after the death of William Billingsley. Lewis Dillwyn kept notes of his experimental porcelain body and glaze changes and modifications (Dillwyn, 1815–1817, see Eccles & Rackham, *Analysed Specimens of English Porcelain in the Victoria and Albert Museum Collection,* 1922) but these were not made available for research consultation until 1920, when they were lodged in the Victoria & Albert Museum archive, whence they first appeared in general publication as an Appendix to Eccles & Rackham's *Analysed Specimens of English Porcelain in the Victoria and Albert Museum Collection,* 1922. What is clear is that the empirical alterations to their paste bodies and glazes made by china works proprietors can now be assessed by the analytical interrogation of shards and other porcelain fragments, which reveal both the successful and unsuccessful modifications that were made during the operation of their factories. It could be argued at this point that the shards would represent the residues from the most unsuccessful attempts at manufacturing body modifications, since these by inference would have been not successfully commercially and therefore would have been broken up and destroyed, therefore comprising a significant presence in the waste pits along with the shards and wasters from the mis-fired porcelains then in full commercial production!

Previous authors and ceramic historians have made statements concerning the finite number of different porcelain bodies that each factory undertook to produce during its lifetime: in some cases, historians claim that only one body was made during the lifetime of the factory under study, such as the unique soft paste phosphatic china at Nantgarw, and only three paste types at Swansea, namely a glassy body, a phosphatic duck-egg body and a magnesian "trident" soapstone body. Yet, the diaries of Lewis Dillwyn clearly refer to at least a dozen experiments he conducted with Samuel Walker between 1815 and 1817, several of which he seemed to find eminently successful, such as his highly esteemed duck-egg paste composition and eventually his more robust but highly unsuccessful trident magnesian body that proved to be the downfall of the Swansea China Works. The situation at other factories is probably very similar – only two bodies are claimed for Coalport, a hard paste china and a phosphatic body, and for Caughley again two bodies, namely a hard paste and a soapstone body, yet collectors and ceramics enthusiasts recognise a special Coalport body that has the blue-green translucency of the fabled duck-egg Swansea porcelain. One example is shown in the pair of Coalport plates in Fig. 3.8, with the fractional pattern mark 4/782 dating to about 1843, in *Nantgarw-style*; another is shown in Fig. 3.9, a possibly earlier example, which is in accord with Henry Morris' decoration at Swansea around 1823–1826, just after closure of the Swansea China Works when it is recorded that Morris bought in ceramics from the Coalport and Staffordshire factories to decorate in his own small muffle furnace at Swansea. Morris was an esteemed Swansea artist and his special mark is also recorded on the base of the platter shown in Fig. 3.9. This discovery means that it is now even more critical to appraise the role of analytical science for assisting in the determination of the factory origin for porcelain artefacts. The critical question that can now be posed is: given the analytical data for the soft paste bone ash factories of Coalport, Swansea and Nantgarw from the 1815–1825 decade, would it be possible to discriminate between porcelain products from these three factories on analytical compositional data alone?

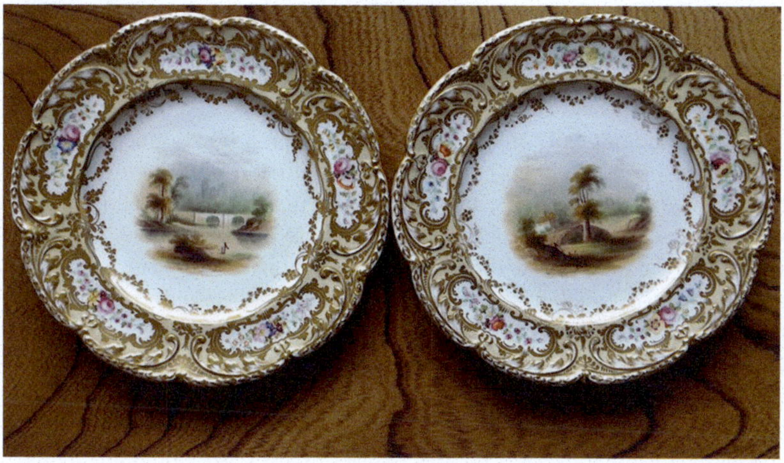

Fig. 3.8 A pair of Coalport plates, *ca.* 1843, heavily gilded and with a Nantgarw-style embossed edge border, depicting rural scenes. Unmarked except for the fractional pattern number 4/782 in red enamel on the base. *Private Collection*

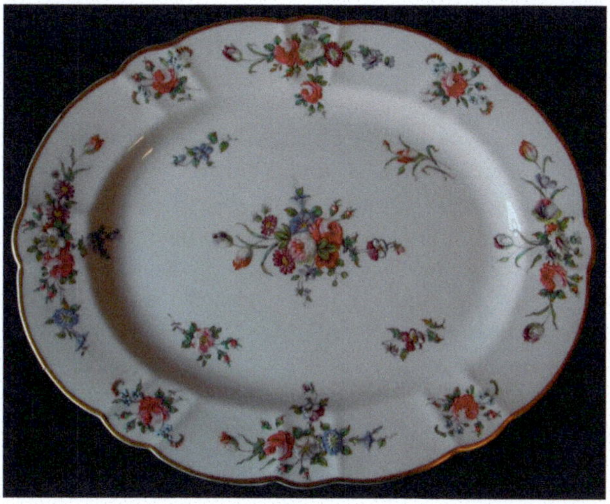

Fig. 3.9 Large oval plate marked SWANSEA. in red enamel capitals, decorated by Henry Morris at Swansea around 1823–1826, currently attributed to possibly Coalport porcelain manufacture. The porcelain has a beautiful duck-egg translucency and this could reflect the suggestion that John Rose experimented with the Swansea China Works porcelain china recipes following the immediate closure of the factory in 1820. However, the unusual embossed moulding at the rim is reproduced in Jones and Joseph *Swansea Porcelain* as occurring but rarely on Swansea porcelain but the shape is definitely not consistent with a Swansea origin. Jones and Joseph confirm that the mark is associated with and used exclusively by Henry Morris, who decorated only locally at Swansea after closure of the Swansea China Works in 1820 and not elsewhere. *Private Collection*

Close examination of the elemental oxide composition analytical data in Tables 2.6, 3.2 and 3.3 is very revealing in that although the individual results do show some small differences, the average values and their standard deviations derived from multiple data sets are actually all lying within the expected error limits of each other so that mathematically these differences may not be significant analytically, with perhaps the possible exceptions of silica, alumina, lime and phosphate compositional values which may just lie outside the established norms! The implication is that analytical data alone, although they do generate much valuable and chemically quantitative information about the changes in composition of porcelain bodies in any particular factory, such as the presence of a hard paste Coalport body with very high silica and alumina content and a very low phosphate content, cannot in themselves in all cases be absolutely definitive in respect of the assignment of a porcelain source factory for some production pieces.

Another valuable example of how analytical science has caused some necessary re-appraisal in ceramic circles is the discovery of hard paste porcelain shards in the Nantgarw China Works waste pit, excavated from the same level as the normal soft paste phosphatic porcelain (Williams, *The Nantgarw Pottery and its Products: An Examination of the Site*, 1931) which contains stratigraphic deposits of wasters, shown diagrammatically in Fig. 3.10 from the site excavation report. It would not be conjectural or controversial to say, therefore, that the discovery of an unattributed piece of suspect Nantgarw porcelain which is characteristically hard paste in texture, and may even have been marked NANTGARW C.W. (!), would quite realistically and properly have been dismissed as a fake by most experts hitherto. However, when one then forensically investigates the origin and documentation, if any exists, which specifies that there was only ever one Nantgarw china body *sine qua non*, we must allow for some degree of licence in this respect. This seemingly unquestionable statement of there being only one solitary composition for all Nantgarw porcelain artefacts can be unambiguously traced back to the comment made by Herbert Eccles when he examined the Nantgarw and Swansea china on show at the Centenary Exhibition held in the Glynn Vivian Gallery in Swansea in 1914, which alleged that *in his opinion only one Nantgarw body existed.* This statement was made on the basis of his handling of the range of Nantgarw porcelain exhibits on show there, which may indeed have all been of a unique type of phosphatic porcelain and visually identical, as no analysis was carried out at that time. This opinion has in the course of time been adopted as a rigorous fact – hence, any variant experimental body produced at Nantgarw by Billingsley, Walker and Young during the years 1817–1822 could not be conceived later as a genuine Nantgarw production: this "opinion" has been reinforced by the several unfounded statements from later authors stating that, of course, Billingsley, Walker and Young would just not have had the time available in the short period of factory operation available to them to make any modifications to their porcelain paste. This latter statement which now seems to have been added to the annals of fact associated with the Nantgarw factory, really does not even allow the consideration that Billingsley and Walker especially were avowed experimentalists (as they had already demonstrated at the Worcester China Works under the personal encouragement of Martin Barr of Barr, Flight & Barr) and had spent the previous decade modifying their existing porcelain recipes.

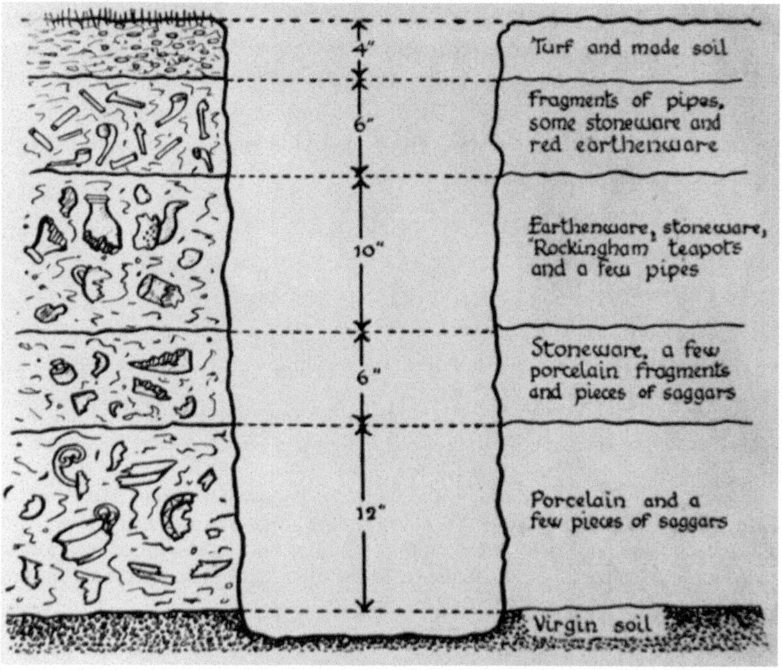

Fig. 3.10 The Nantgarw China Works waste pit discovered by Isaac Williams during his 1931 archaeological excavation carried out at the site: this waste tip lay some 30 feet east of the eastern-most wall of the potting shed. Five distinct levels were identified by the archaeologists in this waste pit – the porcelain shards from the Billingsley/Walker/Young period occupied Level 5 at a depth of approximately 38 inches (~1 metre) below the top surface, dating from 1817–1820, which formed the basis of the modern analytical studies of shards undertaken here and described in this work. The discovery of the actual buildings and kilns ascribed to Billingsley and Walker and still standing was at first surprising because William Chaffers and Llewelyn Jewitt in the mid- to late- nineteenth century had maintained that the site was destroyed after William Weston Young left Nantgarw in 1823 and all useful equipment had been sold off at auction and had been removed to the Coalport China Works by John Rose

Would it not have been possible in the dire financial straits facing Nantgarw in the latter part of their operation arising from the appallingly high local kiln losses of 90% of their Nantgarw output that they might at least have tried to do something about the potential improvement in the robustness of their porcelain body as their commercial non-viability was evident? This is precisely the situation that Lewis Dillwyn found himself in during 1815–1817, when he tried to minimise the loss in the Swansea kilns which was causing him such a problem in meeting the demands of his clientele – and that was only about 70% in comparison with the 90% experienced by his neighbours at Nantgarw! This is fact as Dillwyn's notebooks attest – and the Swansea soapstone high-magnesian paste which he adopted as a result for the replacement for his esteemed duck-egg porcelain, although more robust still failed to save the Swansea China Works (Dillwyn *Note Books*, see Eccles & Rackham, *Analysed Specimens of English Porcelain in the Victoria and Albert Museum Collection*, 1922).

There is obviously a need to examine more closely the compositional data derived from porcelain analyses to form an assessment of its usage in the determination of differences between porcelains and its meaningful adoption as a forensic device for their unambiguous analytical discrimination whilst still retaining evidential statements from early documentation to provide the necessary holistic refinement and support where appropriate.

References

D. Barker, W. Horton, The development of the Coalport China works and a description of ceramic finds. Post Medieval Archaeol. **33**, 3–93 (1999)

P. Colomban, H.G.M. Edwards, C. Fountain, Raman spectroscopic and SEM/EDXS analysis of Nantgarw soft paste porcelain. *Journal of the European Ceramic Society*, submitted for publication, (2020)

L.W. Dillwyn, *Notes on the Experimental Production of Swansea Porcelain Bodies and Glazes*. Made by Lewis Weston Dillwyn with Samuel Walker at the Swansea China Works Between 1815 and 1817. Presented to the Library of the Victoria &Albert Museum, South Kensington, London by John Campbell in 1920. Reproduced in Eccles & Rackham, Analysed Specimens of English Porcelain, 1922, see reference

H. Eccles, B. Rackham, *Analysed Specimens of English Porcelain in the Victoria and Albert Museum Collection* (Victorian and Albert Museum, London, 1922)

H.G.M. Edwards, *Nantgarw Porcelain: The Pursuit of Perfection,* Penrose Antiques Ltd. Short Guides, Series Editor: M.D. Denyer (Penrose Antiques Ltd., Thornton, 2017a). ISBN: 978-0-244-90654-2

H.G.M. Edwards, *Swansea Porcelain: The Duck-Egg Translucent Vision of Lewis Dillwyn* (Penrose Antiques Ltd., Short Guides, Thornton, 2017b). ISBN:9780244325787

H.G.M. Edwards, *Derby Porcelain: The Golden Years, 1780–1830,* (Penrose Antiques Ltd. Short Guides, Thornton, 2017c)

H.G.M. Edwards, *Swansea Porcelain: The Duck Egg Translucent Vision of Lewis Weston Dillwyn* (Penrose Antiques Ltd., Short Guides, Thornton, 2017d)

H.G.M. Edwards, *Swansea and Nantgarw Porcelains: A Scientific Reappraisal* (Springer, Dordrecht, 2017e)

H.G.M. Edwards, *Nantgarw and Swansea Porcelains: An Analytical Perspective* (Springer, Dordrecht, 2018)

H.G.M. Edwards, *Porcelain to Silica Bricks: The Extreme Ceramics of William Weston Young (1776–1847)* (Springer, Dordrecht, 2019)

H.G.M. Edwards, M.C.T. Denyer, *William Billingsley The Enigmatic Porcelain Artist, Decorator and Manufacturer* (Penrose Antiques Ltd., Short Guides, Neopubli, Berlin, 2016). ISBN: 978-3-7418-6

L. Jewitt, *The Ceramic Art of Great Britain from the Prehistoric Times Down to the Present Day*, vol I and II (Virtue & Co. Ltd., London, 1878)

W.D. John, G.J. Coombes, K. Coombes, *The Nantgarw Porcelain Album* (The Ceramic Book Company, Newport, Gwent, 1975)

A.E. Jones, Sir L. Joseph, *Swansea Porcelain: Shapes and Decoration* (D. Brown and Sons, Ltd., Cowbridge, 1988)

K.S. Meager, *The Swansea and Nantgarw Potteries: Catalogue of the Collection of Welsh Pottery and Porcelain on Exhibition at the Glynn Vivian Art Gallery, Swansea* (Swansea Corporation, Swansea, 1949)

J.V. Owen, Antique Porcelain 101: A primer on the chemical analysis and interpretation of eighteenth century British wares, in *Ceramics in America*, ed. by R. Hunter (Chipstone Foundation, USA, 2002), pp. 39–61

J.V. Owen, M.L. Morrison, Sagged phosphatic Nantgarw porcelain (ca. 1813–1820): Casualty of overfiring or a fertile paste? Geoarchaeology **14**, 313–332 (1999)

J.V. Owen, J. Sandon, A rose by another name: A geochemical comparison of Caughley (c.1772-99), Coalport (John Rose & Co., c.1799-1837), and rival porcelains based on sherds from the factory sites. Post Mediaeval Archaeol. **37**, 79–89 (2003)

J.V. Owen, J.O. Wilstead, R.W. Williams, T.E. Day, A tale of two cities: Compositional characteristics of some Nantgarw and Swansea porcelains and their implications for kiln wastage. J. Archaeol. Sci. **25**, 359–375 (1998)

Sir A.H. Church, *English Porcelain: A Handbook to the China Made in England During the 18th Century as Illustrated by Specimens Chiefly in the National Collection*, A South Kensington Museum Handbook, Chapman & Hall Ltd., London, 1885 and (1894)

Sir H.T. de la Beche, T. Reeks, F.W. Rudler, *Catalogue of Specimens in the Museum of Practical Geology, Jermyn Street, London, Illustrative of the Composition and Manufacture of British Pottery and Porcelain from the Occupation of Britain by the Romans to the Present Time*, George Eyre and W. Spottiswoode, 3rd edn (Publishers for the HMSO, London, 1876)

M.S. Tite, M. Bimson, A technological study of English porcelains. Archaeometry **33**, 3–27 (1991)

I.J. Williams, *The Nantgarw Pottery and Its Products: An Examination of the Site* (The National Museum of Wales and the Press Board of the University of Wales, Cardiff, 1932)

Chapter 4
Analytical Studies of Porcelains: Correlation with the Holistic Information About the Eighteenth and Nineteenth Century Factories

Abstract A survey is given of the major eighteenth and early nineteenth century English and Welsh porcelain manufactories with details of their foundation, their founders and, where appropriate, their production formulations and recorded recipe changes. This listing is prefaced by a statement of the analytical chemical information that can be derived from the determination of the compositions of porcelain bodies. The distinction between bone china and phosphatic porcelains is considered. The influence of the Royal Society of London upon early English porcelain manufacture is recounted: the importance of the experiments conducted by John Dwight of Fulham and the chemical analyses of the early Burghley House jars.

Keywords Composition of porcelains · Analytical information · Eighteenth and nineteenth century English and Welsh factories · Foundation of the early factories · Factory founders · Factory histories · Royal Society of London · Bone China · Phosphatic porcelain · John Dwight · Burghley House jars

4.1 Chemical Analytical Information

In this chapter we shall review the chemical analytical techniques and the information that can be derived from their interrogation of porcelains: this will be relevant to any discussion that follows concerning the forensic application of analytical chemistry to porcelain characterisation and for the evaluation of its potential for the discrimination between factory products. The earliest analyses, as have already been outlined above, involved the wet chemical digestion of significantly large samples of finished, decorated porcelain wares, some of which were perfect and others which were already damaged and were accompanied with, or without, restoration. Examples include the work of Sir Arthur Church on the Victoria and Albert Museum Collection specimens, many comprising rather rare figures and cabinet pieces from the Lady Charlotte Schreiber bequest (*English Porcelain,* 1894). In these analyses it is nowhere stated that damage would have resulted from the acquisition of

H. G. M. Edwards, *18th and 19th Century Porcelain Analysis*,
https://doi.org/10.1007/978-3-030-42192-2_4

analytical samples but this would have been *de rigueur* and an essential requirement at that time for the digestive wet chemistry analytical methods employed to determine the quantitative chemical composition of the bodies and glazes. Following on from Sir Arthur Church's work, the analyses of Herbert Eccles and Bernard Rackham (*Analysed Specimens of English Porcelain in the Victoria and Albert Museum Collection,* 1922) gave a comprehensive survey of several important eighteenth and nineteenth century factories which have been described fully earlier. Again, extensive damage would have been caused to the specimens selected from the Victoria and Albert Museum Collection. Maurice Hillis (*Welsh Ceramics in Context, II*, 2005) has investigated the damaged specimens of Welsh porcelains from Swansea and Nantgarw used in the Eccles and Rackham study, which were preserved and are still available for inspection in the Victoria and Albert Museum archive, and pictures of some of these which are reproduced in his article are revealing for the extent of the destructive sampling used for the analytical studies. Other specimens used in the studies are reproduced in black and white photographic prints in the Eccles and Rackham booklet and it can be seen that these often comprise odd saucers and plates, and include specimens with riveted restoration on other pieces such as the Bow mug of questionable authenticity and attribution that we have discussed earlier in Chap. 2.

The protocol adopted towards analytical specimen acquisition and the attitude towards sacrificing a part or whole of a good example of factory finished porcelain in a museum collection or archive for the determination of analytical data is rather different today than it was a 100 years ago, and with relatively few exceptions the general and mandatory requirement facing the analyst now is for non-destructive sampling of the specimen: this means that technically it is now deemed undesirable, if not impossible, to drill a small hole or to remove a sliver of material from a documentary porcelain piece for destructive analysis, even though the quantities required to effect the analysis are now much smaller than those that were necessary a century or more ago. Also, the growth of modern instrumental techniques for the detection of material compositions, especially when coupled with a high powered microscope to examine a sample "footprint" of an area few square microns or less, hence involving only a few mg (10^{-3} g), microgram (10^{-6} g) or even pg (10^{-12} g) of material, has provided the analyst with several techniques that can produce qualitative and quantitative information about the composition of a ceramic specimen and the additional advantage of their combined use *in tandem* affords a new scope for achieving the maximum amount of relevant analytical data from extremely small quantities of a specimen. Actually, this proliferation of additional experimental data from a ceramic specimen can occasionally cause some specific problems in data interpretation as the information has been obtained from only square or cubic micron-sized regions of the material, which may not be representative of the whole specimen in bulk, especially for heterogeneous ceramic artefacts. Hence, localised regions or areas which are particularly rich in a specific mineral or chemical compound can distort the presence quantitatively of this material when factored up into the bulk specimen composition. The reader of most modern research papers in the analysis of ceramics will appreciate clearly the demonstration of this aspect of specimen sampling, and many analysts will now present the data from their microscopic studies separately from the bulk compositions, however these may have otherwise have been determined. Hillis (2001) has also outlined the information that is available for ceramic raw materials, bodies and glazes which can be correlated with analytical data and their interpretation of the compositional recipes. This micro sampling aspect is of concern in several other areas in art involving the scientific analysis of precious specimens at the arts/science interface: a classic example is provided by the *Shroud of Turin*, long believed to be the burial vestment of Jesus Christ following His Crucifixion and burial in the tomb of Joseph of Arimathea in Golgotha, from which small specimens of linen were carefully excised and subjected to radiocarbon analysis and C^{14} dating in three independent accelerator mass spectrometry laboratories in Oxford, Zurich and Tucson in 1989. The date of 1260–1390 CE obtained from the radiocarbon dating experiments (Damon et al. 1989), with a mean of 1325 ± 65 years, and a two sigma confidence interval of 95% for the linen specimens, clearly places the *Shroud* temporally in the mediaeval period and for many condemned the *Shroud* forever as a mediaeval fake (Gove, *Relic, Icon or Hoax: Carbon-Dating the Turin Shroud,* 1996). However, concerns have since emerged about the particular region of the cloth that was chosen for the sample excision, to avoid the important areas containing the body image and blood flow marks, with conservative accusations from forensic scientists that it was not truly representative of the whole specimen (Rogers, *A Chemist's Perspective on the Shroud of Turin*, 2008; Wilson, *The Blood and the Shroud: The Passionate Controversy Still Enflaming the World's Most Famous Carbon-Dating Test*, 1998). This illustrates the dilemma facing modern analysts and conservators in the selection of their specimen sampling procedures: basically this can be summarised as follows, how does one fulfil the need for the provision of analytical information about an important specimen without destroying what can be perceived to be the critical areas of the object under consideration? The ideal conservatorial requirement of micro sampling a region of the object which would not normally be seen on display itself poses a fundamental question for the analyst: would this microsample then be considered as a fair representation of the whole specimen? If not, then there would be some serious implications for the eventual acceptance of the analytical information and interpretation derived from it. For instance, in the ceramics field, some areas of a rare and finished porcelain specimen which may justifiably be considered to be rather negatively for micro-sampling might be regions near already damaged areas which may themselves have already received some unrecorded restoration historically, regions at or near obvious blemishes in the porcelain body (where the reason for the blemish occurrence is unknown and could perhaps be related possibly to an inhomogeneous mixture of body paste locally), and areas near a footrim or underneath the object which could have been subjected to "sagging", distortion or a pooling of excess glaze components (which again could be subject to an inhomogeneity of composition). These factors cause very real problems for analytical scientists and conservators nowadays who wish to obtain compositional information from finished porcelain specimens in public and private collections and impinge upon the current interest being generated in the application of analytical techniques which do not require invasive sampling procedures and thereby allow the interrogation of multi-replicate micro samples over different parts of a porcelain specimen without causing any damage, however small. This concept will be considered later, when the analytical data from a recommended combined analytical techniques approach is applied to ceramic porcelains.

Without a doubt, the most useful and commonly encountered techniques for ceramics analysis cited today are Scanning Electron Microscopy/Energy Dispersive X-Ray Spectroscopic Analysis (SEM/EDAXS), X-Ray Diffraction Analysis (XRD) and X-Ray Fluorescence Spectroscopy (XRF). All of these are essentially elemental detectors; of alternative application are the molecular spectroscopic techniques of infrared and Raman spectroscopy (IRS and RS) which detect molecular compounds. The major difference in the information provided by these two types of technique is that the elemental techniques (SEM/EDAXS and XRF) give the metal and nonmetal determinations directly for translation into elemental oxide percentages whereas the molecular spectroscopic techniques (IRS and RS) provide a determination of the actual chemical compounds or molecular ions present in the sample, which therefore determines the locations of the elemental oxides in their chemically coordinated entities. Thus, both types of technique are complementary in their output information and there is little overlap to be encountered in the emergent analytical data (Edwards, *Nantgarw and Swansea Porcelains: An Analytical Perspective,* 2018b). XRD is also an elemental technique but provides the diffractograms characteristic of specific minerals and crystalline inorganic compounds, which gives an additional dimension to the analysis through the capability of identifying the actual mineral components present in the fired porcelain body and in this respect XRD can act as a potential bridge between the data derived from the elemental and molecular analytical techniques. For example, the SEM/EDAXS experiment will detect the key elements calcium, phosphorus, oxygen, silica, sodium, potassium, lead, aluminium and magnesium in a porcelain specimen (as will the XRF experiment) whereas the IRS and RS experiments will both identify the way in which these elements are compounded together, to form hydroxyapatite, whitlockite, albite, quartz, cristobalite, tridymite, calcite and olivine, for example, as will the XRD data. Hence, the elemental oxide composition is best derived from the SEM/EDAXS experiment whilst the chemical species which indicate the high-temperature reaction sequences that have occurred in the kiln will be identified from XRD, IRS and RS and this information will additionally inform the actual temperature range at which the kiln firing process was operated in the factory when applied to the knowledge of high temperature chemical behaviour and reactivity of the different mineral components comprising the porcelain paste.

The identification of key mineral chemical species in the fired porcelain specimens has provided analysts with extremely useful and novel diagnostic information about the type of porcelain being analysed: for example, in their classic analytical work on eighteenth and nineteenth century English and Welsh porcelains, Tite and Bimson (1991) studied ten factories, namely Chelsea, Longton Hall, Bow, Lowestoft, Nantgarw, Worcester, Vauxhall, Liverpool, Bristol and Plymouth, using both the SEM/EDAXS and XRD elemental techniques. They noted the presence or absence of the following minerals in the XRD data: anorthite $CaAl_2Si_2O_8$, diopside $CaMgSi_2O_6$, enstatite $MgSiO_3$, mullite Al_2SiO_3, whitlockite *beta*-$Ca_3(PO_4)_2$, wollastonite $CaSiO_3$, quartz SiO_2 and the two high temperature phases of quartz, tridymite and cristobalite, which are also both represented formulaically as SiO_2. They were hence able to propose an analytical chemical diagnostic classification of different porcelain types based upon the unique presence of these minerals in the fired porcelain paste, which translates as follows: *hard paste* porcelain, mullite and tridymite; *bone ash* porcelain, whitlockite; *glassy* porcelain, wollastonite; *magnesian (soapstone)* porcelain, enstatite and diopside; *bone china,* anorthite. Of course as mentioned earlier, the critical parameter in the detection of early glassy porcelain and its later variants is the presence of lead oxide derived from the flint glass cullet added (which could contain up to 60% lead oxide), so even small amounts of glass powder or frit additive would give a positive indication of this component being present as it is highly unlikely to have been introduced accidentally from other raw materials. However, a survey of the additives used in early experimental trials for perfecting porcelain body compositions reveals that it was also quite common to use crown glass, or soda glass, a lower melting and less dense glass, which could also be sourced as cullet or frit from glass manufacture, and this, of course, would not give a positive indicator in the analyses for the presence of lead oxide and could result in a mis-interpretation of the derived analytical data.

Even though there was no detailed knowledge available to manufacturers of the chemistry which occurred at the high temperatures in the kiln firing of the ceramic pastes in the manufacture of porcelains in the eighteenth and early nineteenth centuries, china factory proprietors were aware empirically that the addition of certain materials (known as alkaline fluxes) aided the fusion process in the kiln and assisted in the formation of a homogeneous vitreous body (Shaw, *The Chemistry of the Several Natural and Artificial Heterogeneous Compounds Used in the Manufacture of Porcelain, Glass and Pottery,* 1837). Thus, the addition of lime, calcite, and borax to the soda ash, pearl ash and potash already present in the recipe formulations assisted in the melting process at a lower kiln temperature, which gave a better overall control of the process and prevented the high temperature "sagging" of the porcelain pieces when the kiln temperature exceeded that required for ceramic fusion. Pure silica sand does not melt until about 1600 °C, but in admixture with clays and alkaline fluxes, such as those of calcium, sodium and potassium described above, the melt temperature can be decreased to the more workable range of 1200–1300 °C: a characteristic of the hard paste feldspathic porcelain as developed in China at the Jingdezhen kilns was the operational high firing temperature of 1400 °C, which was achievable using the increased air downdraught from their tall kiln chimneys, whereas the manufacturers of the English soft paste porcelains managed to decrease this to 1200 °C or so for their kilns, thereby providing a big saving on fuel costs as well as providing their better control of the processes taking place in the kiln at these elevated temperatures. At the Nantgarw China Works, Billingsley and Walker's formulation body paste recipe required a higher temperature than most other soft paste porcelains and their kiln was believed to operate at a higher firing temperature range of between 1380 and 1420 °C, which was more compatible with the Chinese hard paste porcelain operation (Morton Nance, *The Pottery and Porcelain of Nantgarw and Swansea*, 1942; John, *William Billingsley*, 1968; Edwards and Denyer, *William Billingsley*, 2016). In the kiln at these high temperatures, the temperature control was absolutely vital as a steady high temperature was required to be maintained for about 3 or 4 days at the extreme firing temperature when reached, which was preceded and then followed by the prescribed °C/minute rate increases and decreases of temperature for the heating and cooling events, the whole process taking about 1 week from start to finish. Confirmation of this high kiln temperature achieved at Nantgarw is provided in the paper of Owen and Morrison (1999) in which they analysed "sagged" Nantgarw shards and they have identified materials in specific excavated biscuit porcelain wasters showing evidence of such "sagging" which they calculated could only have been obtained at temperatures in excess of 1430 °C. In a related enterprise, William Weston Young in the early 1820s operated his trial experiments on refractory silica bricks using the high purity *Dinas rock* silica obtained from a localised geological source in the Vale of Neath with only a relatively small amount of calcareous flux included (about 1% or less), and he was able to use the high temperature porcelain biscuit kiln at Nantgarw adequately for this purpose in his trial experiments. It is interesting that upon leaving Nantgarw in 1822/1823 in a continuation of these refractory experiments, he also found that alternative china kilns in factories elsewhere were not able to achieve the high temperatures necessary for the fusion of the silica in his recipe to be undertaken, such as those at the Swansea Pottery and the Minton China Works in Staffordshire (Edwards, *Porcelain to Silica Bricks: The Extreme Ceramics of William Weston Young, 1776–1847,* 2019). Recent studies on shards found during the archaeological excavation of the Nantgarw China Works site in the 1990s indicate that William Weston Young or his successors could also possibly have made a type of hard paste, siliceous porcelain there after the departure of William Billingsley and Samuel Walker in the period 1820–1822, and this, of course, would also have required an effective high temperature kiln to be in operation, seemingly after the official cessation of manufacture of porcelain at the Nantgarw China Works had already occurred and erroneously that the kilns had been removed (Jewitt, *The Ceramic Art of Great Britain from the Prehistoric Times Down to the Present Day,* 1878; Colomban et al. 2020).

A comparison of the derivative chemical information available from the combined analytical studies of porcelains has been undertaken by Edwards (*Nantgarw and Swansea Porcelains: An Analytical Perspective,* 2018b), where the complementarity of the elemental and molecular structural data are demonstrated specifically for the two Welsh porcelain factories of Swansea and Nantgarw. It is clear that whereas the quantitative compositional information about the recipes are best derived from the SEM/EDAXS measurements undertaken on porcelain shards, the molecular data obtained from the same shards using RS and also non-destructively from perfect, finished pieces of Nantgarw and Swansea porcelain provide a new approach to the determination of their prospective factory of origin using a combination of analytical techniques. In this way, the examination of several perfect pieces of Swansea and Nantgarw porcelain showed the presence of key molecular spectral features which enabled assignments to be made to a particular china factory: the RS spectral identification of key features in Swansea glassy, duck-egg and trident soapstone type porcelains facilitated the rapid and totally non-destructive interrogation of unmarked pieces and assisted in their spectroscopic categorisation (Edwards, Chapter 6, in *Nantgarw and Swansea Porcelains: An Analytical Perspective,* 2018b). However, as will be reviewed later, the scientific evidence does not always seem to be compatible with an expressed expert opinion on the factory of origin based upon a wealth of detail including texture, translucency of the paste, the surface glaze, quality of decoration, pattern, and the size and shape of the piece. The ability to use perfect pieces of these rare porcelains without detriment to their integrity or substance, with neither preliminary chemical nor mechanical alterations being necessary to facilitate the observation of the spectral data, meant that this is a novel method which lends itself to serious consideration for future applications in the forensic assignment of the attribution of unknown or suspect porcelains. A notable case where RS has suggested a potential source of origin for an unknown and extremely rare piece of porcelain has been published recently (Edwards et al., Chapter 4 in *Raman Spectroscopy in Archaeology and Art History*, 2019a) wherein a very rare Regency period spill vase decorated with the figure of a ballet dancer was attributed to the Davenport porcelain factory on the basis of the analytical spectral data. A case study of a porcelain table top which upon RS analysis matched that of a marked factory porcelain plate has confirmed its attribution to the Rockingham China Works for the period 1836–1842 (Edwards et al. 2004). Owen (2011) has considered the information available from elemental analytical data derived from ceramics and has cautioned against some of the pitfalls which could occur because of the perceived difficulties in interpretation of these data, several of which have been highlighted above. Hitherto, some case studies of the RS of porcelains have set a firm base for the establishment of the technique in this regard: these include the molecular spectroscopic analysis of an important Rockingham porcelain-topped table (Edwards et al. 2004) (Fig. 4.1), Ming porcelain shards (de Waal 2004a, b; Carter et al. 2017) (Fig. 4.2), Chinese ceramics (Zuo et al. 1999) and early French porcelains (Colomban 2001, 2013; Colomban and Treppoz 2001). In contrast, the SEM/EDAXS technique has been applied to the determination of the composition of many more exemplars of porcelains, such as Nantgarw, Swansea, Derby, Worcester, Liverpool, Chelsea, Bow, Limehouse, Pomona, Bovey Tracey, Coalport and Caughley as outlined in Table 4.1. Several of these examples will cover the earliest attempts at the manufacture of porcelain from the small and relatively recently discovered manufacturing ventures at Pomona (Barker and Halfpenny, *Unearthing Staffordshire*, 1990) and Bovey Tracey (Adams and Thomas, *A Potworks in Devonshire: The History and Products of the Bovey Tracey Potteries, 1750–1836*, 1996) to the better known products of the long-established and major china works at Worcester and Derby.

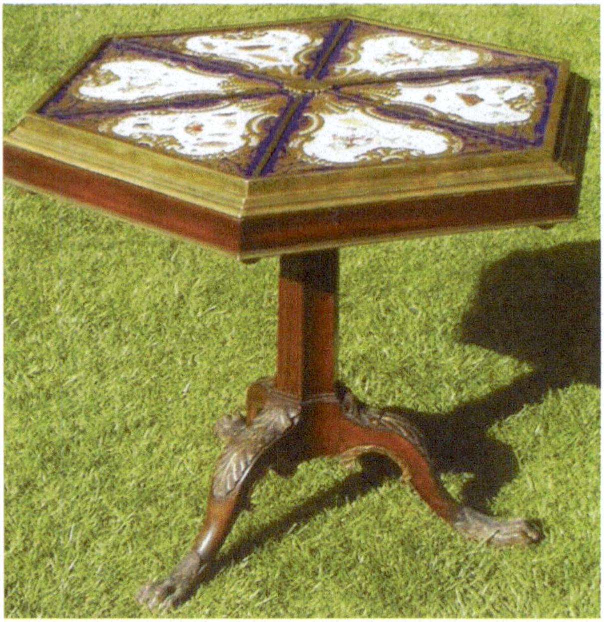

Fig. 4.1 An eighteenth century mahogany tripod table inlaid with six triangular Rockingham porcelain panels bound in brass and highly decorated with fine floral painting, *ca.* 1830–1840, from the sale of effects at the Earl Fitzwilliam's estate at Wentworth Castle, South Yorkshire; the Earl was an enthusiastic patron of the Rockingham China Works in Swinton, South Yorkshire, and commissioned several unique items in porcelain from the factory. The table porcelain was analysed using non-destructive Raman spectroscopy, which provided a match for the porcelain body with a Royal Rockingham period marked plate from *ca.* 1835–1840 and with similar enamelled pigments, therefore placing it firmly as a unique Rockingham porcelain item. *Reproduced courtesy of Bryan Bowden, Esq*

4.2 Historical Summaries of Some Early Porcelain Factories

Background information to the analytical data presented in Table 4.1 is given below in an alphabetical array rather than in any chronological arrangement for the holistic evaluation of the products of each factory cited in the Table, which will assist in the categorisation of their classification of porcelain types manufactured there and to evaluate accordingly if any porcelain body changes were recorded over time:

- ***Bovey Tracey:*** Lord William Courtenay, Earl of Devon, acquired the china clay and lignite mines at Blue Waters near Bovey Tracey in Devon in 1750 and he encouraged several experienced potters to join him there to manufacture ceramics using these readily available local raw materials (Adams and Thomas, *A Potworks in Devonshire: The History and Products of the Bovey Tracey Potteries, 1750–1836*, 1996). Although the readily available local supplies of Cornish stone and china clay were advantageous geographically for the foundation of a porcelain factory to reduce transportation costs, a problem encountered early on was

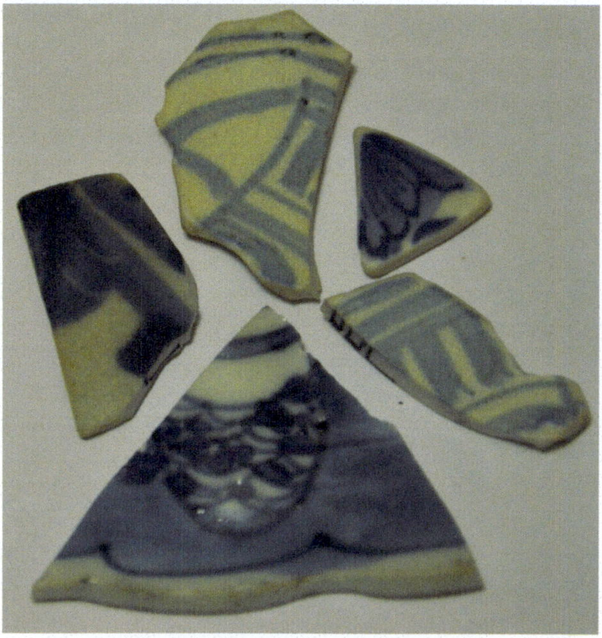

Fig. 4.2 Five Ming period (Wanli 1368–1644) underglaze blue and white porcelain shards from the shipwrecks of two large Portuguese carracks which foundered off the Cape of Good Hope, South Africa with cargoes of Ming porcelain destined for Europe; these comprise two of fifteen known Portuguese wrecks in the area which sank there before 1650. From the top, the shards are numbered 1–5: shards 1–4 were excavated from the wreck of the *Santa Maria Madre de Deus* which sank in Morgan's Bay, Eastern Cape, in 1643, and shard 5 is from the *Santo Espirito* which sank in Bonza Bay, East London, in 1608. *Shards acquired from the J.A. van Tilburg Collection at the University of Pretoria, South Africa*

the poor quality of the brown coal/lignite mined locally: the Dean of Exeter commented at the time that *"The Bovey Coal was not of a heat intense enough to answer the purpose"*, which is particularly relevant for the attainment of the necessarily high and stable temperatures required for the operation of a kiln involved in the manufacture of hard paste porcelain as identified in the discussion above. After this inauspicious start, the continued manufacture of porcelain at Bovey Tracey was minimal and they thereafter concentrated upon the production of salt-glazed earthenwares, but bankruptcy finally set in in 1752. Then in 1766, Nicholas Crisp moved to Bovey Heathfield from the Vauxhall manufactory in London following its closure and his own bankruptcy there in 1763/1764 (Owen et al. 2000), bringing with him some key ceramics workers and he was instrumental in forming the new Indeo Pottery at Indeo House in Bovey Tracey the same year. This lasted until about 1774 when it again fell into disuse until William Ellis, a local potter revived the Indeo Pottery which produced various ceramics and earthenwares until 1836. During this time another local pottery was set up locally in 1801 by the Honeychurch family, called the Folly Pottery, which

Table 4.1 SEM/EDAXS determinations of the elemental oxide compositions of selected eighteenth and nineteenth century English and Welsh porcelains

Factory	Period	Type	Analyst(s)	Specimens	SiO_2	Al_2O_3	CaO	P_2O_5	K_2O	Na_2O	MgO	Fe_2O_3	PbO
Bovey Tracey	1767–1774	Hard paste	Owen et al. (2000)	1	63.0	30.6	0.3	0.3	3.8	1.1	0.3	0.5	–
Bow	1750–1755	Phosphatic	Tite and Bimson (1991)	1	45.6	8.7	23.6	18.6	1.1	0.8	0.6	0.5	–
	1755–1760			2	50.9	5.6	23.4	15.5	0.7	0.6	0.5	0.2	0.2
	1765–1770			1	49.2	5.6	24.5	16.2	0.6	0.5	0.3	0.3	0.4
	1750–1760	Phosphatic	Owen and Day (1994a, b)	10	43.7	2.6	28.5	20.6	0.3	0.2	0.5	0.1	–
	1745–1749	Hard paste	Ramsay and Ramsay (2008)	1	64.3	21.7	5.6	–	2.4	4.1	1.7	–	–
Bristol	1775	Hard paste	Tite and Bimson (1991)	1	70.8	24.0	0.5	–	3.0	0.7	0.5	0.5	–
Caughley	1770–1799	Magnesian	Owen and Sandon (2003)	12	76.0	4.9	1.6	0.3	2.6	1.6	9.9	0.3	3.4
	1790–1799	Hard paste		8	74.6	19.7	0.6	0.3	2.9	1.4	0.2	0.2	–
Chelsea	1745–1749	Glassy	Tite and Bimson (1991)	1	71.7	4.3	13.1	–	4.3	0.5	0.3	0.4	5.1
	1755–1760	Glassy		1	70.3	4.1	18.8	–	3.5	0.5	0.4	0.2	1.9
	1760–1765	Phosphatic		1	46.4	11.3	21.8	17.2	1.4	0.6	0.5	0.3	–
Coalport	1795–1820	Phosphatic	Owen and Sandon (2003)	2	46.6	12.7	20.9	16.7	1.6	0.9	0.5	0.2	–
	1815–1825			1	44.4	13.4	21.4	16.9	2.0	1.2	0.5	0.3	–
	1820–1830			5	45.6	12.6	21.2	17.3	1.6	1.0	0.5	0.2	–
	Post 1830			7	43.7	14.2	21.4	17.1	1.9	0.8	0.5	0.2	–
	Indeterminate			2	44.1	12.4	21.8	18.1	1.6	1.0	0.5	0.2	–
	1795–1820	Hard paste		4	73.5	18.9	0.3	1.3	2.6	1.3	0.2	0.3	–
	1820–1830			2	74.8	19.0	0.3	0.8	2.5	1.2	0.2	0.3	–
	Indeterminate			11	75.3	18.9	0.2	0.5	2.7	1.3	0.2	0.2	–

Derby	1750–1770	Glassy	Owen and Barkla (1997)	6	72.3	3.6	5.2	1.5	2.4	0.2	0.4	0.2	13.8
	1810–1840	Bone china		1	38.5	15.1	23.0	17.8	2.0	1.1	0.5	0.3	1.6
	1770–1796	Phosphatic		6	46.7	10.0	21.8	16.5	1.0	0.8	0.6	0.4	1.3
Fulham	1675–1690	Silicaceous	Tite et al. (1986)	1	76.9	18.0	0.3	–	2.7	0.4	0.3	0.6	–
Isleworth	1766–1778	Phosphatic	Freestone et al. (2003)	12	40.5	7.4	23.4	17.0	2.0	0.7	0.6	0.4	3.4
			Owen (2011)	22	41.3	8.2	21.6	18.8	2.6	1.1	0.6	0.5	3.6
Limehouse	1744–1747	Silicaceous	Owen (2000a, b)	1	76.1	16.0	0.5	0.1	1.3	0.7	0.2	0.5	3.8
	1744–1747		Freestone (1993)	4	78.1	16.9	0.4	0.2	1.4	0.6	0.2	0.7	–
				4	72.5	10.8	6.2	0.1	3.3	2.5	1.0	0.7	1.3
Liverpool (Chaffers)	1756–1765	Soapstone	Tite and Bimson (1991)	1	70.3	4.3	4.8	1.6	2.9	2.2	11.9	0.7	1.3
Liverpool (Reid, Brownlow Hill)	1755–1776	Hard paste	Owen and Hillis (2003)	)1	82.6	8.3	5.2	–	2.0	0.5	0.3	0.4	0.5
Liverpool (Gilbody)	1755–1761	Phosphatic	Owen et al. (1998)	4	57.2	8.5	17.9	13.7	1.1	0.5	0.5	0.3	–
Liverpool (Pennington, Copperas)	1770–1779	Phosphatic	Owen et al. (1998)	5	49.3	11.2	20.5	14.9	1.6	0.8	0.6	0.5	–
Liverpool (Christian/ Pennington, Shaws Brow)	1765–1799	MgO/PbO	Owen et al. (1998)	4	75.6	2.8	1.9	0.2	2.1	1.1	8.4	0.4	7.4
Longton Hall	1755–1760	Glassy	Tite and Bimson (1991)	2	70.1	3.0	14.1	1.0	2.8	0.4	0.3	0.2	8.1
Lowestoft	1770–1775	Phosphatic	Tite and Bimson (1991)	2	44.4	7.8	23.6	19.8	1.1	0.9	0.7	0.4	–
	1757–1799	Phosphatic	Owen (2001)	12	42.5	7.1	27.2	20.4	0.8	0.9	0.7	0.3	–

(continued)

Table 4.1 (continued)

Factory	Period	Type	Analyst(s)	Specimens	SiO_2	Al_2O_3	CaO	P_2O_5	K_2O	Na_2O	MgO	Fe_2O_3	PbO
Nantgarw	1817–1820	Phosphatic	Tite and Bimson (1991)	3	44.5	13.0	0.6	16.7	2.3	0.8	0.6	0.2	–
			Owen and Morrison (1999)	7	43.7	12.8	22.0	17.4	2.2	0.5	0.5	0.2	–
		Siliceous		3	80.5	9.2	0.5	0.5	5.5	1.5	2.0	0.1	–
			Owen et al. (1998)	8	43.5	12.7	23.3	17.6	2.1	0.4	0.4	0.2	–
		Intermediate		1	70.8	8.9	9.2	7.5	2.3	0.2	0.2	0.2	–
		Siliceous		1	80.3	9.1	0.6	0.5	5.6	1.8	1.9	0.1	–
Pomona	1744–1747	Siliceous	Bemrose (1973)	1	75.0	15.5	0.8	0.1	2.4	0.6	0.8	1.8	–
				1	80.0	10.5	5.5	0.3	2.0	1.3	1.4	0.7	1.1
			Freestone (1993)	1	75.6	9.2	4.8	0.3	2.9	1.0	1.2	0.5	2.5
Plymouth	1768–1770	Hard paste	Tite and Bimson (1991)	1	72.3	23.1	0.7	–	3.0	0.7	0.3	0.5	–
		Hard paste	Wood and Cowell (2002)	1	63.4	30.8	0.5	–	1.7	3.6	–	–	–
Rockingham	1820–1826	Hard paste	Cox and Cox (2001)	2	73.3	19.9	0.9	0.2	3.2	1.7	0.3	0.4	–
	1826–1842	Phosphatic		3	56.4	12.2	15.7	10.7	2.3	1.0	0.5	0.7	–
	1826–1842	Bone china		1	31.2	13.2	29.2	22.2	1.2	0.8	0.7	1.0	–
	Pre-1825	Bone china		1	44.0	16.1	17.3	13.9	2.0	1.0	0.6	0.5	5.0
	1826–1842	Bone china		12	35.2	16.2	25.7	19.0	1.7	1.0	0.7	0.4	–

Swansea	1815–1819	Phosphatic	Owen et al.[a] (1998)	1	45.2	20.0	16.7	13.0	2.8	1.4	0.4	0.3	–
			Owen and Sandon[b]	1	52.8	19.9	13.0	9.8	2.6	1.2	0.5	0.2	–
		Siliceous	Owen et al. (1998)	7	71.6	23.9	0.3	0.2	1.7	0.6	0.2	0.7	–
Vauxhall	1753–1764	Soapstone	Tite and Bimson (1991)	2	75.6	4.0	5.6	–	3.7	1.2	9.1	0.5	0.4
Worcester	1751–1752[c]	Early WH Siliceous	Owen (1997)	2	76.9	18.5	0.2	–	1.5	0.6	0.3	0.6	–
			Owen (1997)	4	74.5	3.4	6.2	3.6	2.8	1.3	6.7	0.5	1.2
			Phosphatic Owen (1997)	1	75.0	3.8	5.0	2.9	3.8	0.8	0.7	0.4	7.6
	1752–1770 Wall		Owen (1997)	6	71.0	3.8	1.7	0.3	3.4	1.5	12.2	0.4	5.8
				9	73.1	3.1	2.0	0.4	3.2	1.2	10.2	0.4	6.4
	1774–1783 Davis			2	69.8	2.9	2.2	0.4	3.5	1.3	10.3	0.4	9.6
	1783–1792	Flight		6	72.8	6.0	1.2	0.4	3.9	1.0	8.4	0.4	5.8
	1792–1804	Flight and Barr		8	74.9	7.4	0.8	0.4	4.1	1.5	6.7	0.2	4.0
	1804–1813	BarrFB		2	73.9	6.6	0.5	0.3	4.2	1.1	7.0	0.2	6.0
	1813–1840	FlightBB		1	74.4	7.0	0.6	0.3	4.7	0.6	5.4	0.3	6.6
	1760	Soapstone	Tite and Bimson (1991)	1	74.1	3.8	1.9	4.9	2.8	1.0	10.8	0.3	4.9
	1765–1770			1	72.8	3.3	1.2	7.2	3.3	1.0	10.7	0.3	7.2
	1765			1	72.6	3.5	1.6	6.5	3.7	0.8	10.9	–	6.5

(continued)

Table 4.1 (continued)

Factory	Period	Type	Analyst(s)	Specimens	SiO_2	Al_2O_3	CaO	P_2O_5	K_2O	Na_2O	MgO	Fe_2O_3	PbO
Comparison porcelains													
Chinese	Yuan Yingqing	Hard paste	Tite et al. (1984)	6	75.6	18.2	0.4	0.3	3.1	1.4	0.2	0.9	–
Japanese		Hard paste	Ramsay and Ramsay	85	74.4	18.7	0.2	–	4.2	1.0	0.2	1.0	–
Chinese		Kaolinized porcelain stone	Tite et al. (1984)	1	77.0	16.0	–	0.3	3.0	2.2	0.2	0.7	–

[a]Swansea, duck-egg porcelain, *Biddulph* service, finished and decorated, gilded platter
[b]Swansea, duck-egg porcelain, finished and decorated, gilded dinner plate
[c]Worcester, 1751–1752, early Dr Wall period, Warmstry House site

itself became bankrupt in 1835. Later restarts saw the production of earthenwares at this site until 1885 and then carried on into the twentieth century. With regard to the production of porcelain at Bovey Tracey, therefore, we must really consider just two short periods, firstly, a very limited production between 1750 and 1752, followed by the Crisp revival at the Indeo Pottery which took place between about 1767 until 1776, or thereabouts. The analysis of Bovey Tracey porcelain shards by Owen et al. in 2000 (Owen et al., *Geoarchaeology,* 2000) concerns this later period of the Crisp production of hard paste porcelain and their analytical results of 63% silica, 31% alumina, with 5% combined soda and potash and low percentages of lime, magnesia and a trace of phosphorus pentoxide confirm this type of porcelain production. It is interesting to speculate on the origin of the recipe for this hard paste porcelain but comparison of these results with the analytical data for a Plymouth shard (Wood and Cowell 2002) reveals that the two are almost identical in elemental oxide percentages: for silica, alumina, lime, combined soda and potash, Bovey Tracey gives 63%, 31%, 0.3% and 5%, respectively, whereas Plymouth gives 63%, 31%, 0.5% and 5% respectively – analytical data that are so close in value that it is tempting to infer that the analysed shards from the Bovey Tracey site originated at the same source factory (Plymouth) or alternatively perhaps that identical (or indeed very similar) recipes and formulations were used at both sites. Both sites, Bovey Tracey and Plymouth, were quite local to each other and there could well have been an exchange of information and specific knowledge between the workforce at each site, especially towards the end of the Plymouth factory's existence in 1770, which would still have been in the temporal frame of Nicholas Crisp's ongoing production of porcelain at Bovey Tracey! What this mean analytically, of course, is that an unknown specimen of porcelain form either site is analytically not differentiated through the elemental oxide percentages alone.

- ***Bow:*** One of the first English porcelain factories, along with its neighbour Chelsea, Bow porcelains (Adams and Redstone, *Bow Porcelain*, 1981) have been the subject of several experimental analytical studies in which the analyses have not always concurred with established and avowed expert ceramic opinion. Tite and Bimson (1991) analysed four specimens of Bow shards from three production periods, namely, 1750–1755, 1755–1760 and 1765–1770: all were phosphatic in type with an average percentage of phosphorus pentoxide of 16.2%, representing a bone ash content in the original recipe of approximately 38–40%. Corresponding percentages of silica of 49.2%, alumina of 6.4%, magnesia of 0.4%, lime of 23.8% and a combined soda and potash content of 1.2% are supportive of an assignment to a predominantly phosphatic and soft paste porcelain body. On the other hand, some recent studies of first patent Bow porcelains by Ramsay and Ramsay (2008) give a totally different body composition with 60.4% silica, 26.8% alumina, 6.3% lime, and 6.5% combined soda and potash, with zero percent phosphorus pentoxide, which is more typical of a true or hard paste porcelain. It is clear that from the first patent Bow body registered by Edward Heylyn and Thomas Frye in 1745 (Hurlbutt, *Bow Porcelain,* 1926), several major body variants were being experimentally evaluated and this complex story has been unravelled in a very recent and comprehensively detailed scholastic work published by Ramsay and Ramsay (*The Evolution and Compositional Development of English Porcelains from the 16th Century to Lund's Bristol c. 1750 and Worcester c. 1752 – the Golden Chain.*, 2017). In their ground-breaking research on Bow porcelains, Ramsay and Ramsay have also analysed data pertaining to the mysterious early English "*Factory A*" porcelains, for which it has long been postulated belonged perhaps to a completely unassigned and unrecognised porcelain factory from the mid-eighteenth century but for which a cogent scientific argument based upon analytical data has now been presented for their re-assignment to early Bow factory porcelain body variants. An important point emerges from this particular study in that, unlike Derby or Rockingham porcelains, where the successful recipes seemed to have been maintained unchanged or at least very little changed, over relatively larger periods of time, other factories of an admittedly earlier foundation experimented with several body variants which could possibly then have been retained in parallel and a concurrent commercial production established over a significant production period. Forensically, therefore, it might then be foreseen to be rather difficult to establish a characteristic set of analytical data which would unambiguously define a factory porcelain body with a single set of unique characteristic elemental oxide compositions, which could perhaps be utilised to identify the potential origin for an unknown piece, unless examples of these body variants had been well-characterised analytically separately for known periods of production at the factory (Freestone 1996).

- ***Bristol:*** see Plymouth/Bristol below.
- ***Caughley:*** Founded in 1775 near Broseley in Shropshire by Ambrose Gallimore and Thomas Turner, who was the brother-in-law of Josiah Spode, a high quality soft paste porcelain was produced here until 1799, mainly functional in scope and specialising in the production of useful articles such as tea wares and dinner sets in underglaze blue chinoiserie patterns. Some 151 patterns have been recorded for Caughley china in the researches of the Caughley China Society. The site was selected in the first place for its excellent access to the transportation network afforded by its proximity to the River Severn and the presence there already of an earthenwares industry. The arrival of Thomas Turner catalysed the change in emphasis from the factory production of earthenware to porcelain and this was enhanced by the recruitment from the Worcester China Works of Robert Hancock, an esteemed engraver of chinoiserie patterns. Also known as the Salopian China Manufactory, Caughley became famous for its unusual design and production of porcelain table wares and useful articles such as scalloped dishes, pickle trays, asparagus servers, radish plates, masked jugs and pounce pots (Godden, *Caughley and Worcester Porcelains,* 1775–1800, 1969b). A set back occurred in 1783 when some key members of the workforce left Caughley to join Josiah Spode at his china works in Etruria, Stoke-on-Trent, Staffordshire, including Thomas Minton, who later was instrumental in the establishment of his own eponymous china manufactory in Staffordshire, which later became one of the most successful china and glazed earthenware factories in the mid- to late-

nineteenth century. Around this time the Salopian China Manufactory set up a warehouse in Lincoln's Inn Fields in London, which also became a later outlet for the sale of Spode china. In 1783, Robert Chamberlain founded his own china factory in King Street, Worcester, and he became an important client of Caughley for the purchase of china from them in the white for decoration and sale from his own establishment, until he eventually produced his own porcelain in 1788, thereafter known as "Chamberlain's Worcester" porcelain. It has been suggested that Robert Chamberlain had become increasingly more disillusioned with what he perceived was a fall-off in quality of the porcelain biscuit being supplied to him at Worcester in large quantities by the Salopian China Manufactory, this prompting his venture into porcelain manufacture of his own. It would be interesting to verify analytically if this allegation was true from a change in chemical composition between early and later Caughley shards which may reflect the decreased quality perceived by Chamberlain and which caused him so much anguish and result in his embarkation into porcelain production on his own. Caughley seemed to have been a nursery for aspiring porcelain manufacturers, as John Rose also started there and then decamped to a neighbouring pottery site at Jackfield, where he commenced to manufacture porcelain in competition with Caughley in collaboration with Thomas Rose and with Messrs. Anstice and Horton, calling his new venture the Coalport China Works (Barrett, *Caughley and Coalport Porcelains*, 1951). This erosion of key staff and growth of nearby competitors, coupled with Thomas Turner's "*indifferent state of health*", must have had a toll on Turner's ability to keep the Caughley manufactory going as in 1798/1799 he eventually sold out the Caughley operation to John Rose at Coalport. Initially, Rose continued to maintain the production of porcelain at Caughley in the white and he oversaw its transportation across the River Severn to the neighbouring Coalport China Works site for decoration but a particularly serious accident in October, 1799 resulted in the death of 29 of the workforce engaged in transporting china from Caughley to Coalport on the River Severn, when an overloaded barge carrying china capsized. As a result, John Rose thereafter ran down the Caughley factory operation and used the production hardware to enlarge his Coalport China Works base. Thomas Turner then disappeared from the ceramics scene and died in 1809.

- ***Chelsea:*** The establishment of the Chelsea China Works by Nicholas Sprimont in 1745 gives Chelsea the credit of being one of the first English china manufactories, although there is documentary evidence that suggests the first factory was actually that at Pomona at Newcastle-under-Lyme in Staffordshire, which was founded in 1743/1744 (Bemrose 1973), closely followed by those at Bovey Tracey in Devonshire and at Bow, in London, both founded around 1744/1745. Early Chelsea products undoubtedly had a glassy porcelain body and the use of glass cullet was certainly prevalent in their recipe composition: analytically, this is not as easy to define as with later factories as it seems that the incorporation of lead-free crown glass (or soda glass) was the norm at Chelsea, perhaps in admixture occasionally with a lead-rich flint glass additive, and this would of course render only a rather small presence of lead oxide to be detectable analytically in the body. This can be seen in Table 4.1, where the analytical data for Chelsea porcelain of Tite and Bimson (1991) are revealed: the earliest Chelsea shards from the marked triangle (*ca.* 1745–1749) and red anchor mark (ca. 1755–1760) periods both show lead oxide percentages of 5.1% and 1.9% only, along with high silica and low alumina percentages and higher percentages of alkaline oxide potash and lime cumulatively amounting to about 18–20% in total. In contrast, however, a shift in Chelsea body composition seems to have occurred in the later gold anchor period between 1760 and 1770, since this body is totally different in composition to the earlier glassy variety and analysis now records some 46% silica, 11% alumina, 22% lime and 17% phosphorus pentoxide – the latter being equivalent to a significantly high 44% bone ash raw material component. A similar situation can be described for the Vauxhall manufactory (ca. 1753–1764), where the analysis of shards (Tite and Bimson 1991) showed that very small lead oxide percentages of approximately 0.3% could not be correlated with the recorded use of ground glass cullet in the paste – which can now therefore be confirmed as originating from a crown glass component rich in potash but also being lead-free. The Vauxhall porcelain differs from Chelsea in that it has a significant percentage of magnesia in the shards, which approximates to about 25–30% of soapstone in the original recipe, although the silica percentage is still high at approximately 75%.

- ***Coalport:*** John Rose formed the Coalport China Works in 1795 on the site of the Jackfield earthenware ceramics factory neighbouring the Caughley (or Salopian) China Manufactory, which was situated just a mile upstream on the River Severn, where he had been apprenticed to Thomas Turner until he left in 1793 to set up at Jackfield. In 1792 the Coalport Canal had opened between Coalport and the River Severn, which further improved the transportation network for the importation of raw materials and export of the finished porcelain goods. Financial support for the venture was forthcoming from Edward Blakeway. As noted above for Caughley, Rose acquired the Salopian China Manufactory in 1799 and for most of that year he continued using Caughley as a producer of biscuit china for decoration at his works in Coalport, until the fatal accident involving the transportation of this china between the two sites resulted in his re-assessment of the situation and his enlargement of the main site operation at Coalport using the Caughley site hardware (Godden, *Coalport and Coalbrookdale Porcelains*, 1970). A spin-off operation was started by Thomas Rose in 1800 who set up a small china works nearby on the opposite bank of the Coalport Canal with William Reynolds and Robert Horton; the former died in 1803 and opened the way for Robert Anstice to join as partner, until John Rose acquired this site too for his expanding business at Coalport in 1814. Many ceramic historians have claimed that John Rose then later purchased the equipment and materials from the Nantgarw and Swansea china manufactories in 1819 and 1820, on the basis of a seemingly false assertion being made by Llewellyn Jewitt (Jewitt, *The Ceramic Art of Great Britain from the Prehistoric Times Down to the Present Day,* 1878). Recent documentation and archaeological excavations have shown Jewitt's statement to be a rather suspect if not totally incorrect assumption (Edwards, *Porcelain to Silica Bricks: The Extreme Ceramics of William Weston*

Young, 1776–1847, 2019). The reality may be that Rose did purchase several items at the closing-down auctions of the Nantgarw and Swansea China Works in 1823 and 1824 but to infer that this involved additionally detailed recipe formulations, kilns and manufacturing equipment must be now considered to be rather tenuous. William Weston Young and Thomas Pardoe certainly operated the Nantgarw China Works well into 1823, until the death of Thomas Pardoe in June of that year, and it has recently been shown (Edwards et al. 2019a, b) that a hard paste porcelain was also being made at the Nantgarw China Works site, probably by William Weston Young or his Pardoe family successors, which it would not have been possible to achieve if the site had been cleared as Jewitt has proposed (Jewitt, *The Ceramic Art of Great Britain from the Prehistoric Times Down to the Present Day,* 1878). In fact, the archaeological excavation of the Nantgarw site by Isaac Williams in 1931–1932 (Williams, *The Nantgarw Pottery and its Products: An Examination of the Site*, 1932) proves conclusively that this notion is clearly absurd and it is now realised that Llewellyn Jewitt had probably never visited the Nantgarw China Works site to record his observations directly and that his hitherto assumed factual statement was based upon a false presumption, perhaps arising from a misreading of the auction sales reports. Despite this controversial situation, an interesting monogram has surfaced from the Coalport China Works archive comprising an ampersand containing intertwined C, S and N letters, representing the "acquisition" of the Caughley, Swansea and Nantgarw China Works by John Rose (Fig. 4.3): this only has appeared on Coalport porcelains from the 1840s, after the death of John Rose, and the current thinking is that this perhaps was a marketing device being utilised by the Coalport China Works at that time to announce the superiority of their translucent china over that of contemporary competitors by advertising their "incorporation" of the three specified highly esteemed porcelain manufactories which were still held in very high public regard. In 1799, Rose certainly did acquire the Caughley (Salopian) enterprise but he never took over at either the Swansea China Works or at the Nantgarw

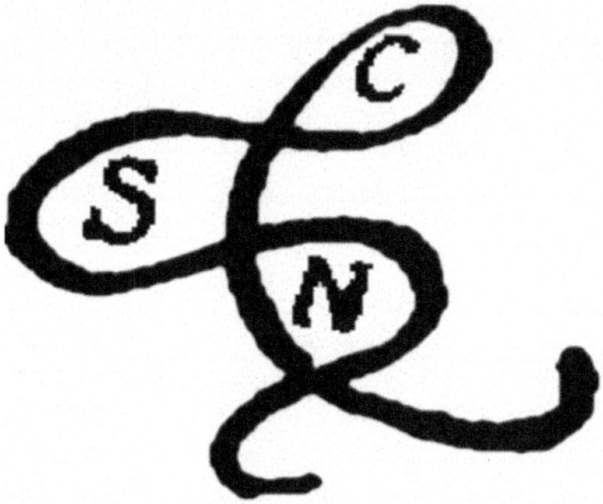

Fig. 4.3 Ampersand used by the Coalport China Works *ca.* 1860 and thereafter containing the letters C, S and N, claimed to represent the takeover of the Caughley, Swansea and Nantgarw factories by John Rose in 1790 and 1820–1823, respectively. Used by John Rose's successors at Coalport after his death in 1848

China Works, although he did enlist the services of William Billingsley and Samuel Walker from Nantgarw into his employment at Coalport in 1820. In 1820, Coalport received a prestigious accolade and gold medal from the Society of Arts for their new "*feldspar porcelain*" which was glazed with a lead-free glaze, so removing the perceived problematic toxicity of the industrial use of lead compounds and the premature deaths and disablement of the young workforce employed in the glazing sheds of the ceramics factories: much Coalport porcelain flatware from this period shows a proud announcement of this award in the form of a backstamp in puce enamel (Messenger, *Coalport, 1795–1926: An Introduction to the History and Porcelain of John Rose and Company*, 1995). In 1830 the Coalport China Works initiated a novel method of applying a transfer printed blue outline onto china being enamelled to assist the painters and thereby increase production output – this became known affectionately as "clobbering" and was widely adopted by other factories later in the nineteenth century. John Rose died in 1841 and the china business was continued by his nephew and William Pugh, until Pugh died in 1875 and the factory was closed down. It reopened in 1880 and was reinstated as the Coalport China Works but fell into further financial difficulties again in the 1920s, when it was subsumed by the Cauldon Potteries in Shelton, Staffordshire, and eventually by the Wedgwood Group in 1967, who have retained the Coalport name on their china products to the present time.

- ***Davenport:*** In 1785, John Davenport was working as a potter for Thomas Wolfe at Stoke-on-Trent and in 1794 he acquired his own pottery at Longport in Staffordshire, manufacturing creamware and blue and white transfer earthenware (Lockett, *Davenport Pottery and Porcelain, 1794–1887,* 1972). In 1806, he saw evidence of the phenomenal commercial success of Spode's bone china products and he rapidly devolved into making porcelain of a high quality: the Prince of Wales, later Prince Regent and King George IV, was an esteemed client and ordered several services from the Davenport factory around this time. John Davenport retired in 1830, leaving his sons William and Henry to carry on the business. Henry died in 1835 and the business then became known as William Davenport and company, until William died in 1869, the business continuing with William's sons until 1887. It seems that between 1806 and 1815 a high quality soft paste porcelain was made, gradually evolving into a bone china of exceptionally fine quality. Lockett and Godden (*Davenport China, Earthenware and Glass, 1794–1887*, 1989) state that "*Little is known of pre-1830 Davenport porcelain except that they were of very high quality and could compete in decoration with the finest Derby and Swansea*". Edwards (*Swansea and Nantgarw Porcelains: A Scientific Reappraisal*, 2017) has commented that a spill vase of exceptionally fine translucency in a private collection which had been previously assigned to Davenport *ca.* 1815–1820 is actually of Nantgarw porcelain, the decoration and translucency being so fine as to render this possible mis-attribution to be made. It has also been mentioned above that a very rare and fine quality porcelain spill vase of hitherto unknown origin and of a very clear white translucency was firmly assigned to Davenport porcelain of *ca.* 1815–1820 from the non-destructive analytical results of a Raman molecular spectroscopic (RS) study (Edwards et al., *Raman Spectroscopy in Archaeology and Art History,* 2019a).

- ***Derby:*** The Derby china factory was one of the first to be established in England by Andre Planche, an émigré from the French porcelain manufactories, with a formal foundation date of 1748, and the acquisition by its then proprietor, William Duesbury, of the Chelsea factory in 1770 gave Derby an impetus which rapidly saw it become, along with Worcester, one of the most highly esteemed English china manufactories in the late eighteenth century (Twitchett, *Derby Porcelain, 1748–1848,* 2002; Edwards, *Derby Porcelain: The Golden Years, 1770–1830*, 2018a). Owen and Barkla (1997) have analysed shards of Derby porcelain and have deduced that three types of porcelain were made there during the period 1770–1830: the very first was a glassy porcelain, rich in lead oxide, which would have been compatible with their acquisition of the Chelsea factory in 1770 and is sometimes referred to as "Chelsea-Derby". The second of these is the standard phosphatic porcelain body which contains approximately 40% bone ash (Anderson, *Derby Porcelain and the Early English Fine Ceramic Industry*, 2000). The third is a bone china which contains a similar amount of bone ash but a correspondingly reduced percentage of silica content and a proportionally higher percentage of alumina. It can be reasonably proposed that this bone china variety would have been produced sometime between 1795 and about 1810, when the dramatic success of Josiah Spode's bone china was being marketed so well at the Spode China Works in Staffordshire. The analysis (Tite and Bimson 1991) of a modern Spode-type Staffordshire bone china made by the Hammersley china works in 1955 reportedly gives the following percentage composition: silica 34%, alumina 15%, lime 27%, phosphorus pentoxide 21%, magnesia 0.6%, potash 1.8% and soda 0.8%, which are remarkably similar to the Derby bone china shards from some 200 years previously, although a small amount of lead oxide was recorded in the Derby analyses which correlates with the use of a flint glass cullet additive by William Duesbury (Anderson, *Derby Porcelain and the Early English Fine Ceramic Industry,* 2000). It also seems very reasonable to presume, therefore, that Robert Bloor, who adopted the proprietorship of the Derby China Works following the Duesbury and Kean era, did not modify the original phosphatic body recipe - although the later Bloor Derby glaze composition (*ca.* 1810–1848) was clearly different from that used during the Duesbury era as the evidence of crazing at the interface between the later glaze and body is universally acknowledged to be recognisable as a "Bloor Derby" characteristic.
- ***Fulham:*** see later.
- ***Isleworth:*** The smallest of the "early five" porcelain manufactories based in London, Isleworth joins the other three of Chelsea, Limehouse and Vauxhall in being located on the banks of the River Thames at the site now occupied by Nazareth House in Twickenham (Massey 2003; Godden 2008). The remaining London factory at Bow was situated on the banks of the River Lea, a tributary of the River Thames. The Isleworth Pottery was established by Joseph Shore from Worcester in 1756 and commenced making creamware and earthenware slip but

started manufacturing porcelain in 1766 until 1787. Shore died in 1768 and the business was taken over by his son William and a partner Benjamin Quarman: it continued in operation making earthenware until 1831. Archaeological excavations have been carried out on the site and the most recent of these was in 2016, establishing the remains of a biscuit kiln and glost kiln, with many wasters being excavated (Howard 1998, 2001). Two important analytical investigations have been carried out on Isleworth shards by Ian Freestone et al. (2003) and by Victor Owen in Panes et al. (2012), the latter attempting to differentiate scientifically between Bow and Isleworth porcelains, which is not a simple task analytically.

- **Limehouse:** The Limehouse Porcelain Manufactory was among the first of the English porcelain manufactories, being founded just a year after Bow and Chelsea in 1745, by William Ball, John Wilson and Joseph Wilson within 2 miles of the Bow manufactory at Fore Street near Duke Shore, at the Limehouse reach on the River Thames in London (conforming to the location of the present-day Narrow Street) (Tyler et al., in *Digging for Early Porcelain*, 2000). In 1990 the site was excavated archaeologically and remains were found of a kiln and a large quantity of glazed and unglazed china wasters, amounting altogether to 1402 shards (Potter, in *The Limehouse Porcelain Manufactory: Excavations at 108–116 Narrow Street, London, 1990,* 1998). Jay and Cashion (2013) report a recent Raman spectroscopic analytical study of these wasters and confirm that the body is comprised mainly of silica and alumina with magnesia and lead oxide also present, and with a silicaceous and lead glaze where appropriate. Generally, the porcelain quality was not very good, which contributed to the closure of the factory in 1748, but much interest has been stimulated in Limehouse porcelain from analytical studies of the shards by Freestone (1993), Owen (in *The Limehouse Porcelain Manufactory: Excavations at 108–116, Narrow Street, London,1990,* 2000b) and by Ramsay, Daniels and Ramsay (*Limehouse Porcelain: Are Limehouse Porcelains in Fact All Limehouse? Evidence from Archaeology, Science, and Historical Documents,* 2015). From the analytical data obtained from the relevant factory shards, Ramsay, Daniels and Ramsay (2015) have identified a time-line pathway that existed from Bow-Limehouse-Lund's Bristol-Worcester which spanned just 8 years in time between 1744 and 1752. In a scholastic piece of detective work, the analytical data from the two separate and independent studies of Ramsay et al. and Owen reveals that there were up to three distinct phases of porcelain production at Limehouse. Ramsay et al. (*Limehouse Porcelain: Are Limehouse Porcelains in Fact All Limehouse? Evidence from Archaeology, Science, and Historical Documents*, 2015) could ascribe the following chronology to their shards – 1745-early 1746: a silica/alumina body with a silica/alumina/lime glaze, early 1746 – June 1747: a silica/alumina/lime body with a moderate lead oxide glaze, and from June 1747 until early 1748: a magnesia/phosphate body with a moderate lead oxide glaze. All of these were compositionally derived from Bow porcelain. Freestone (1993) defined two bodies, both silicaceous and aluminaceous, namely, two parts of alkaline lime/bottle glass and one part of a silica/alumina body similar in composition to the Bow first patent body, followed by a second group with lower silica levels but including higher percentages of lime, potash and soda. Both bodies had a low-lead content glaze. Finally, Owen (*The Limehouse Porcelain Manufactory: Excavations at 108–116, Narrow Street, London,1990,* 2000b) also found two types of porcelain body, the earliest comprising mainly silica and alumina with a low lime content and a lime-alkaline glaze and the later period shards with a similar body but with a lead oxide glaze. Bernard Watney (in *Limehouse Ware Revealed,* 1993) confirmed the presence of a magnesian-phosphatic body in his collection of Limehouse porcelain shards. Ian Freestone (1993) noted that glass frit had been added to the china clays in the formulation and that this suggested a link between the three factories of Pomona, Bow and Limehouse. The poorer quality of the porcelain body at Limehouse has been attributed to the inability of the proprietors to source the premier grade china clay which their Bow rivals obtained from the Cherokee "*unaker*" deposits, and as a result were forced to use the inferior sedimentary ball clays probably sourced from Cornwall. The eventual demise of the Limehouse operation can be ascribed to, firstly, a lack of demand for their porcelain product, which was deemed to be inferior to Bow (Elliott 1929), and, secondly, the drying up of their source of soapstone deposits from the mine at Kynance Cove in Cornwall which became worked out in 1747. The intriguing link between Bow, Limehouse and then Lund's Bristol and eventually Worcester proposed by Ross Ramsay is extremely relevant to the wider consideration of transferability of knowledge and technologies from one factory to another at this time, particularly when they were close neighbours to each other. The hiring of personnel from a rival factory was quite common as we have seen earlier for Caughley and Coalport and this was also found to be the case here between Bow and Limehouse; it appears that most staff movement between factories occurred around Michaelmas (November) annually in the ceramics industry when contracts of employment were renegotiated. It is even more revealing when we consider the compositions of the formulations for Bow, Limehouse and Lund's Bristol: these comprise three ceramic types of soft paste porcelain: a high clay silica/ball clay/saltpetre composition at Limehouse, which was switched to the magnesian bodies in June 1747, then comprising a magnesia/lead oxide porcelain made from soapstone, silica and lead oxide glass frit, followed by a magnesia/phosphate/lead oxide porcelain made from soapstone, bone ash, silica and lead oxide flint glass frit. The composition of Lund's soft paste Bristol porcelain produced between 1749 until late 1750 was a magnesia/phosphate/lead glass body with a moderate lead oxide glaze followed by a late 1750 until early 1751 body variant with a magnesia/lead glass body plus a moderate lead oxide glaze, which clearly pointed to information being acquired from Bow and Limehouse. It is also known from correspondence that Edward Heylyn, Thomas Frye, William Ball and Benjamin Lund frequented each other's factories and it can be surmised that the transfer of empirical compositional and technological information about their experimental products would have been undertaken in a supportive way by these four leaders in English porcelain experimentation in its earliest days. One unexpected aspect to emerge from this in-depth and comprehensive research into the Limehouse factory and the accompanying analyses of its shards is that there is clear evidence for an active English synthetic porcelain activity some 35 years prior to that reported at Meissen in 1709, which involved the personage of John Dwight, whose critical role in china production in the seventeenth century will be discussed further below in the context of the influence of key members of the prestigious Royal Society of London!

- ***Liverpool:*** Liverpool was home to a group of porcelain manufactories between 1754 and 1804 which built upon the production of tin-glazed delftware made there from about 1710 (Brown and Lockett, *Made in Liverpool: Liverpool Pottery and Porcelain*, 1700–1850, 1993): the porcelains made in Liverpool were mainly functional and decorated in underglaze blue with the then fashionable chinoiserie designs (Spero, *Liverpool Porcelain*, 2006). The porcelain factories and their founders (Hillis, *Liverpool Porcelain*, 2011) can be summarised briefly and chronologically as follows: Richard Chaffers (1754–1765) at Shaw's Brow, who was licensed to mine soapstone in Cornwall in 1756 for his soapstone bodied paste; Philip Christian took over at Shaw's Brow in 1765 upon the death of Chaffers, from 1765 until 1778; Samuel Gilbody operated a factory converting an earthenware product to porcelain as next door neighbour to Chaffers on Shaw's Brow from 1754–1761, concluding with his bankruptcy in 1760; William Reid opened his factory at Brownlow Hill in 1754 which also reached bankruptcy in 1761 – this was taken over by William Ball in 1761 who then sold the business on to Thomas Lewis in 1763, who then leased the factory to James Pennington and Company the same year; James, John and Seth Pennington continued to produce porcelain at the Brownlow Hill site until 1767, then they moved to Park Lane until 1773. John Pennington opened two more porcelain factories at Copperas Hill (1770–1779) and Folly Lane (1779–1786), the latter being continued by his widow Jane Pennington until 1794 after John Pennington died in 1786. Afterwards, the porcelain production at the various Liverpool factories went into decline until their eventual closure at the turn of the century. The porcelain type associated with the Liverpool factories is generally soft paste, but clearly the recipes had a soapstone component, as the analytical data in Table 4.1 from Tite and Bimson's work (1991) for, admittedly, only a single shard from Chaffers' site dated to the period 1756–1765 indicates a very significant magnesia content of 11.9%, representing almost a 35% soapstone raw material content in the recipe formulation. Interestingly, this same shard has a phosphorus pentoxide percentage of 1.6%, low lime and alumina contents of 4.8% and 4.3%, respectively, and a high silica percentage of 70.3%. The Liverpool porcelain manufacturing story does not end there as Thomas Wolfe opened up a factory at Islington which operated between 1790 and 1795, making a hybrid hard paste porcelain there similar to New Hall in Staffordshire and in 1796, Samuel Worthington diversified his established earthenware factory at Herculaneum into the manufacture of some porcelain, importing workers from Staffordshire to fulfil his need for skills and expertise (Brown and Lockett, *Made in Liverpool: Liverpool Pottery and Porcelain*, 1700–1850, 1993; Watney, *Liverpool Porcelain of the 18th Century*, 1997).

- ***Longton Hall:*** An early English soft paste porcelain factory in Staffordshire founded in 1749 by the "Three Williams", namely William Littler, William Jenkins and William Nicklin, built a reputation for its decorative figures with overglaze blue enamel ("*Littler's blue*") produced between 1754 and 1757, often described as "snowmen" on account of the overly thick glaze which has caused the enamels to run and streak, simulating the effects of melting snow. Generally, the thickly potted and rather heavily bodied functional wares that were the major production of Longton Hall mostly sold locally despite the opening of a London outlet in 1758, which was unsuccessful and closed within the same year (Watney, *Longton Hall Porcelain*, 1957). A notable employee as a decorator and enameller in the early years was William Duesbury, who later left to help set up and eventually run the Derby China Works in the early years, during which time Derby acquired the Chelsea manufacturing enterprise. The business passed to Robert Charlesworth in 1755, who was responsible for closing the operation in 1760. During that time, William Littler moved to establish a new pottery at West Pans in Scotland, where he continued decorating Longton Hall ceramics. For analyses see Owen and Day (1998).
- ***Lowestoft:*** From 1757 until 1802, functional and useful soft paste porcelain wares were made at Lowestoft in Suffolk: extensive copying of the shapes from the Bow and Worcester porcelain factories took place over three identifiable periods, namely Early 1756–1761, Middle 1761–1768 and Late 1768–1802 (Godden, *The Illustrated Guide to Lowestoft Porcelain*, 1969a; Godden, *Lowestoft Porcelain*, 1999b). Lowestoft claims to be the third porcelain early soft paste manufactory in England for longevity next to Worcester and Derby. The first advertisement for its porcelain wares appeared in 1760 and these were sold mostly to a local market clientele although in 1770 a warehouse shop was opened in Cheapside, London (Spencer, *Early Lowestoft: Study of the Early History and Products of the Lowestoft Porcelain Manufactory*, 1981) and it appears that an export market also existed (Owen 2001). The Early and Middle periods concentrated upon underglaze blue and white decorations in chinoiserie styles and polychrome enamels did not appear until after 1768. The actual foundation date of the Lowestoft porcelain manufactory is unclear (Brooks, *China Returns to Lowestoft: The Story of Lowestoft Porcelain,* 2001) but it seems that the names of Philip Walker, Obed Alred, John Richman and Robert Brown were all associated with early Lowestoft: Robert Brown, a chemist, is mentioned as being the manager of the factory in 1770 and after his death in 1771, the business passed to his son, also Robert Brown, who carries on until the closure of the factory. Godden (*Lowestoft Porcelain*, 1999b) has noted that some former employees at Lowestoft were formally recorded as working in the Worcester China Works in 1800, before the formal closure of the Lowestoft operation in 1802. As early as 1795, Robert Brown commented that the competition from the Staffordshire bone china factories resulted in the running down of the china works at Lowestoft, so a dispersal of the skilled workforce would have been expected to have occurred soon afterwards. In Table 4.1, two Lowestoft shards from the Late Period, 1770–1775, were analysed by Tite and Bimson (1991), who determined that they were clearly of the phosphatic type with significant percentages of phosphorus

pentoxide around 20% with medium silica and low alumina percentages of 44% and 8%, respectively, and with no addition detected of flint glass cullet and soapstone. The compositional distinction between Lowestoft and Bow porcelains has been investigated by Owen and Day (1994b) using analytical data from excavated shards.

- ***Minton:*** The Minton China Works was founded by Thomas Minton in 1793 in Stoke-on-Trent and they quickly established an enviable international reputation for the manufacture of a range of fine ceramics in classical shapes. They have been described as Europe's leading manufacturer of ceramics in the Victorian era, supplying earthenware, stone china, tiles and architectural ceramics for public buildings which can still be seen today in the Houses of Parliament, London, and the US Capitol, Washington. Originally trained as an engraver with Thomas Turner in the Caughley (Salopian) China Works in the 1780s, Thomas Minton started manufacturing earthenwares in his business called Thomas Minton and Sons, then he set up to make bone china in partnership with Joseph Poulson in Poulson's neighbouring pottery in 1798 (Godden, *Minton Pottery and Porcelain of the First Period, 1793–1850*, 1968). When Poulson died in 1808, Minton carried on the venture in bone china manufacture until 1816, then ceased until he built a new "china pottery" in 1824. As a profitable sideline, Thomas Minton with Poulson, Wedgwood and Adams and the proprietors of the New Hall China Works set up the Hendra Company at St Dennis, Cornwall, to mine and export china clay and Cornish stone minerals at an attractively favourable rate to themselves. Early Minton porcelain is rather restrained in Regency style and relied upon geometric patterns which carried on into the architectural ceramic tile designs for which they became famous later in the century. Upon the death of Thomas Minton in 1836, his son Herbert succeeded to the business and he developed their architectural ceramics work in association with leading architects such as Augustus Pugin. His launch of the famed *Palissy* earthenware at the Great Exhibition of 1851, a colourful, lead-glazed majolica, became a staple brand-name of Mintons and was later copied by many others.
- ***Nantgarw:*** The Nantgarw China Works was founded by William Billingsley and Samuel Walker upon their departure from the Royal Worcester China Works, owned by Barr, Flight and Barr, after the death of Martin Barr in November 1813. From their arrival in Worcester from Brampton-in-Torksey in 1808, Billingsley and Walker had engaged in rather clandestine experiments with the encouragement of Martin Barr to refine the Worcester body using a special experimental reverberatory kiln and new paste recipes, much experimentation being carried on cloaked in the highest secrecy at night. Just before he died, Martin Barr realised that Billingsley and Walker wished to avail themselves of the opportunity to manufacture their new porcelain from their experimental ideas and he presented them with an honorarium of £200 to effect this: after the death of their mentor in 1813, the Royal Worcester China Works passed into a new ownership consortium of Flight, Barr and Barr and the new owners did not wish to proceed with the re-conformation of the Worcester production line to accommodate the new bodies proposed by Billingsley and Walker in the light of their ongoing experiments

there under Martin Barr's direction. In this unsupportive atmosphere, therefore, Billingsley and Walker left Worcester in early 1814 and took out a lease on a property in the small, isolated Welsh village of Nantgarw (Welsh: *a rough stream*) alongside the newly constructed Glamorgan Canal which had opened in 1793 between the deep-water Cardiff Docks and the coal mining and iron-working communities of the Rhondda, Pontypridd and Merthyr. The Glamorgan Canal would facilitate the easy transportation inwards of raw materials such as local coal, lime and bone ash, and others such as china clay, sand and flints from Cornwall and East Anglia via Cardiff docks, whilst at the same time providing good and safe transportation outwards of the finished, delicate porcelain artefacts to London and elsewhere. In September 1814, William Billingsley, Samuel Walker and their financial backer, William Weston Young, submitted a request for financial support for their fledgling china works in the form of a *Memorial* to the British government which proved unsuccessful (Morton Nance, *The Pottery and Porcelain of Swansea and Nantgarw,* 1942; Edwards, *Porcelain to Silica Bricks: The Extreme Ceramics of William Weston Young,* 1776–1847, 2019; Edwards, *Nantgarw and Swansea Porcelains: An Analytical Perspective*, 2018b). The chairman of the commission, Sir Joseph Banks, alerted Lewis Weston Dillwyn of Swansea to the exceptional quality of the porcelain being produced by his neighbours at Nantgarw from the specimens submitted in support of the application and to the need of the applicants for a sponsor. As a direct result of the refusal of official financial support, later in 1814 the Nantgarw enterprise (Phase I) was closed down and Billingsley and Walker, but not Young, were recruited by Lews Weston Dillwyn, to launch his Swansea China Works for the manufacture of porcelain at Swansea, which led directly to Dillwyn experimenting with various porcelain bodies from 1815 to 1817 (Dillwyn, *Notes,* see Appendix in Eccles and Rackham, *Analysed Specimens of English Porcelain in the Victoria and Albert Museum Collection,* 1922, and in Edwards, *Nantgarw and Swansea Porcelains: An Analytical Perspective*, 2018b). The resulting Swansea duck-egg china body was held in very high esteem by the clientele but efforts designed to improve its robustness by the incorporation of soapstone into the recipe by Dillwyn, to form his rather gritty "trident" paste said to have a pigskin-like texture, which was not favoured by a perfectionist such as William Billingsley, and subsequently led to the departure of first Billingsley and then Walker from Swansea. Billingsley and Walker then re-established the Nantgarw China Works (Phase II) with further financial support from William Weston Young and ten of his local business associate investors in 1817. The phenomenal success of Nantgarw china in the London market was paradoxically the downfall of the Nantgarw China Works, which could not meet the insatiable demand of its London clientele because of an unsustainable and unacceptably high 90% wastage in kiln firing of the biscuit porcelain through "sagging" and ineffective kiln control at the high temperatures involved in the production process of 1420 °C. Billingsley and Walker then left Nantgarw, now on the verge of bankruptcy, for their final employment at the Coalport China Works of John Rose in 1820, where William Billingsley died in 1828: In the next two decades Samuel Walker then disappeared from the ceramics scene, until he re-surfaced in Liverpool *en route* for emigration to the USA to set up the Temperance Hill Pottery in New York State in 1847, where it is believed that a small quantity of experimental porcelain was manufactured alongside a brown-glazed *Brameld*-type "Rockingham" earthenware. This story has been recounted in detail to amplify the long established opinion that only one porcelain body was made at Nantgarw (Morton Nance, *The Pottery and Porcelain of Swansea and Nantgarw*, 1942), when in fact very recent analytical evidence is now appearing that suggests strongly that a hard paste porcelain type was being made there in limited quantities at an unknown date, but quite possibly by William Weston Young between 1821 and 1823, in addition to the phosphatic bone china made by Billingsley and Walker during their tenure of the Nantgarw China Works between 1817 and 1820 (W.W. Young *Diaries* 1843). Table 4.1 reflects this additional information from the analytical data, where analytical data on Nantgarw shards clearly indicates a range of compositions exists with a significantly varying phosphate content. Of especial mention here is the statement made by Owen and Morrison (1999) in their paper that they had been presented with a glazed decorated, i.e. finished, Nantgarw porcelain plate, presumably already damaged, which matches their analytical data from two phosphatic unsagged shards, except that the silica content was 10% more and the phosphate was depleted by the same amount, namely 10%; these are figures that the authors did not consider remarkable but are evidential in that some variation in the Nantgarw recipe formulations could probably have been applied evidently at some stage during its short history. In 1831, William Henry Pardoe reopened the Nantgarw site for earthenware production and it seems that he could have embarked there upon the production of small quantities of porcelain very briefly over the space of only a year or two in the mid- to late-1850s (Edwards, *Porcelain to Silica Bricks: The Extreme Ceramics of William Weston Young,* 1776–1847, 2019) as an auction sale report from 1858 advertises quantities of china tea sets and dinner services from Pardoe's Nantgarw China Works due to the closure of manufacturing operations there. A return to earthenware and clay pipe production continued at the Nantgarw site in the 1860s until the close of the nineteenth century and the last family proprietor, Percival Pardoe, ran the earthenware business at the factory until his death in 1931.

- ***New Hall:*** Founded by a group of six local potters at Shelton, Staffordshire, namely Messrs. Hollins, Keeling, Warburton, Turner, Clowes and Bagnall, who bought out Richard Champion's patent for a hybrid hard paste porcelain from Plymouth and Bristol in 1781, since Champion's china export business was suffering badly from the American War of Independence and subsequently the secession of the emerging United States of America. The operation commenced firstly at Tunstall at Keeling's existing pottery site but within the first year a disagreement between the partners resulted in the premature departure of Keeling and Turner, the remainder then relocating to Shelton Hall, a pottery works belonging to Hollins, Warburton and Company and Thomas Palmer near the centre of Hanley, Stoke-upon-Trent. Because of several locations nearby called Shelton, the new business was renamed Shelton New Hall, and finally just New

Hall. Initially, New Hall manufactured a type of hard paste porcelain, which was not based upon Champion's recipe as first believed, but rather modified with the adoption of a two-stage firing process more akin to the soft paste porcelain firing requirements (Holgate, *New Hall and its Imitators*, 1973; Godden, *New Hall Porcelains,* 1999a). John Turner relocated to establish his own pottery at Lane End which was carried on by his sons William and John. John Daniel joined the company in 1801, which then diversified into the manufacture of bone china before 1815, in emulation of the success of Josiah Spode – a company catalogue published in 1812 already mentions the production of "real china", which is believed to refer to this bone china variant rather than to the Chinese hard paste porcelain variety. This appears to have not been too successful a switch, perhaps, when viewed in competition with the Spode by now well-established bone china production and the New Hall operation eventually closed down in 1835.

- ***Pinxton:*** The Pinxton China Works was established in the village of the same name in Nottinghamshire in 1796 on the banks of the Cromford Canal in premises rented from the Rev. D'Ewes Coke by his son, John Coke of Brookhill Hall. A local businessman, John Coke engaged William Billingsley, who was at that time the prestigious and esteemed head of the china enamelling and decorating studio at the Derby China Works, as a partner in this enterprise (Gent, *The Patterns and Shapes of Pinxton Porcelain,* 1796–1813, 1996; Sheppard, *Pinxton Porcelain, 1795–1813,* 1996). It seems that John Coke attempted to interest William Duesbury, proprietor of the Derby China Works, in the possibility of opening another china factory at Pinxton but Duesbury was not interested. It is quite probable that William Billingsley, who had succeeded his mentor Edward Withers in his position at Derby in 1790, was yearning to produce a new translucent porcelain body which would surpass any that had yet been created commercially and which would be a suitable canvas for his exquisite enamelling skills on porcelain. Since his apprenticeship at the Derby China Works some years before, William Billingsley had experimented in this direction with Zachariah Boreman in his home in Derby using a muffle furnace and he probably felt that the time was ripe for his experimental ideas to reach a production scale-up (John, *William Billingsley*, 1968). However, the situation at Pinxton did not seem to fulfil his ambition completely and Billingsley did not achieve his objective there, so he left Pinxton and John Coke's employment in April 1799 to set up a china decorating business, first in Mansfield and then at Brampton-in-Torksey (Torksey), purchasing porcelain from his previous employers at Derby and at Pinxton for this purpose. It appears that Billingsley still had the idea of perfecting a recipe for a highly translucent porcelain as there is evidence from the porcelain shards found at Brampton-in-Torksey pointing to some experimental production work having been carried out there. A catalyst in rekindling this venture in porcelain manufacture at Brampton-in-Torksey would have been the collaboration at Torksey between William Billingsley and Samuel Walker, the son of a neighbouring farmer, who was adept in kiln design engineering and construction, so giving Billingsley a practical input in this direction which would be maintained for the remaining 20 years of Billingsley's life. The dream of creating porcelain was clearly responsible for the move of Walker and Billingsley from Torksey to the Worcester China Works in 1808, where they had a rather clandestine arrangement with the proprietor Martin Barr to experiment with the preparation of new porcelain body pastes in experimental reverberatory kilns, this work being carried out at night, and would have been supplementary to Billingsley's assumed advisory role in the enamelling workshop at Worcester. In the meantime, John Coke took on Henry Bankes as a partner in the Pinxton China Works in 1801, after the departure of William Billingsley, but he ceased to have interest in the Pinxton China Works after his marriage in April 1806, preferring to concentrate on his coal mining business instead, called Coke and Co. Ltd. John Cutts, now the sole remaining proprietor at Pinxton, finally managed the Pinxton China Works up until its eventual closure in 1813. Subsequently, the Pinxton China Works site, having become a coal-mining enterprise following its closure, was demolished in 1934 and became a scrap-metal dealership, so nothing now remains of the original porcelain factory and it is presumed that any archaeological context would have been severely compromised by the later intrusive industrial and mining activities which were carried on at the same site.

- ***Plymouth/Bristol:*** The very first English hard paste porcelain manufactory was established at Plymouth in 1768 by William Cookworthy, a Quaker pharmacist, who had the idea that his recently found "Cornish stone" could be an acceptable replacement for the mysterious *petuntse* or China stone which had been revealed by Father Francois Xavier d'Entrecolles to be a prime ingredient along with china clay for the hard paste porcelain which was then being imported in vast quantities from Jingdezhen, China (McKenna, *Cookworthy's Plymouth and Bristol Porcelain,* 1947; Adams, *True Porcelain Pioneers: The Cookworthy Chimera, 1670–1782*, 2016). Cookworthy knew that these two vital components, namely Cornish stone and china clay, were accessible near St Austell in Cornwall and he immediately patented their use in the manufacture of porcelain in 1768 (Penderill-Church, *William Cookworthy, 1705–1780*, 1972). The Plymouth manufactory was set up initially with a consortium of businessmen led by William Cookworthy and involving Thomas Pitt, Baron Camelford. In 1773 the soapstone patent held by Cookworthy was sold to Richard Champion. The factory closed in Plymouth in 1770 and relocated to Bristol, where it continued to operate until 1781 under the management of Richard Champion, who sold it on to the fledgling New Hall porcelain factory in Staffordshire along with his soapstone patent acquired from Cookworthy in 1773, which actually expired the following year in 1782. Cookworthy died in 1780, which perhaps precipitated the demise of the Bristol manufactory (Penderill-Church, *William Cookworthy (1705–1780): A Study of the Pioneer of True Porcelain Manufactured in England,* 1972) At Bristol by 1770, Champion had acquired the moulds and technical expertise from the recently defunct Longton Hall manufactory and he had also hired some of the former workforce from the Worcester porcelain factory to work there. Throughout their correlated existence Cookworthy's Plymouth and Champion's Bristol factories manufactured hard paste porcelains as evidenced by the analytical data from the shards presented in Table 4.1: Tite and Bimson's analyses (Tite and Bimson, 1991) gave almost identical results for the Plymouth shard (1768–1770) and the Bristol shard (*ca.* 1775) with approximately 71% silica and 23% alumina content for each and a combined soda and potash percentage of 4%. Wood and Cowell's data for a Plymouth shard (Wood and Cowell, 2002) give a rather different data set with 63% silica and 31% alumina content with 5% combined soda and potash. It should be recognised, however, that a second manufactory was opened in Bristol from 1750–1752 by Benjamin Lund and William Miller, who manufactured soft paste porcelain there for a very short time before they merged with the Worcester China Works of Dr. John Wall which opened in 1751 (Jones 2007).

- ***Ridgway:*** Job and George Ridgway established a ceramics factory with William Smith (who died in 1798) in Hanley, Stoke-on-Trent, in 1794; Job Ridgway died in 1814 and the business thereafter changed its name frequently in a rather complex fashion. Godden (Godden, *Ridgway Porcelain*, 1972b; Godden, *The Illustrated Guide to Ridgway Porcelains,* 1972a) lists no fewer than 16 potters with the name Ridgway associated with a group of factories in Hanley. Initially producing earthenwares and stonewares, Ridgway was inspired by the success of Spode and Worcester and started producing bone china in 1808. Another factory was started by John and William Ridgway in Shelton and then became separated as the Cauldon Place Works in 1830. John Ridgway in particular became known for his production of very high quality porcelain, being appointed "Royal Potter" to Queen Victoria after the Great Exhibition of 1851 and he incorporated this accolade into his factory mark. William Ridgway concentrated upon the production of a large quantity of ceramics for export to America, comprising mainly transfer patterned earthenwares. From 1972, Ridgway porcelains became part of the Royal Doulton Group.
- ***Rockingham***: The authoritative work of Cox and Cox (*Rockingham Porcelain, 1745–1842*, 2001) forms the basis for the analytical data shown in Table 4.1; Cox and Cox (2001) state that between 1820 and 1826 Thomas Brameld and his brothers, George and John Wager Brameld experimented in the creation of porcelain bodies in Yorkshire with widely different compositions and of six specimens analysed by Alwyn Cox two were found to be clearly of a hard paste composition, three phosphatic porcelain and one bone china. The bone china had 100% more bone ash than the phosphatic analogue, the former having just in excess of 52% bone ash in its formulation, which was a very significant amount of bone ash even for a phosphatic porcelain. According to Llewellyn Jewitt (*The Ceramic Art of Great Britain from the Prehistoric Times Down to the Present,* 1878) Rockingham actually reported four recipes for porcelain bodies based upon calcined bone ash, Cornish stone, flints and flint glass components. It is interesting that Thomas Brameld recommended the use of fish bones for his calcined bone ash, then sheep and then horse bones if fish bones were not available, unlike William Billingsley at Nantgarw and William Duesbury at Derby who strongly favoured the use of ox bones for this calcined bone ash porcelain component, which they regarded as superior to the others (Anderson, *Derby Porcelain and the Early English Fine Ceramic Industry,* 2000; Edwards, *Nantgarw and*

Swansea Porcelains: An Analytical Perspective, 2018b). Cox and Cox (*Rockingham Porcelain, 1748–1847*, 2001, p. 173) comment on the surprising discovery of a hard paste porcelain in the Rockingham portfolio of porcelain bodies which they analysed; analytical data for this hard paste body report some 73% silica and 20% alumina content with other components being in minor abundance, with a total alkaline flux of some 6%. The earliest specimen analysed was dated pre-1825 and is illustrated in Cox and Cox's book (*Rockingham Porcelain,* 2001, p. 174), which apparently approaches the composition of a bone china with a phosphorus pentoxide percentage of 14%, equivalent to a raw material component of 46% calcined bone ash. At the height of the Rockingham factory porcelain production between 1826 and 1842, it appears that Brameld had already selected a satisfactory bone china composition by 1826, which then remained unchanged over this period of production; 12 specimens of Rockingham porcelain from this period showed a constant percentage elemental oxide composition and a calcined bone ash percentage of 46%. It is interesting that at this stage of production, Brameld's bone ash was apparently obtained from the recommended ox bones source, which apparently brought him into line with most other manufacturers! It is quite possible that his supplies of fish bones could not satisfy his requirement for a calcined bone ash raw material to match his increased production levels. The process of limited experimental trials followed by the selection of the best body available, called "*Another Porcelain Body*" by Brameld (Cox 1978/1979) from those experiments, which then was adopted for the factory production for the next 16 years without any further experimental change, to seek something perhaps a little better is often a desirable experimental route to follow, but this does not always seem to be the case. For example, over a similar period, between 1821 and 1836, the highly successful Spode factory made no fewer than nine changes to its feldspathic porcelain body in the quest for the search for a better body (Whiter, *Spode,* 1970; Cox and Cox, *Rockingham Porcelain,* 2001). Brameld's final chosen recipe was for 1 part blue clay (ball clay from Cornwall), 1.5 parts Cornish china clay and 1 part Cornish stone (sourced from St Stephen's China and Stone Company, St Austell, Cornwall and the Meledor China Clay Company, St Columb, Cornwall, both priced at 70s per ton shipped to Hull Docks), 2.5 parts calcined bones (giving a bone ash component composition of 46%, and believed to have been sourced from Staffordshire) and also an unspecified amount of flint raw material sourced from Ramsgate, Sandwich and Shoreham on the Kent and Sussex coast. Brameld did raise a note of caution in referring to his observation of the higher level of iron oxide impurity in the ball clay raw material which was detrimental to the colour of the final porcelain body after firing.

- ***Spode:*** Josiah Spode (1733–1797) was apprenticed as a potter to Thomas Whieldon between 1749 and 1754 in his Staffordshire pottery and then went on to employment with William Balls in Stoke-on-Trent in 1754. He rented a small pottery in Stoke-on-Trent in 1767 and started his own "Spode" pottery in 1776 making first cream-coloured earthenware, pearlware, black basalt, caneware and jasperware and then finally bone china in the early 1790s. He perfected the underglaze blue transfer patterns used on fine earthenware for which he became famous in 1783/1784, launching the "Italian" or "Caramanian" range which is still being made today. He created a formulaic and experimental recipe for bone china in 1770 and this was marketed very effectively by his sons Josiah Spode II and Samuel Spode in the 1790s – so successful was this enterprise that the name "bone china" as an industry standard remains synonymous today with Spode. He recruited two skilled engravers, Thomas Lucas and James Richer, from Coalport to perfect his transfer printing process on large items of tableware: Thomas Minton trained as an engraver at Caughley and then supplied Spode with copper plate engravings until he opened his own factory in 1796. The basis for Spode's transfer printing process was the use of special flexible, thin tissue paper and water soluble gum which could be applied in cut sections to large and intricate ceramic shapes which could be floated off after pigmenting to leave a true pattern ready for glazing in a glost kiln (Whiter, *Spode: A History of the Family, Factory and Wares from 1733–1833,* 1970). Spode's "bone china", which was adopted very soon by his rivals at Minton and Davenport, was a mixture of bone ash, china stone and kaolin and a critical part of the recipe preparation for the body involved the elimination of a preliminary stage of fritting the bone ash separately and then re-grinding it with additional components for a separate firing stage. The formulation was 6 parts bone ash (44%), 4 parts Cornish stone (30%) and 3.5 parts kaolin (26%) which were all ground together and fired in a single stage process which better approximated to the ancient Chinese method of manufacture of true hard paste porcelain rather than the two-stage process better suited to and adopted for the English soft paste and phosphatic porcelains. The use of soapstone and bone ash together in such high proportions was revolutionary and conferred upon Spode china a very high translucency and thinness of potting which could compete effectively with the soft paste porcelains of their competitor factories. Two further advances made by Spode were their "stone china" introduced in 1813, which was robust, unglazed, of a light texture and of a reasonably translucency, and "felspar porcelain", introduced in 1821, which involved the substitution of feldspar for Cornish stone in an otherwise standard bone china body. This latter "felspar porcelain" actually closely resembles Mason's "ironstone china" which was introduced by Charles Mason (Josiah Spode II's nephew) in 1813. The Spode China Works became Copeland and Garrett in 1833 until 1847 and thereafter became W.T. Copeland and Sons into the twentieth century, where it is still traded as Spode to this day.

- ***Swansea:*** in common with many early nineteenth century porcelain manufactories which had an historical precedent in earthenware production, the Swansea China Works founded by Lewis Weston Dillwyn in 1813 started as the Cambrian Pottery in 1762, with William Coles and then George Haynes, manufacturing earthenware and useful articles for mainly local consumption. Dillwyn's venture into porcelain was at first unsuccessful when he recruited two people in 1812 from Coalport who were not able to make the porcelain he needed, but the venture accelerated rapidly when he took on William Billingsley and Samuel Walker in October 1814, arguably the finest china decorator and ceramic kiln designer and manager in Europe at that time (Edwards, *Swansea Porcelain: The Duck Egg Translucent Vision of Lewis Weston Dillwyn,* 2017; Jones and Joseph, *Swansea Porcelain; Shapes and Decoration*, 1988). Between 1815 and 1817, Dillwyn, in collaboration with Walker undertook a series of experimental trials in which he systematically varied the composition and ingredients of his basic recipe to improve upon a rather heavily potted Swansea "glassy" porcelain being manufactured at that time and thereby discovered his much esteemed duck-egg china, so-called for its beautiful blue-green translucency to visible light (Edwards, *Swansea Porcelain: The Duck Egg Translucent Vision of Lewis Weston Dillwyn*, 2017; Meager, *The Swansea and Nantgarw Potteries*, 1949; Morton Nance, *The Pottery and Porcelain of Swansea and Nantgarw*, 1942), This duck-egg porcelain was highly appreciated not only for its beautiful translucency but for the superb decoration and gilding with locally employed artists of the highest calibre, including Billingsley himself. Problems in kiln wastage verging between 70% and 75% necessitated a further attempt to improve the robustness of the duck-egg china, but the incorporation of steatite and proportional reduction in the bone ash and china clay components resulted in an inferior although much stronger body, which unfortunately cost Dillwyn his market edge and clientele support so that his china business faced imminent closure in 1820. This story is important because, unlike that of Rockingham, Dillwyn did not remain with his successful body and preservation of a constant recipe composition but he constantly strove to achieve an even better porcelain body. His experimental notes from 1815 to 1817 detail some dozen or so attempts at creating variants of Swansea porcelain which are summarised in his Notebook (Dillwyn, Notebook, see Eccles and Rackham, *Analysed Specimens of English Porcelain in the Victoria and Albert Museum Collection,* 1922) and are reproduced in Eccles and Rackham (*Analysed Specimens of English Porcelain in the Victoria and Albert Museum Collection*, 1922) and in Edwards (*Nantgarw and Swansea Porcelains: An Analytical Perspective*, 2018b). In Table 4.1 are presented the analytical data for Swansea porcelains which include the work of Owen et al. (1998) and Owen and Sandon (2003): in the former paper, a finished, decorated specimen of a Swansea duck-egg china platter from the *Biddulph* service (Royal Institution of South Wales, Swansea) was analysed and in the latter a finished and decorated Swansea duck-egg china dinner plate was analysed. It is indeed intriguing that although both specimens are clearly of the finest type of duck-egg Swansea porcelain, they have significantly different analytical compositional data and we need to explore the possible reasons for this discrepancy. For example, the Swansea duck-egg dinner plate composition is significantly different to that of the other duck-egg platter from the *Biddulph* service in that it is 14% enriched in silica, but depleted in phosphate to the extent of 33% – these figures just do not correlate with the historical conclusions that when Lewis Dillwyn had found his "perfect" recipe for duck-egg porcelain he stayed with that for the duration of his manufacturing process until he made the fateful decision to move onto a soapstone body, called the "trident" body, for increased robustness. Confirmatory evidence for the change in recipe between these two Swansea examples is also provided by analytical data for their glaze analyses: the *Biddulph* service specimen porcelain glaze contains only half the amount of lead as its analogous Swansea dinner plate, namely 13% compared with 25%, respectively, yet both are deemed to be standard examples of Dillwyn's duck-egg porcelain by collectors.

- ***Vauxhall:*** The Vauxhall porcelain manufactory was founded in Lambeth, London, by Nicholas Crisp, a jeweller of St Paul's Churchyard, and John Sanders, a Delft pottery manufacturer, on the site of Sanders' pottery in 1752 to produce a soapstone based porcelain in competition with the Chinese hard paste porcelains then being imported in large quantities. Initially the Vauxhall factory produced figurines and decorative wares (Massey 2014), and later in 1754 functional and useful tea wares, in blue and white and polychromatic enamels, but they could not compete economically with the Chinese imports or with the quality of the production by their neighbours at Chelsea. John Sanders died in 1758 and soon thereafter, Nicholas Crisp ran into financial difficulties and became bankrupt in 1763 (Owen et al. 2000), the factory being finally sold off in May 1764 (Massey and Spero, *Ceramics of Vauxhall, 18th Century Pottery and Porcelain,* 2007). Little was known of the Vauxhall porcelain manufactory until excavations commenced on the factory site in 1988, yielding shards and evidence of the kiln operations; an interesting discovery was the green translucency of the Vauxhall porcelain body, attributed to the extensive use of soapstone in the paste which was modelled upon the composition used in the Worcester porcelain manufactory. However, Worcester porcelain does not have this green translucency and although Swansea duck-egg translucent porcelain does have this property it does not contain soapstone in its formulation! Points such as this ned further investigation, but it is possible that charge transfer complexes involving iron (III) impurities in the Vauxhall raw materials at source are responsible. Also, the Swansea soaprock porcelain variant known as the "trident ware", does have a significant soapstone additive in its recipe but does not have a green translucency, but rather has a soft peach to muddy brown translucency. We should conclude, therefore, that the explanation of the greenish translucency of Vauxhall porcelain being due to its soapstone component is erroneous and that this must be attributable to another source, perhaps the presence of small amounts of rare earth metallic ions in the china clays or perhaps the presence of iron (III) compounds as stated above. Some possible origins of this green translucency in Swansea duck-egg porcelain have been considered in detail by Edwards (*Swansea and Nantgarw Porcelains: A Scientific Reappraisal,* 2017; *Swansea Porcelain: The Duck-Egg Translucent Vision of Lewis Weston Dillwyn,* 2017) and it seems that the formation of chemical charge-transfer complexes involving the rare-earth impurities in the particular Cornish clay used in the recipe formulations are responsible for this very attractive porcelain translucency. Table 4.1 gives Tite and Bimson's (1991) results from the analysis of Vauxhall shards from 1753 to 1764, which clearly demonstrate the soapstone porcelain type produced there with an average silica percentage of 75%, alumina of 4%, potash of 4%, magnesia of 9%, lime of 6% and even a small presence of lead oxide, implying the addition of flint glass cullet in the recipe, as was practised at the neighbouring Chelsea china works.

- ***Worcester:*** Worcester, founded by Dr. John Wall and William Davis in 1751, is only one of two of the early English porcelain manufactories to survive into the twenty-first century as the Royal Worcester Porcelain Company, now part of the Portmeirion Group since 2009, having been granted the Royal warrant through the creation of a special armorial service for the Duke of Clarence, later King William IV, in 1789 (Jones, *Porcelain in Worcester, 1751–1951: An Illustrated Social History,* 1993; Jones, *Origins of Worcester Porcelain: Local Ingenuity and the Pathways from Staffordshire, Stourbridge, Bow Limehouse and Bristol,* 2019). The other survivor is the Royal Crown Derby China Works, which claims a foundation pre-1750 (actually listed as 1748) from Andre Planche, but the first factory was closed in 1848, reopening a few years later at a different site under the present title until the present day. Dr. John Wall and William Davis initially carried out experiments in creating a soaprock porcelain body in Davis' apothecary's shop in Broad Street, Worcester, but early attempts at porcelain production were not successful: through an association with Lund and Miller at Bristol they then realised that the Bristol manufactory would be ripe for their acquisition and with 13 other local businessmen and investors they raised £4500 to launch the Worcester Tonquin Manufactory and leased Warmstry House on the banks of the River Severn on the 16th May, 1751, the deeds being signed by all 15 partners on the 4th June 1751, each under an agreement to forfeit a £4000 penalty if they incautiously revealed the secrets of porcelain manufacture (Rissik Marshall, *Coloured Worcester Porcelain of the First Period, 1751–1783,* 1954; Sandon, *The Illustrated Guide to Worcester Porcelain, 1751–1793,* 1969). Richard Holdship, a partner, was prominent in the process of buying out the Bristol manufactory of Lund and Miller in early 1752 (Jones 2007) and he personally acquired from Benjamin Lund, a fellow Quaker, the rights to mine 20 tons of soaprock per annum from Cornwall. The transfer of technology and knowledge from the Bristol manufactory provided the effective basis for the launch of the new Worcester china manufacturing enterprise, which succeeded in making mainly functional blue and white decorated items, whose soapstone body proved resistant to cracking under boiling water, unlike other rival porcelains (Barrett, *Worcester Porcelain and Lund's Bristol,* 1966). Agents were opened in London at Aldersgate in 1754 and in Worcester at Samuel Bradley's shop in the High Street. A landmark event was the production of printed scenes onto porcelain from engraved copper plates in the mid-1750s, which was heralded by the move to Worcester of the celebrated engraver Robert Hancock in 1756. In 1767, Dr. Wall engaged James Giles, the noted enameller, to decorate Worcester polychrome porcelain in his atelier in London. In 1774 Dr. John Wall retired and William Davis took over the running of the porcelain manufactory. In 1775, Thomas Turner departed Worcester to set up his rival porcelain manufactory at Caughley, taking with him many of the skilled workforce personnel. In 1783, Messrs. Flight and Barr assumed control of the Worcester manufactory, which continued thereafter under different combinations of the Flight and Barr family names until 1840 (Sandon, *Flight and Barr Worcester Porcelain, 1783–1840,* 1993). The analytical figures for early Worcester porcelain shards given in

Table 4.1 (Tite and Bimson 1991) clearly indicate the soapstone character of the earliest porcelain body with an average magnesia content of 10.8%, representing some 33% soapstone raw material content in the recipe formulation. Very high silica percentages averaging 73% and low alumina of 3.5% and lime of 1.5% are also typical of this paste – but an interesting figure relates to the high lead oxide content of greater than 6%, which must imply a significant addition of flint glass cullet to the paste, verging towards a glassy type of porcelain body in description. Owen (1997) has concluded from the analytical data obtained from the earliest Worcester shards that the claims by Dr. Wall and Davis in the initial setting up of their china manufactory at Warmstry House that they had invented a truly different porcelain composition were substantiated, although there was some evidence that possibly these early shards had in fact originated from the Plymouth and Bristol manufactories which were subsumed into the Worcester enterprise in 1752 after the closure of these combined factories in 1751.

4.2.1 Inclusion of Data for Chinese and Japanese Porcelains

Finally, in Table 4.1, entries have also been made for comparison purposes of two batches of analyses carried out for Chinese and Japanese porcelains which, of course, can be categorised as "true" porcelains of the hard paste variety. The Chinese porcelains were sourced from the Yuan Dynasty (eleventh to the fourteenth centuries) and are known as Yingqing china predominantly from Jingdezhen (Tite et al. 1984), the technical ancestor of all later Chinese hard paste porcelain. These were fired normally between 1250 and 1300 °C but required a glazing procedure at the relatively high temperature of 1220 °C to provide the finished product. In contrast, the origin of the Japanese porcelain shards was not described in detail in the paper. As a comparison, the analytical data for the composition of the Chinese porcelain stone or *petuntse* raw material is also included.

Although summary descriptions of five factories have been provided above in this listing of early English and Welsh porcelain manufactories, namely Davenport, Minton, New Hall, Ridgway and Spode, it is generally accepted that these five comprise the major producers of English bone china in the early- to mid-nineteenth century, starting with the eponymous Spode Bone China around 1795, which laid the foundation for the standard English bone china body during the next 20 years or so. At the time of writing, the analysis of Pinxton porcelain shards does not seem to have been accomplished using SEM/EDAXS experiments: the future interest in this approach could apply to an assessment of the paste compositions being trialled by William Billingsley and John Coke at their fledgling works in Pinxton in 1796 immediately following Billingsley's departure from William Duesbury's Derby China Works. It is recorded that the kiln wastage upon firing in Pinxton was disproportionately large, as was echoed in Nantgarw some years later, and it appears that the phosphatic recipe was perhaps not compatible with commercial success in surviving the high firing temperatures in the biscuit kiln. It is appreciated, however, that the archaeological contextual integrity of the Pinxton china works site buildings may have been severely compromised by later coal mine workings and scrap metal recovery operations, but it is quite possible that a waste pit for shards could still exist which has not yet been disturbed by other industrial surface activity and an analytical investigation of any Billingsley/Coke period shards and porcelain fragments from the Pinxton factory site would provide some interesting information about paste and glaze composition in the period between Billingsley leaving Derby in 1795 and joining Worcester in 1808, and particularly for the narrower period between 1795 and 1799 when he was active there in the manufacture of porcelain.

4.3 A Terminological Issue: Bone China *Versus* Phosphatic Porcelain

During our survey of the early English and Welsh porcelain factories accomplished thus far it will have been apparent that the phrases "*bone china*" and "*phosphatic porcelain*" have been used by most previous authors to describe a ceramic factory's output and it is not obviously clear what these terms can actually mean, especially when we shall come to consider the categorisation of porcelains in a scientific manner. Earlier, the simple classification of porcelains into hard paste and soft paste seemed adequate enough but this view was soon regarded as being far too simplistic and too generic because of the intricate modifications that were made to the paste formulations in often empirical ways. Eccles and Rackham (*Analysed Specimens of English Porcelain in the Victoria and Albert Museum Collection*, 1922) introduced yet another category of porcelain, which they termed "*hybrid*" porcelain, to account for certain products and bodies which sat firmly between the two extremes quoted above. Clearly, the original terminology is not satisfactory, too all-embracing and requires some fine definitive tuning: a further complexity arises when the older literature is scanned and the terms "bone china", "phosphatic porcelain" and even another described as "bone ash porcelain" are used by authors and ceramic historians to describe the products of different ceramics factories. It is hence appropriate to now revisit the terminology and to attempt to clarify this issue before we progress onto a detailed chemical interpretation of the analytical chemical elemental oxide data and its role in classifying porcelain types.

Porcelain has been defined as a vitreous ceramic material which has been subjected to high temperatures during which key elemental oxides and compounds used in its formulation have been converted into a translucent body shape, which may or may not be glazed; and unglazed porcelain is called "biscuit". Bone china is defined as a porcelain compound containing bone ash, a feldspathic material and kaolin, producing a ware with a translucent body which contains a minimum of 30% bone ash derived from calcined animal bones. Bone china is said to be the strongest and most robust category of porcelain ceramics, possessing a high mechanical strength and fracture resistance which contribute to its fine potting capabilities and translucency. Phosphatic porcelain, also known as bone ash porcelain from the calcium phosphate ($Ca_3(PO_4)_2$, whitlockite) contained therein, is described similarly and is also categorised as a soft paste porcelain: Eccles and Rackham in their original studies (*Analysed Specimens of English Porcelain in the Victoria and Albert Museum Collection*, 1922) were rather ubiquitous in their descriptors of the porcelain products of the 23 factories they listed – restricting their categories to hard paste, such as Chinese, Plymouth, Bristol and New Hall porcelains, soft paste such as Bow, Nantgarw and Swansea porcelains, and hybrid such as Coalport and Worcester porcelain, reserving the category of bone china exclusively for Spode. Glassy porcelains comprised a different category and it was recognised that many of the early factories made identifiable products which contained glass frit additives, such as Chelsea, Swansea and Worcester. The problem is, however, that many soft paste porcelains, and the so-called hybrids and bone china manufactories all contain calcium phosphate sourced from calcined bone ash. The addition of bone ash to porcelain seems to be an exclusively English invention, first credited to the Bow China Works of Thomas Frye and Edward Heylyn and to Frye's patent of 1748 for its use in china manufacture at Bow, whose site was located conveniently near the slaughterhouses and cattle markets of East London which could provide a readily available source of animal bones for calcination. Early Bow porcelain contained up to 45% calcined bone ash and they termed their product "*fine porcelain*". Many authors still persist in crediting the use of bone ash in soft paste porcelain and to the invention of the resulting "bone china" to Josiah Spode in 1790, but chronologically this cannot be upheld today, although there is no doubt that Spode was the most successful eighteenth century entrepreneur in the marketing of this particular ceramic body, which he called "*Staffordshire bone porcelain*" or "*Stoke china*". So successful was this product that by 1815 many English porcelain factories had converted to the adoption of its formula recipe and it then became known as the "industry standard"; those factories that survived into the twentieth century all manufactured bone china, for example, Worcester, Spode, Derby, Minton and Coalport, although not exclusively and often this was undertaken alongside the manufacture of other porcelain types.

The confusion in the correct terminology applied for bone china, phosphatic soft paste porcelains and bone ash china cannot be resolved simply into a differential compositional formulation for those ceramics that have a bone ash component or not in the recipe but is rather more subtly dependent upon the initial process adopted in the manufacture for firing the body paste. Whereas soft paste phosphatic porcelains usually involve a *two-stage* manufacturing process, whereby the calcined bone ash is mixed with sand and kaolin and heated to form a "frit" which is then cooled and ground finely for admixture with water and other raw material components such as more kaolin, talc, lime, potash and glass cullet if required, for reworking at a higher firing temperature in a kiln to make the final biscuit body, the preparative procedure used for bone china is essentially different. Josiah Spode used a *one-stage* process of firing his components together: firstly, he calcined his crushed cattle bones at 1250 °C, then he mixed his finely ground bone ash with kaolin and feldspar, a traditional formulation being 25% kaolin, 25% Cornish stone and 50%

bone ash, in a single firing at 1200 °C to give his bone china biscuit body, comprising the minerals anorthite, $CaAl_2Si_2O_8$, and whitlockite, beta-tricalcium phosphate, $Ca_3(PO_4)_2$ (Dodd and Murfin, *Dictionary of Ceramics,* 1994/5; Freestone 1999; Dinsdale, *Pottery Science: Materials, Products and Processes,* 1986). This single stage firing process for bone china therefore actually mimics closely that of the true or hard paste porcelain as perfected by the Chinese more than it does soft paste porcelains, which have a multi-stage firing process, at least one of these being in the preparatory operations for the raw materials which resulted in a frit which was then subjected to further treatment, mixing with further additives and kiln firing.

The result of a rather loose terminology applied historically is, therefore, that some rather quaint and outdated concepts are revealed in the earlier literature of what constitutes bone china and soft paste phosphatic porcelains and these should not be perpetuated but amended or corrected in future publications. The defining mantra is: all bone china is soft paste phosphatic porcelain but not all soft paste phosphatic porcelain is bone china, and it must be stated that the critical criterion for the definition of bone china is the adoption of a single stage operation for its primary body firing process directly from the raw materials formulation rather than a two-stage process as found in the production of other soft paste porcelains. In this the apparent threshold of a phosphatic content of 30% calcined bone ash in the raw material components of the porcelain paste is therefore rather immaterial.

4.4 Historical Issues for Modern Analytical Interpretation

The collection of historical facts in Sect. 4.2 relating to the foundation of porcelain factories and their ongoing business for 21 different English and 2 Welsh porcelains, which can be broadly classified as 13 soft paste, 6 bone china and 4 hard paste porcelains, has raised several common themes and some potential issues for the interpretation of modern analytical data from their shards and surviving ceramic fragments which must be considered. These can be summarised below:

- Many of the porcelain manufactories were created from sites that were already established as earthenware or salt-glazed pottery works, such as the Coalport/Jackfield Pottery, Swansea/Cambrian Pottery and Rockingham/Brameld Pottery. Others were totally novel porcelain start-up foundations, such as Nantgarw, Chelsea and Caughley.
- Of the 23 china factories listed in the above summary many started as family concerns and were inherited businesses.
- Most of the china factories did not survive for an extended period and a significant number did not last even as long as 10 years after their foundation, the most common cause for their closure being bankruptcy for the commercially unsuccessful business or the death of the founder, leaving no one with the required skills or acumen to take over the business. Examples include Nantgarw, Swansea, and Limehouse. In other cases, neighbouring factories took over the failing busi-

ness and acquired thereby the hardware and specific technological knowledge, such as Coalport/Caughley, Bristol/Plymouth, New Hall/Bristol, and Derby/Chelsea.

- Almost every factory at some stage acquired skilled personnel from other factories, often erstwhile competitors, such as Swansea/Nantgarw, Pinxton/Derby, Coalport/Caughley. It was quite a common practice to hire staff annually at Michaelmas in ceramics factories and this also encouraged the transfer of workforce personnel from one factory to another with the distinct possibility of there being a knowledge transfer of the practical details about the manufacturing processes then occurring from one factory to another. This certainly happened at Worcester/Brampton-in-Torksey, Derby/Chelsea, Coalport/Caughley, Coalport/Nantgarw, Swansea/Nantgarw and Pinxton/Derby.
- In only a few cases is it specifically stated that a major change in body composition was undertaken during the lifetime of production at the factory site, for example Limehouse, Swansea and New Hall; this, of course, would be extremely important for the analytical interpretation of compositional data from shards, especially as it places them chronologically with or without an archaeological context.
- The practice of using "grog", pieces of broken ceramic from other factories which could be ground up and used as a body component or "filler", was fairly common amongst porcelain factory proprietors in the eighteenth and early nineteenth centuries; this implies that forensically it must be borne in mind that all the shards found on site may not be from the ceramic process production that took place there and the presence of potential interlopers must be recognised. Of a similar but more serious occurrence is the practice of a factory purchasing the products from another factory to complete their service commission orders on time: this certainly occurred where perhaps production failings or kiln wastage necessitated its adoption for the adequate delivery of a commission on time. An example is the esteemed *Lysaght* dinner-dessert service from the Swansea factory, which included several dessert comports and dishes bought in from Coalport in the white for local decoration and gilding *en suite* at Swansea by the artist Henry Morris (Edwards, *Swansea and Nantgarw Porcelains: A Scientific Reappraisal*, 2017). This was especially prevalent where larger and perhaps more intricate porcelain artefact designs were envisaged, such as centrepieces and ice-pails, and this was certainly the case operating also for the *Gosforth Castle* Swansea service commission.

4.5 The Curious Case of the Fulham Pottery

It became apparent during research of the recent literature concerning the growth of early English porcelain manufactories that several historical misconceptions or ideas had just been accepted as original statements of fact, seemingly without question. It is intriguing that it has always been historically assumed that the English porcelain industry grew spontaneously with the foundation of the Chelsea and Bow manufactories in 1744/1745, closely followed or perhaps even slightly preceded by the short-lived Newcastle-under-Lyme (Pomona) and Bovey Tracey enterprises. The latest historical detective work by Ramsay, Daniels and Ramsay on the Limehouse factory has strongly suggested that an extensive prior knowledge about porcelain ceramic manufacturing already existed in England in the early eighteenth century and that this had itself been generated by a much earlier experimental base extending back into the seventeenth century (Ramsay et al., *The Limehouse Porcelain Factory, Its Output, Antecedents and the Influence of the Royal Society of London on the Evolution of English Porcelains based on Composition and Technology*, 2013; *Limehouse Porcelain: Are Limehouse Porcelains in Fact All Limehouse? Evidence from Archaeology, Science, and Historical Documents*, 2015). Much has been made about the early experiments carried out in Italy and France before finally Ehrenfried von Tschirnhaus and Johann Bottger claimed the first synthesis of a European porcelain in 1709 at Meissen in Saxony – so providing the starting point for all subsequent experimentation and enterprising ventures in commercial European ceramics production. It is perhaps now timely to revisit this perception in the light of some recently researched historical documentation.

The major questions facing the historically accepted origins of the English porcelain industry are: when and where did it all start, and did the early factories arise spontaneously, or had there been a now-forgotten catalyst or stimulus which created a base of information and practical knowledge upon which the early factories could draw? The assumption has always been presumed, quite fairly in retrospect, that English eyes cast upon the European scene in the 1740s would have recorded the advances being made in the area of synthetic ceramics in France and Germany in an at least partially successful competition with the Chinese imports and this would have then provided an impetus for starting work in England in that area. This seems a reasonable proposition, however, with hindsight it is not unreasonable to consider an alternative scenario that there was possibly a real caucus of information already extant upon which the early English porcelain industry could rely which could have provided a self-initiated and positive response when called upon in the early to mid-1700s, matching the growth of an industry that was already blossoming in France and Germany? The potential idea for this train of thought actually arises when one considers the type of porcelain being made in these early English factories in the 1740s and 1750s – which has been analytically determined to be indigenous and actually unlike anything contemporary that was being made overseas in France, Germany or even in China itself around that time. So, what could be the basis for an assumption by historians that the English porcelain industry resulted from knowledge gained piecemeal and divined from its erstwhile competitors abroad and maybe even relying on an initial knowledge transfer westwards form Europe to spur on the ceramics manufacturing industry? The compositional formulations of the European and Chinese porcelains are very different and the types of porcelain made are also very significantly different, which would lead us to believe that any knowledge transfer from Meissen or Sevres to England in particular was not realistic in that the porcelains are actually so very different in their composition as are the firing processes used to achieve their manufacture. A clue arises in the use of bone ash as a component in the early Bow porcelains in England: where did that idea arise, and did it arise spontaneously or from a cachet of previously recorded but perhaps now lost experimental evidence? What has until now been assumed historically is that Thomas Frye patented the idea of using bone ash for his Bow porcelains in 1748 and that it was used thereafter by a significant number of the early English fledgling porcelain factories. An analogous situation arises from the discovery of soapstone/talc in some of the earliest English shards analysed, from whose compositions these can now thereby be re-classified clearly as magnesian soft paste porcelains uniquely: This material was not used at all in the Continental European factories or in Chinese porcelains, so how did its use in porcelain synthesis originate in England, and, therefore, this fact decrees that it most certainly could not have arisen from the recipe knowledge transfer from other by then well-established Continental porcelain manufactories? It has been shown incontrovertibly that the Limehouse Porcelain Manufactory seemed to be the innovator of soapstone ceramics in English porcelain production (Watney, in *Limehouse Ware Revealed,* 1993; Sandon, *Flight, Barr and Barr Worcester Porcelain, 1783–1840*, 1993) and that the exhaustion of its Kynance Cove source mine, which provided the Cornish mineral for this vital component in 1747, was probably the prime contributing factor in the closure of the Limehouse magnesian porcelain manufactory only a few months later in 1748.

Ramsay et al. (*Limehouse Porcelain: Are Limehouse Porcelains in Fact All Limehouse? Evidence from Archaeology, Science, and Historical Documents*, 2015) have discovered that documentation relating to the foundation of the Limehouse Porcelain Manufactory in East London in 1744 strongly indicated that its key personnel had prior information of a technical nature which could not be explained by a knowledge transference from either Meissen or other contemporary French factories. Then, correspondence between members of the Royal Society of London, which had been founded by King Charles II upon his Restoration in 1660 and which numbered among its membership luminaries such as Sir Isaac Newton, Robert Hooke and Robert Boyle, has surfaced which showed that in the mid- to late-1670s active experimentation was already being undertaken on porcelain synthesis in Oxford and in London by Robert Boyle and John Dwight, encouraged by Royal Society support and funding (Daniels, *The Origin and Development of Bow Porcelain, 1730–1747, Including the Participation of the Royal Society, Andrew Duche and the American Contribution,* 2007; Daniels and Ramsay 2009).

4.5.1 The Role of the Royal Society in English Porcelain Manufacture

Hitherto, the role of the Royal Society in English porcelain research and manufacture has been rather ignored, although it was hinted at historically in an earlier ceramics work by Solon (*The Old English Potter,* 1883). Daniels (*The Origin and Development of Bow Porcelain, 1730–1747, Including the Participation of the*

Royal Society, Andrew Duche and the American Contribution, 2007) has commented upon her interpretation of several key documents which demonstrate clearly that in the mid-to late-seventeenth century, only a few years after its foundation in 1662, the Royal Society was actively encouraging and sponsoring research and experimentation in the ceramics area, some 50 years prior to the successful Meissen porcelain manufacturing operation.. These, coupled with the documentary evidence relating to John Dwight's Fulham Pottery put together by Haselgrove and Murray (1979), make a compelling case for re-considering this period in early English porcelain and ceramics experimentation and its potential influence upon a wider European scene.

John Dwight, who was born in Chester in 1637/1638, established a manufactory of salt-glazed earthenware in Fulham at the junction of the New King's Road and Burlington Road, not far from Putney Bridge on the River Thames, which he called the Fulham Pottery. The Fulham Pottery managed to survive the major political changes and events occurring through the early eighteenth century and under the later ownership of William de Morgan became a driving force for architectural and functional pottery and tiles in Victorian England well into the 1880s. Excavations at the site have revealed the presence of a ceramic kiln which has now been preserved on this site. Just prior to the foundation of his pottery at Fulham, John Dwight filed a patent in April, 1671 describing his experiments in the use of clays and minerals to make stoneware and a phrase from this patent is particularly revealing, namely *"… the discovery of the mistery of transparent earthenware commonly known by the name of porcelaine in China."* Critically, this statement alerts the reader to Dwight's patent for the manufacture of a "*transparent porcelain*" rather than just the opaque earthenware at Fulham with which he is normally credited. Sir Arthur Church had knowledge of Dwight's patent and he alleged that Dwight must have achieved something very similar to what we now call porcelain in that ".. *he produced if not porcelain then something distinctly porcelaneous*". This comment was certainly supported by the earlier contemporary statement made by John Houghton FRS in 1693/1694, who maintained that "*Mr Dwight of Fulham uses clay, the same earth China ware is made from and this is made by not lying long in the earth but in the fire …. we make as good China here as any in the world*". Robert Hooke FRS on February 1674 (in Haselgrove and Murray 1979) confirmed the production by Dwight of his English china including some figures and "*Severall little Jars of several colours all exceeding hard as flint, very light and very good shape*" and again on May 16th 1674, he described Dwight's products as "*In glazed with ashes. Very hard and close excessive deer*". The latter comment perhaps refers to Dwight's potential use of a silicaceous- alumina-calcareous (SiAlCa) glaze, with perhaps some lead oxide in addition.

Later, Robert Plot FRS in 1677 noted that Dwight had "*discovered the mystery of the Hessian wares and ways to make an Earth white and transparent as Porcellane*" but he also mentioned that Dwight had experienced problems with his glazing of this white *porcellane* which was suspected by Ramsay and Ramsay ("*The Evolution and Compositional Development of English Porcelains from the 16th Century to Lund's Bristol c. 1750 and Worcester c. 1752 – the Golden Chain*", 2017) to arise from the silica -alumina -calcareous glaze composition which required very high firing temperatures which Dwight was unable to achieve in his kiln. This was also believed by Tite et al. (1986), who attributed the defects in his manufactured products to the adoption of an intractable SiAlCa glaze. It has been suggested that Dwight may have acquired some information indirectly from Xavier d'Entrecolles about the recipe for Chinese porcelain, but of course this is just not tenable as d'Entrecolles did not inform his Jesuit hierarchy in Paris of his discovery of the Chinese manufacturing secret until 1717 by which time Dwight had died in 1708. Furthermore, Spataro et al. (2009) have discovered evidence for Dwight's reversion to a lower firing temperature of a lead glaze, which he then used to recover several articles which had nevertheless successfully survived the preliminary SiAlCa glaze trials.

Houghton wrote again in 1694, bemoaning the fact that a restricted importation was in effect for the hard paste porcelain from China of "*curious manufacture*", but he sought solace in the fact that "*Mr Dwight of Fulham has done it, casting thin and firing it at high temperature*". Sir John Lowther FRS in 1698 (Haselgrove and Murray 1979) visited Dwight at his Fulham Pottery and was shown "*20 or 30 varieties of more china like than is in ye world beside, nothing in Germany is like his, nor had he any help from them at setting up, but owes all to his own studies*".

A most telling comment in Daniels' research (Daniels, *The Origin and Development of Bow Porcelain, 1730–1747, Including the Participation of the Royal Society, Andrew Duche and the American Contribution,* 2007) is a note that refers to a visit to the Fulham Manufactory in 1675 by Ehrenfriede von Tschirnhaus, who later "*transferred to the Meissen Factory*" in Saxony, where history now tells us that European porcelain was invented! Von Tschirnhaus (1651–1708) arrived in London from Paris in May, 1675 with an introductory letter from Spinoza and met with influential member scientists of the Royal Society, most probably including Dwight, Hooke and Boyle, where they could well have discussed their novel experiments on refractory porcelains which then stimulated von Tschirnhaus' recorded visit to the Fulham Pottery to view the experimental products for himself. On his return to Paris, von Tschirnhaus immediately commenced his own experiments on porcelain synthesis and eventually he set up with Bottger at Meissen under the patronage of Augustus, the Elector of Saxony. It could be argued, therefore, that Dwight's successes at Fulham initiated and stimulated those experiments of von Tschirnhaus made after his visit there in 1675 which culminated in the historical labelling of the first European synthetic porcelain at Meissen in 1709. It must be said that this is in disagreement with the opinion of Honey (1939), who whilst recognising Dwight's contribution to early porcelain manufacture in England, controversially and rather curiously states that it would have had no influence at all on Continental European porcelain manufacture, whilst still applauding the influence later English factories such as Bow had upon their counterparts in France, especially at Vincennes! This must surely come under the category of a mixed ceramic metaphor!

The role of the Royal Society in the early manufacturing efforts of John Dwight should not therefore be underestimated; Robert Hooke, a polymath and a leading member of the Royal Society in the seventeenth century, who has been described as "*England's Leonardo*" (Chapman, *England's Leonardo: Robert Hooke, 1635–1703, and the Art of Experiment in Restoration England,* 1996), in a report to the Royal Society on the 5th December 1678, stated that "*John Dwight had manufactured a ceramic as hard as porphyry … an excellent Chinese earth which could endure the greatest fire without vitrification*". At this time Hooke was a well-respected and highly regarded experimentalist and was appointed the Curator of Experiments at the Royal Society, being particularly interested in mechanics and heat, so he was very appropriately qualified to comment upon Dwight's ceramic experiments. Gunther's monumental survey of experimental science at Oxford (*Early Science at Oxford,* 1923–1967), where Hooke was for many years assistant to Robert Boyle, comprised 14 volumes of which no fewer than 5 were devoted to Hooke himself. In 1656, Robert Boyle took up lodgings in what is now University College at the University of Oxford, where he employed Robert Hooke as his assistant, during which time Boyle's Law establishing the relationship between the pressure, volume and temperature of a gas ($pv = nRT$) was formulated. They founded the first purpose-built chemistry laboratory in England off South Parks Road in Oxford and modelled it along similar lines to the Abbot's Kitchen at Glastonbury Abbey: this location was very close to Wadham College, whose Principal was John Wilkins, a catalyst for intellectual meetings there with like-minded natural philosophers and thinkers including Sir Christopher Wren, Robert Boyle and Robert Hooke. The "Abbot's Kitchen" at Oxford is believed to be the first purpose built chemistry laboratory and in his book, *The Sceptical Chymist* (1661), Boyle is credited with the invention of the word "chemistry" to describe that particular branch of natural philosophy. In 1660 this Wadham group, numbering seven from Oxford along with five more from Gresham College, London, founded a wider group which then received its Royal Charter from King Charles II in 1662 as the Royal Society of London. John Dwight received his degree in natural philosophy from Oxford University in 1661 and obtained employment first in Wigan, whilst being mentored by Boyle, and Boyle, Dwight and Hooke became lifelong friends thereafter, Dwight being heir to several of the personal effects of Robert Boyle after his death in 1691.

Robert Hooke was not alone in appreciating Dwight's discovery of translucent porcelain as Sir Hans Sloane also referred to his use of a "*Chinese earthen clay*" for its manufacture. Finally, a Dr. William Sherrard brought supplies of china clay and china stone from Paris to England in 1712 and surely the Fulham Pottery may have been perhaps a rather obvious destination for these specimens with the known work on refractory materials being undertaken there at that time? We can but enquire the reason that this landmark, cutting-edge work of John Dwight has been largely ignored and perhaps the only sensible appraisal of this situation is that there seems to be no documentary and physical evidence of his manufacture of his translucent porcellaneous material commercially surviving today – but he must have made a significant number of specimens for Hooke, Sloane and several other observers to describe its ceramic body properties so accurately in terms of its thin potting, robustness to mechanical shock and heat resistance – the very characteristics for which the Chinese hard paste china then being imported into Europe was so highly esteemed.

A final part of this tale emerges with the analytical case study of the three *Burghley House Jars,* which comprise a refractory porcelain of very unusual composition (Ramsay and Ramsay, *The Evolution and Compositional Development of English Porcelains from the 16th Century to Lund's Bristol c. 1750 and Worcester c. 1752 – the Golden Chain*", 2017; Spataro et al. 2008) comprising a large *Virtues* jar and two smaller versions with lids. Nondestructive chemical analysis in the British Museum Research Laboratories has revealed that their composition categorises them as very early porcelain (Tite et al. 1986) and this has further been credibly attributed to the Fulham manufactory. If this is really the case, then it firmly establishes that not only did John Dwight actually carry out early experiments at Oxford and Fulham on early porcelain variants but that he must have succeeded in making items that were deemed satisfactory for sale or at least presentation pieces for an esteemed clientele … and that, surely, must suggest strongly that the first manufacture of a successful porcelain in Western Europe should be accorded to John Dwight, some 40 years before von Tschirnhaus and Bottger claimed this for themselves at Meissen. Ramsay and Ramsay (*The Evolution and Compositional Development of English Porcelains from the 16th Century to Lund's Bristol c. 1750 and Worcester c. 1752 – the Golden Chain*, 2017) also allude to the critical mentoring role of Robert Boyle in this enterprise with John Dwight as analogous to that adopted between von Tschirnhaus and Bottger – both resulting apparently in the production of what must be early and historically important translucent refractory materials, be these porcelain as we know it today or possibly at least a noteworthy translucent variant.

The valid question may be posed at this point as to the reason there was so much activity and interest being shown by the Royal Society, and in particular prominent natural philosphers such as Boyle and Hooke, in the manufacture of porcelains. It should be remembered that in the 1660s and 1670s, chemistry was only just emerging as a definite and independent science from the alchemical experimental mysteries of the Renaissance, where much synthesis and many chemical transformations were steeped in secrecy, myths and arcane knowledge. An important part of these diverse alchemical processes and procedures involved the working of recipes involving strong acids such as *aqua regia* (a one to three mixture of concentrated nitric and hydrochloric acids) and fused alkalis at elevated temperatures for many days in crucibles, whose resistance to corrosion and attack was paramount. There was much experimentation being carried out on refractory materials for the purpose of the manufacture of alchemical crucibles and this surely established a basis for later work in the perfection of porcellaneous materials for the prosecution of chemical reactions and work at elevated temperatures. It is believed that a tradition in the manufacture of high quality crucibles was already existent in Saxony in the seventeenth century and that this could have well generated the initial experimental work there on porcelains at the Meissen factory in the late 1690s and prompted von Tschirnhaus to move there with the experience of his recent kiln firing knowledge gained in France, where King Louis XIV (*Le Roi Soleil*) was already sponsoring refractory china work at St Cloud near Paris, and prior to that the experiments being undertaken in London at the Fulham Pottery of John Dwight. It is interesting to propose that these early experiments, which were perhaps initially designed on the basis of effecting better alchemical transmutations, actually realised the manufacture of translucent porcelains which chronologically can be detected and recorded many years before the Jesuit priest Father Xavier d'Entrecolles revealed the secret Chinese recipe from Jingdezhen for the manufacture of their hard paste or true porcelain. It is also very clear that the composition of the early English wares from Fulham, Bow, Limehouse and Chelsea differed completely from those being produced on the Continent and this strongly suggests, perhaps surprisingly in view of currently maintained historical ideas, that the production of porcelains in Western Europe developed along two main and parallel fronts, namely the English and the Franco/German themes whose materials and products were compositionally and significantly very different.

4.5.2 The Burghley House Jars

Of special interest to early ceramics research, several papers have appeared relatively recently on the nature and provenancing of the very curious and probably unique *Burghley House Jars* (Ramsay and Ramsay, *The Evolution and Compositional Development of English Porcelains from the 16th Century to Lund's Bristol c. 1750 and Worcester c. 1752 – the Golden Chain*, 2017) mentioned above: they do not seem to fit in with any standard classification of porcelain from their analytical elemental oxide data and it is strongly suspected by several experts that they could possibly be finished and decorated products of the Fulham Pottery and represent specimens of finished John Dwight porcelain dating from the late seventeenth century. As far as contemporary ceramics historians are concerned they are indeed "*strangers in a strange land*" (Robert Heinlein, "*Stranger in a Strange Land*", 1961, with reference to Exodus 2:22). As Ramsay and Ramsay (2017) have so clearly enunciated, the Burghley House *Virtues* jar and one of its smaller companions analysed with a high silica content, an alumina content of ~18% and a potash content of ~5% which is totally different to any Continental glassy porcelain or Chinese true porcelain in circulation at that time (Wesley 2008: Spataro et al. 2008). It, therefore, must be considered to be peculiarly indigenous English in composition: additional significant levels of titania and iron oxide being present point to the use of a Dorset secondary ball clay, which John Dwight was known to have favoured in his recipes at Fulham. The independent analytical data of Wesley (2008) and of Spataro et al. (2008) firmly established that the *Virtues* jar was in fact an early porcelain, predating 1683, and that this was neither a stoneware nor a glassy, vitrified material. Ramsay and Ramsay (2017) associate the *Virtues* porcelain jar with John Dwight and manufactured around 1675, although it appears that this view is not confirmed by the opinions of Tite et al. (1986) and Spataro et al. (2008). Table 4.2 shows the compositional analyses of the Fulham shards and measurements made directly upon the Burghley House *Virtues* jar and its companion small jar: it can be seen that although very similar, the analyses are not precisely identical, but of course in the

Table 4.2 Compositional data for early porcelains from the Fulham Pottery and the Burghley House *Virtues* jars from SEM/EDAXS analyses

Specimen	Type	Date	Analyst(s)	SiO_2	Al_2O_3	CaO	P_2O_5	K_2O	Na_2O	MgO	Fe_2O_3	PbO
Fulham Pottery	Shards	~ 1680	Tite et al. (1986) and Spataro et al. (2008)	76.9	18.0	0.3	–	2.7	0.4	0.3	0.6	–
Burghley House	Virtues Jar	~1675	Spataro et al. (2008)	66.8	18.2	0.6	–	4.5	1.6	1.5	1.1	2.2
	Small Jar	~1675		69.0	18.9	1.4	0.1	5.8	1.4	0.5	0.6	1.7

nature of the ongoing experimental porcelain formulations current at that time it is quite conceivable that the data actually do encompass the same factory of production. Ramsay and Ramsay (2017) argue that this potential early Dwight porcelain recipe was the forerunner of the other early English factories following on later, such as those at Pomona, Limehouse, Bristol and to the earliest Dr. Wall's Worcester, and thereby leading on naturally to Bow and Vauxhall. For a more detailed story relating to the Burghley House jars the reader is referred to the excellent account afforded by Ramsay and Ramsay (*The Evolution and Compositional Development of English Porcelains from the 16th Century to Lund's Bristol c. 1750 and Worcester c. 1752 – the Golden Chain*, 2017) which also clearly puts forward the active role of individual scientists in the Royal Society in creating English porcelain after the death of John Dwight in 1708 in competition with the burgeoning porcelain industries then appearing in France and Saxony.

In summary, it can be stated with some confidence that recent research has shown very conclusively that there was a English porcelain and vitreous ceramics industry operating in the late seventeenth century which occupied the attention of some of the finest intellectuals and experimental scientists, and that porcelain was possibly being successfully made perhaps some 35–40 years prior to the claim by Meissen in 1709. The question as to whether or not this advance was led by John Dwight and his associates at Fulham must still remain unanswered. Expert opinion is still divided on this, with Ramsay and Ramsay (*The Evolution and Compositional Development of English Porcelains from the 16th Century to Lund's Bristol c. 1750 and Worcester c. 1752 – the Golden Chain*, 2017) being supportive of the advent of the production of seventeenth century porcelain in England but Spataro et al. (2008) alternatively casting some doubt upon this. Of course, neither the early English nor the German porcelains can be regarded now as being in the category of "true" porcelains as befits the Chinese formulations and recipes, which were very different from them compositionally, for their export wares had not yet reached Europe in large quantities, and indeed did not do so until later in the first quarter of the eighteenth century.

References

B. Adams, *True Porcelain Pioneers: The Cookworthy Chimera, 1670–1782* (B. Adams, Dawlish, 2016)

E. Adams, D. Redstone, *Bow Porcelain* (Faber and Faber, London, 1981)

B. Adams, A. Thomas, *A Potworks in Devonshire: The History and Products of the Bovey Tracey Potteries, 1750–1836* (Sayce Publications, Bovey Tracey, 1996)

J.A. Anderson, *Derby Porcelain and the Early English Fine Ceramic Industry*, Ph.D. Thesis, University of Leicester, UK, October, 2000

D. Barker, P. Halfpenny, *Unearthing Staffordshire* (City of Stoke-on-Trent Museum and Art Gallery, Stoke-on-Trent, 1990)

F.A. Barrett, *Caughley and Coalport Porcelains* (F. Lewis, London, 1951)

F.A. Barrett, *Worcester Porcelain and Lund's Bristol* (Faber and Faber, London, 1966)

P. Bemrose, The Pomona Potworks, Newcastle, Staffordshire. Trans. Engl. Ceram Circ. **9**, 1–18 (1973)

R. Boyle, *The Sceptical Chymist or Chymico-Physical Doubts and Paradoxes Touching the Spagyrist's Principles Commonly Called Hypoststical as They Are Wont to Be Proposed and Defended by the Generality of Alchymists* (J. Cadwell for J. Crooke at *The Ship* in St Paul's Churchyard, London, 1661)

C.J. Brooks, *China Returns to Lowestoft: The Story of Lowestoft Porcelain* (Jack Rose Old Lowestoft Society Publishing, Lowestoft, 2001)

E.M. Brown, T.A. Lockett, *Made in Liverpool: Liverpool Pottery and Porcelain, 1700–1850* (National Museums and Galleries on Merseyside, Liverpool, 1993)

E.A. Carter, M.L. Wood, D. de Waal, H.G.M. Edwards, Porcelain Shards from Portuguese Wrecks: Raman Spectroscopic Analysis of Marine Archaeological Ceramics. Herit. Sci. **5**, 17 (2017). https://doi.org/10.1186/s40494-017-0130-9

A. Chapman, *England's Leonardo: Robert Hooke, 1635–1703, and the Art of Experiment in Restoration England*. Proc. R. Inst. G. B. **67**, 39–275 (1996)

Sir A.H. Church, *English Porcelain: A Handbook to the China Made in England During the 18th Century as Ilustrated by Specimens Chiefly in the National Collection* (A South Kensington Museum Handbook, Chapman and Hall Ltd., London, 1885 and 1894)

P. Colomban, Differentiation of Antique Porcelains, Celadons and Faiences from the Raman Spectra of their Bodies, Glazes and Paintings. Asian. Chem. Lett. **5**, 133 (2001)

P. Colomban, The destructive/non-destructive identification of enameled pottery, glass artifacts and associated pigments – A brief overview. Arts **2**, 77–110 (2013)

P. Colomban, F. Treppoz, Identification and differentiation of ancient and modern European porcelains by Raman macro- and micro-spectroscopy. J. Raman. Spectrosc. **32**, 93–102 (2001)

P. Colomban, H.G.M. Edwards, C. Fountain, Raman Spectroscopic and SEM/EDXS analysis of Nantgarw soft paste porcelain. J. Eur. Ceram. Soc. Submitted for publication, (2020)

A. Cox, The analysis of Rockingham porcelain in the light of Thomas Brameld's notebook and comparison with some other examples pof 19th century bone china, J. North. Ceram. Soc., **3**, pp. 29–35, 1978/1979

A. Cox, A. Cox, *Rockingham Porcelain, 1745–1842* (Antique Collectors Club, Woodbridge, 2001)

P.E. Damon, D.J. Donahue, B.H. Gore, A.L. Hatheway, A.J.T. Jull, T.W. Linick, P.J. Sercel, L.J. Toolin, C.R. Bronk, E.T. Hall, R.E.M. Hedges, R. Housley, I.A. Law, C. Perry, G. Bonani, S. Trumbore, W. Woelfli, J.C. Ambers, S.G.E. Bowman, M.N. Leese, M.J. Tite, Radiocarbon dating of the Shroud of Turin. Nature **337**(6208), 611–615 (1989)

P. Daniels, *The Origin and Development of Bow Porcelain, 1730–1747, Including the Participation of the Royal Society, Andrew Duche and the American Contribution* (Resurgat Publishers, Oxford, 2007)

P. Daniels, W.R.H. Ramsay, Bow porcelain. New primary source documents and evidence pertaining to the early years of the manufactory between 1730–1745 and the John Campbell's letter to Arthur Dobbs. South. Inst. Technology Journal of Applied Research. **1**, 1–30 (2009)

D. de Waal, Raman investigation of ceramics from 16th and 17th century Portuguese shipwrecks. J. Raman Spectrosc. **35**, 646–649 (2004a)

D. de Waal, Raman identification of the pigment in blue and white Ming porcelain shards. Asian. Chem. Lett. **8**, 57–65 (2004b)

L.W. Dillwyn, Notes on the Experimental Production of Swansea Porcelain Bodies and Glazes. Made by Lewis Weston Dillwyn with Samuel Walker at the Swansea China Works Between 1815 and 1817. Presented to the Library of the Victoria and Albert Museum, South Kensington, London by John Campbell in 1920. Reproduced in Eccles and Rackham, Analysed Specimens of English Porcelain, 1922, see reference

A. Dinsdale, *Pottery Science: Materials, Products and Processes* (Ellis Horwood, Chichester, 1986)

A. Dodd, D. Murfin, *Dictionary of Ceramics*, 3rd edn. (The British Pottery Manufacturers, Federation, Institute of Mineralogy, London, 1994/5)

H. Eccles, B. Rackham, *Analysed Specimens of English Porcelain in the Victoria and Albert Museum Collection* (Victorian and Albert Museum, London, 1922)

H.G.M. Edwards, *Swansea Porcelain: The Duck-Egg Translucent Vision of Lewis Dillwyn* (Penrose Antiques Ltd. Short Guides, Thornton, 2017). ISBN: 9780244325787

H.G.M. Edwards, *Derby Porcelain: The Golden Years, 1770–1830* (Penrose Antiques Ltd. Short Guides, Thornton, 2018a)

H.G.M. Edwards, "Raman Spectroscopic Studies of Swansea and Nantgarw Porcelains", Chapter 6, in *Nantgarw and Swansea Porcelains: An Analytical Perspective*, (Springer International Publishing AG/Springer Nature, Cham, 2018b)

H.G.M. Edwards, M.C.T. Denyer, *William Billingsley the Enigmatic Porcelain Artist, Decorator and Manufacturer* (Penrose Antiques Ltd. Short Guides, Neopubli, Berlin, 2016). ISBN: 978-3-7418-6802-3

H.G.M. Edwards, P. Colomban, B. Bowden, Raman spectroscopic analysis of an English soft paste porcelain plaque-mounted table. J. Raman. Spectrosc. **35**, 656–661 (2004)

H.G.M. Edwards, A.P.H. Surtees, R. Telford, "Dancing on Eggshells: A Holistic Analytical Study of a Ballet Dancer on Regency Porcelain", Chapter 4, in *Raman Spectroscopy in Archaeology and Art History*, ed. by P. Vandenabeele, H. G. M. Edwards, (Royal Society of Chemistry Publishing, Cambridge, 2019a)

H.G.M. Edwards, P. Colomban, C. Fountain, *Raman spectroscopic and SEM/EDAXS analyses of Nantgarw porcelain* (in preparation for publication, 2019b)

W. Elliott, Soft paste Bristol porcelain and the intimate relationship of the factory at Lowdin's glasshouse with Limehouse and Worcester. Trans. Engl. Porc. Circ. **2**, 6–23 (1929)

I.C. Freestone, A technical study of Limehouse, in *Limehouse Ware Revisited*, ed. by D. Drakard, (English Ceramic Circle, London, 1993), pp. 68–73

I.C. Freestone, A-marked porcelain: Some recent scientific work. Trans. Engl. Ceram Circ. **16**, 76–84 (1996)

I.C. Freestone, *The science of early English porcelain*, Sixth Conference and Exhibition of the European Ceramics Society, vol **1** (Brighton, 1999), pp. 11–17

I.C. Freestone, L. Joyner, R. Howard, The composition of porcelain from the Isleworth manufactory. Trans. Engl. Ceram Circ. **18** (2003)

N.D. Gent, *The Patterns and Shapes of the Pinxton China Factory, 1796–1813* (N.D. Gent Publications, Pinxton, 1996)

G.A. Godden, *Minton Pottery and Porcelain of the First Period, 1793–1850* (Barries and Jenkins, London, 1968)

G.A. Godden, *The Illustrated Guide to Lowestoft Porcelain* (Barrie and Jenkins, London, 1969a)

G.A. Godden, *Caughley and Worcester Porcelains, 1775–1800* (Barrie and Jenkins, London, 1969b)

G.A. Godden, *Coalport and Coalbrookdale Porcelains* (Barrie and Jenkins, London, 1970)

G.A. Godden, *The Illustrated Guide to Ridgway Porcelains* (Barrie and Jenkins, London, 1972a)

G.A. Godden, *Ridgway Porcelain* (Antiques Collectors Club, Woodbridge, 1972b)

G.A. Godden, *New Hall Porcelains* (Antique Collectors Club Art Books, Woodbridge, 1999a)

G.A. Godden, *Lowestoft Porcelain* (Ainsworth and Nelson, Ponsharden, 1999b)

G.A. Godden, Isleworth porcelain in Victorian collections. Trans. Engl. Ceram Circ. **20**, 75–90 (2008)

H.E. Gove, *Relic, Icon or Hoax: Carbon-Dating the Turin Shroud* (Institute of Physics Publishing, Bristol/Philadelphia, 1996)

R. Gunther, *Early Science at Oxford*, vol 14 (Gunther Publication, Oxford, 1923–1967)

D. Haselgrove, J. Murray, John Dwight's Fulham Pottery, 1672–1978: A collection of documentary sources. J. Ceram. Hist. Stoke-on-Trent City Mus. **11** (1979)

R. Heinlein, *Stranger in a Strange Land* (Ace Publishers/Grosset and Dunlap, New York, 1961)

M. Hillis, An introduction to ceramic raw materials, bodies and glazes. North. Ceram. Soc. J. **18**, 77–111 (2001)

M. Hillis, The Development of Welsh Porcelain Bodies, in *Welsh Ceramics in Context, Part II*, ed. by J. Gray, (Royal Institution of South Wales, Swansea, 2005), pp. 170–192

M. Hillis, *Liverpool Porcelain, 1756–1804* (M. Hillis Publications, Liverpool, 2011)

D. Holgate, *New Hall and its Imitators* (Faber and Faber, London, 1973)

W.B. Honey, The relations between English and continental porcelain. Trans. Engl. Ceram Circ. **7**, 88–98 (1939)

R. Howard, Isleworth pottery: Recognition at last? Trans. Engl. Ceram Circ. **16**, 345–368 (1998)

R. Howard, Report on the archaeological evaluation of the Isleworth Pottery site. Trans. Engl. Ceram Circ. **17**, 467–469 (2001)

F. Hurlbutt, *Bow Porcelain* (G. Bell and Sons, London, 1926)

W. Jay, J.D. Cashion, Raman spectroscopy of Limehouse porcelain sherds supported by Mossbauer spectroscopy and scanning electron microscopy. J. Raman Spectrosc. **44**, 1718–1732 (2013)

L. Jewitt, *The Ceramic Art of Great Britain from the Prehistoric Times Down to the Present Day*, vol I and II (Virtue and Co. Ltd., London, 1878)

W.D. John, *William Billingsley* (Ceramic Book Co., Newport, 1968)

R. Jones, *Porcelain in Worcester 1751–1951: An Illustrated Social History* (Parkbarn Publishing, Guildford, 1993)

R. Jones, The origins of Lund's Bristol porcelain and the site of the Bristol manufactory. North. Ceram. J. **23**, 83–104 (2007)

R. Jones, *The Origins of Worcester Porcelain: Local Ingenuity and the Pathways Form Staffordshire, Stourbridge, Bow, Limehouse and Bristol* (Parkbarn Publishing, Guildford, 2019)

A.E. Jones, Sir L. Joseph, *Swansea Porcelain: Shapes and Decoration* (D. Brown and Sons, Ltd., Cowbridge, 1988)

T.A. Lockett, *Davenport Pottery and Porcelain, 1794–1887* (David and Charles, Newton Abbot, 1972)

T.A. Lockett, G.A. Godden, *Davenport China, Earthenware and Glass, 1794–1887* (Barrie and Jenkins, London, 1989)

R. Massey, The Isleworth Pottery insurance policies, 1765–1800. Trans. Engl. Ceram Circ. **18**, 295–299 (2003)

R. Massey, Vauxhall figures. Trans. Engl. Ceram Circ. **25**, 1–21 (2014)

M.F. Massey, S. Spero, *Ceramics of Vauxhall, 18th Century Pottery and Porcelain* (English Ceramic Circle, London, 2007)

K.S. Meager, *Swansea and Nantgarw Potteries: Catalogue of the Collection of Welsh Pottery and Porcelain on Exhibition at the Glynn Vivian Art Gallery, Swansea* (Swansea Corporation, Swansea, 1949)

M.F. Messenger, *Coalport, 1795–1926: An Introduction to the History and Porcelain of John Rose and Company* (Antiques Collectors Club, Woodbridge, 1995)

E. Morton Nance, *The Pottery and Porcelain of Swansea and Nantgarw* (B.T. Batsford Ltd., London, 1942)

J.V. Owen, Quantification of early Worcester porcelain recipes and the distinction between Dr Wall- and Flight-period wares. J. Archaeol. Sci. **24**, 301–310 (1997)

J.V. Owen, A preliminary assessment of the geochemistry of porcelain sherds from the Limehouse factory site, in *Limehouse Ware Revisited*, ed. by K. Tyler, R. Stephenson, (MoLAS Monographs, No. 6, Museum of London, London, 2000a), pp. 61–63

J.V. Owen, A preliminary assessment of the geochemistry of porcelain sherds from the Limehouse site, in *The Limehouse Porcelain Manufactory: Excavations at 108–116, Narrow Street, London,1990*, ed. by K. Tyler, R. Stephenson, (Museum of London Archaeology Service, Monograph No. 6, London, 2000b)

J.V. Owen, Provenience of 18th century British porcelain shards from Sites 3B and 4E, Fortress of Louisburg, Nova Scotia: Constraints from mineralogy, bulk paste and glaze compositions. Hist. Archaeol. **35**, 108–121 (2001)

J.V. Owen, What analytical data can and cannot tell us about 18th century fine ceramics. Trans. Engl. Ceram Circ. **22**, 215–230 (2011)

J.V. Owen, R. Barkla, Compositional characteristics of 18th century Derby porcelains: Recipe changes, phase transformations and melt fertility. J. Archaeol. Sci. **24**, 127–140 (1997)

J.V. Owen, T.E. Day, Estimation of the bulk composition of fine-grained media from microchemical and backscatter image analysis: Application to biscuit wasters from the Bow factory site, London. Archaeometry **36**, 217–226 (1994a)

J.V. Owen, T.E. Day, Eighteenth century phosphatic porcelains: Bow and Lowestoft – further confirmation of their compositional distinction. Trans. Engl. Ceram Circ. **16**, 342–344 (1994b)

J.V. Owen, T.E. Day, Assessing and correcting the effects of the chemical weathering of potsherds: A case study using soft paste porcelain wasters from the Longton Hall (Staffordshire) factory site. Geoarchaeology **13**, 265–268 (1998)

J.V. Owen, M. Hillis, From London to Liverpool: Evidence for a Limehouse-Reid porcelain connection based on the analysis of sherds from the Brownlow Hill (ca. 1755–1776) factory site. Geoarchaeology **18**, 851–882 (2003)

J.V. Owen, M.L. Morrison, Sagged phosphatic Nantgarw porcelain (ca. 1813–1820): Casualty of overfiring or a fertile paste? Geoarchaeology **14**, 313–332 (1999)

J.V. Owen, J. Sandon, A rose by another name: A geochemical comparison of Caughley (c.1772–1799), Coalport (John Rose and Co., c.1799–1837), and rival porcelains based on sherds from the factory sites. Post-Mediev. Archaeol. **37**, 79–89 (2003)

J.V. Owen, J.O. Wilstead, R.W. Williams, T.E. Day, A tale of two cities: Compositional characteristics of some Nantgarw and Swansea porcelains and their implications for kiln wastage. J. Archaeol. Sci. **25**, 359–375 (1998)

J.V. Owen, B. Adams, R. Stephenson, Nicholas Crisp's "*Porcellien*": A petrological comparison of sherds from the Vauxhall (London., *ca.* 1751–1764) and Indeo Pottery (Bovey Tracey, Devonshire, *ca.* 1767–1774) factory sites. Geoarchaeology **15**, 43–78 (2000)

N. Panes, R. Howard, J.V. Owen, Attribution enhanced – Isleworth porcelain re-examined. Trans. Engl. Ceram Circ., 89–116 (2012)

J. Penderill-Church, *William Cookworthy, 1705–1780: A Study of the Pioneer of True Porcelain Manufacture in England* (D. Bradford Barton Publishers, Truro, 1972)

J. Potter, The Limehouse story, in *Digging for Early Porcelain*, ed. by D. Barker, S. Cole, (City Museum and Art Gallery, Stoke-on-Trent, 1998), pp. 40–53

W.R.H. Ramsay, E.G. Ramsay, A case for the production of the earliest commercial hard paste porcelains in the English-speaking world by Edward Heylyn and Thomas Frye in about 1743. Proc. Roy. Soc. Victoria **120**, 236–256 (2008)

W.R.H. Ramsay, E.G. Ramsay, *The Evolution and Compositional Development of English Porcelains from the 16th Century to Lund's Bristol c. 1750 and Worcester c. 1752 – the Golden Chain* (Invercargill Press, Invercargill, 2017)

W.R.H. Ramsay, P. Daniels, E.G. Ramsay, *The Limehouse Porcelain Factory, Its Output, Antecedents and the Influence of the Royal Society of London on the Evolution of English Porcelains based on Composition and Technology* (Invercargill Press, Invercargill, 2013)

W.R.H. Ramsay, P. Daniels, E.G. Ramsay, *Limehouse Porcelain: Are Limehouse Porcelains in Fact All Limehouse? Evidence from Archaeology, Science, and Historical Documents* (Resurgat Publishers, Oxford, 2015)

H. Rissik Marshall, *Coloured Worcester Porcelain of the First Period, 1751–1783* (Ceramic Book Company, Newport, 1954)

R.N. Rogers, *A Chemist's Perspective on the Shroud of Turin* (Barrie M. Schwortz/Florissant Publishing, Florissant, 2008)

H. Sandon, *The Illustrated Guide to Worcester Porcelain, 1751–1793* (Praeger, New York, 1969)

H. Sandon, *Flight, Barr and Barr Worcester Porcelain, 1783–1840* (Antiques Collectors Club Distribution, Woodbridge, 1993)

S. Shaw, *The Chemistry of the Several Natural and Artificial Heterogeneous Compounds Used in Manufacturing Porcelain, Glass and Pottery* (Scott, Greenwood and Son, London, 1837). Re-issued in its original form in 1900, 713 pp, 1900

C.B. Sheppard, *Pinxton Porcelain, 1795–1813, and the Porcelains of Mansfield and Brampton-in-Torksey* (C.B. Sheppard Publications, Alfreton, 1996)

L.M. Solon, *Art of the Old English Potter* (Bemrose and Son, London, 1883), pp. 32–35

M. Spataro, M. Bimson, N. Meeks, Part3: A scientific investigation of three pieces of porcelain presumed seventeenth century, from Burghley House. Trans. Engl. Ceram Circ. **20**, 187–194 (2008)

M. Spataro, N. Meeks, M. Bimson, A. Dawson, J. Ambers, Early Porcelain in 17th century England: Non-destructive examination of two jars from Burghley House. Br. Museum. Res. Bull. **3**, 37–48 (2009)

C. Spencer, *Early Lowestoft: A Study of the Early History and Products of the Lowstoft Porcelain Manufactory* (Ainsworth and Nelson, Ponsharden, 1981)

S. Spero, *Liverpool Porcelain, 1755–1799* (Simon Spero Publishing, London, 2006)

M.S. Tite, M. Bimson, A technological study of English porcelains. Archaeometry, **33**, 3–27 (1991)

M.S. Tite, I.C. Freestone, M. Bimson, A technological study of Chinese porcelain of the Yuan Dynasty. Archaeometry **26**, 139–154 (1984)

M.S. Tite, M. Bimson, I.C. Freestone, *A technological study of Fulham stoneware* (Proceedings of the 24th International Archaometry Symposium, Washington, DC, 1986), pp. 95–104

J. Twitchett, *Derby Porcelain, 1748–1848* (The Antique Collector's Club, Woodbridge, 2002)

K. Tyler, R. Stephenson, J.V. Owen, C. Philpotts, *The Limehouse Porcelain Manufactory: Excavations at 108–116 Narrow Street, London, 1990* (Museum of London Archaeological Service, Monograph No.6, London, 2000)

B.M. Watney, *Longton Hall Porcelain* (Faber and Faber, London, 1957)

B.M. Watney, *"Limehouse, its Relationship to Newcastle-Under-Lyme (Pomona) and Other Manufactories", in Limehouse Ware Revealed, Ed* (D. Dakard, English Ceramic Circle, London, 1993)

B.M. Watney, *Liverpool Porcelain of the 18th Century* (R. Dennis, London, 1997)

M. Wesley, Part 1: The Burghley House Buckingham porcelains as documentary objects. Trans. Engl. Ceram Circ. **20**, 169–182 (2008)

L. Whiter, *Spode: A History of the Family, Factory and Wares from 1733–1833* (Barrie and Jenkins, London, 1970)

I.J. Williams, *The Nantgarw Pottery and its Products: An Examination of the Site* (The National Museum of Wales and the Press Board of the University of Wales, Cardiff, 1932)

I. Wilson, *The Blood and the Shroud: The Passionate Controversy Still Enflaming the World's Most Famous Carbon-Dating Test* (Weidenfeld and Nicholson, London, 1998)

N. Wood, M. Cowell, William Cookworthy's Plymouth porcelain and its origins in Jingdezhen practice, in *Proceedings of the International Symposium on Ancient Ceramics, Shanghai*, ed. by G. Taoxi, K. Jishu, (China Shanghai Institute of Ceramics, Chinese Academy of Sciences, Shanghai, 2002), pp. 326–336

W. W. Young, *The Diaries of William Weston Young, 1776–1847 (1802–1843)*, vol 30, West Glamorgan Archive Service, Swansea, SA1 3SN. https://arcgiveshub.jisc.sc.uk/data/gb216-d/dxch/ddxch/i/hub

J. Zuo, C. Xu, C. Wang, Z. Yushi, Identification of the pigment on painted pottery from the Xishan site by Raman microscopy. J. Raman. Spectrosc. **30**, 1053–1055 (1999)

Chapter 5
Analytical Compositional Data and the Interpretation of the Data Acquired from Elemental Oxide Determinations.

Abstract A detailed and critical comparative survey is made of the published elemental oxide analytical data which has been obtained for many of the English and Welsh porcleian manufactories discussed earlier in the text. The use of replicate specimens or sampling to establish the experimental errors in the data is discussed and the pitfalls awaiting incomplete data interpretation are highlighted. The selection of multiple component analytical data is merited and several experimental protocols are suggested for the evaluation of porcelain type compositions and the establishment of a factory attribution, including the phosphorus pentoxide, magnesia and lead oxide protocols.

Keywords Elemental oxide analytical data · Data interpretation · Experimental errors in data determination · Protocols for factory attribution using experimental data

The elemental oxide compositional data for the eighteenth and nineteenth century porcelain manufactories under consideration here has been gathered in Table 4.1: this comprises analytical data acquired from two basic techniques, namely early wet chemical digestion and the later scanning electron microscopy with electron dispersive X-ray analysis, SEM/EDAXS. Both analytical techniques can be considered destructive of sample in the sense that the specimen is subjected to chemical and/or mechanical pretreatment before the analyses are carried out. Whereas the earlier analytical wet chemical technique was undertaken using finished porcelain samples, which may have been already damaged so the further damage created by the removal of a significant quantity of the specimen for analysis was not considered, resulting in a serious invasion of the integrity of the specimens, the dictats of modern analysis require that normally only broken shards are taken, for which further fragmentation or damage is tolerably accepted. Later, the application of molecular spectroscopic techniques which are non-invasive of the specimen and which can be applied to perfect and finished pieces of porcelain will be reviewed, but the molecular information retrieved from such analyses is essentially rather different to that which is obtained from the elemental SEM/EDAXS experiments that we have considered

H. G. M. Edwards, *18th and 19th Century Porcelain Analysis*,
https://doi.org/10.1007/978-3-030-42192-2_5

thus far in that the quantitative analytical estimation of the elemental oxide percentages is not possible but rather the molecular composition of the finished, fired porcelain is interrogated. Together, *in tandem* these techniques can reveal much information not only about the recipe formulations and changes made therein temporally but also the molecular conversions that have taken place in the firing process which reveal details of the technology being used to make the ceramics.

The first point to be made from a consideration of the information collected in Table 4.1 is that it is very diverse in both the number and type of sample being reported: in many cases only one specimen has been analysed from a particular factory and rarely does this exceed ten specimens. Hence, the data derived from each determination are quoted to different accuracies. For example, the citation of a silica percentage as 63.0% for a single Bovey Tracey hard paste porcelain shard does not imply that it is ten times or more accurate than the corresponding value derived from 12 Coalport soft paste shards which spans the range 76 +/− 2% within one standard deviation of experimental error: the latter figure is mathematically averaged over each determination and is therefore considered to give a more reliable and accurate figure than a single specimen result. In many cases, of course, it is just not possible for the analysts to acquire multiple shards for their analyses: even for the Limehouse Pottery Manufactory archaeological excavation carried out in 1990, where some 1402 porcelain shards were discovered, relatively few of these have been actually analysed for their elemental oxide compositional data. In a recent study, Colomban et al. (2020) have analysed porcelain shards from a 1990s archaeological excavation at the Nantgarw China Works site and some 80 shards were provided from the museum archive for analysis, weighing approximately 4 kg, of which 8 were subjected to a combined molecular spectroscopic and SEM/EDAXS elemental compositional analysis, some of which were glazed, some of which were biscuit unglazed porcelain and three of which more rarely were glazed and decorated with enamels (Colomban et al. 2020). The latter three were the most interesting analytically because shards of finished porcelain are so rarely found in site investigations as the final stage firing of the underglaze enamels is accomplished at much lower kiln temperatures in the "glost kiln" and the resultant loss of the artefact is correspondingly little encountered compared with losses experienced at the first firing stage with its attendant kiln high temperatures. Another "hidden" parameter which needs to be realised here is that the experimental determinations themselves are subject to error: Tite et al. (1984) specify the following errors collectively for their SEM/EDAXS measurements as silica 1%, alumina and potash 1–2%, iron oxide, lime and soda 5%, and magnesia and phosphate 10–20%. Clearly, such values need to be incorporated into the interpretative conclusions of the analytical data, particularly where the discrimination between factory products of different phases of operation or between different factories is being attempted and especially where the differences in the percentage compositions between the body paste determinations may only be estimated at about 1–2%.

Secondly, a very important deduction can be made from an internal comparative consideration of the elemental oxide percentages for each individual factory, where the range of values is seen to be often rather wide, which seemingly implies that even within a particular paste type the composition of the body has changed significantly. The origin of this could be that there has been a physical change in the relative compositions of each component of the raw materials or it could even reflect an imprecision in the making up of the paste itself or in the stages involving the component "frits" in a multi-stage process. A contributory factor to this imprecision would be the indeterminate change in hydration of raw materials, such as feldspar, calcined bone ash, lime, potash, soda ash and the primary and secondary china clays: unless each component was stored strictly in controlled surroundings or thermally treated properly before usage a consistent weight for each mineral component could not be achieved and this would introduce additional errors into the recipe formulation. An estimate has been made earlier of the effect this might have had upon the composition and certainly, as a result, several percent difference in the elemental oxide determinations would not be unexpected and untoward.

Thirdly, an important conclusion from the data provided in Table 4.1 is that in many cases, some of which will be considered in greater detail below, it is not possible to discriminate analytically between the porcelains produced by Factory X and Factory Y when one compares just one single elemental oxide determination, such as silica, for example. This is a very important analytical conclusion, because it means that in some cases the analyses do not positively permit an identification to be accomplished using a single parameter, especially for an important and ubiquitous determination such as silica, but we must then look to assessing other contributory data sets: hence, this is probably the most significant and general outcome of Table 4.1 – namely, that there is no analytical determination of a single analyte that can positively discriminate between all factories in the list, but it is recognised that an analytical protocol needs to be established whereby a yes/no answer to the specific presence, relative amount or complete absence of an individual elemental oxide component will eliminate sequentially several possibilities until only the source factory remains by this deduction. For example, such a protocol could well be based upon answers to the following test queries: how much silica is present (high/medium/low), how much alumina is present (high/medium/low), is magnesia present in a significantly large quantity, is lime a major component, are soda ash and potash present in quantity and does their component ratio have a particular significance, what is the percentage of phosphorus pentoxide or phosphate present, is iron oxide present and how much, does the lead oxide content reflect the usage of flint glass frit or cullet as a body component, and finally are there any other additives which can act as additional mineral markers of a potential factory origin, such as barytes found to occur in early Bovey Tracey porcelain, borax and arsenic oxide in Swansea porcelains, and the trace presence of cobalt oxide from smalt addition to remove a yellow colouration caused by high levels of ferric oxide in the china clays as used by Coalport and Swansea (Edwards, *Swansea and Nantgarw Porcelains: A Scientiifc Reappraisal*, 2017). It is instructive at this point to consider what inference can be drawn from the key elemental oxide percentages that are derived from the shards and other porcelain items studied in Table 4.1 to determine the type of porcelain that we are dealing with: three typical protocols for the combination of elemental oxides percentages and the interpretation of the porcelain type derived therefrom are given below, each starting with a "unique" derivative elemental oxide based on a single component raw material:

A. ***The phosphorus pentoxide P_2O_5 protocol:***

High P_2O_5, Medium SiO_2, Low Al_2O_3	Phosphatic
High P_2O_5, Medium SiO_2, Medium Al_2O_3, High CaO	Phosphatic
Medium P_2O_5, Medium SiO_2, High Al_2O_3, Medium CaO	Phosphatic
Low P_2O_5, High PbO, High SiO_2, Low Al_2O_3, Medium CaO	Glassy
Low P_2O_5, High SiO_2, Low Al_2O_3	Soft paste
Low P_2O_5, High MgO, High SiO_2, Medium Al_2O_3	Soaprock
Low P_2O_5, High MgO, High SiO_2, Low Al_2O_3	Soaprock
Low P_2O_5, High SiO_2, High Al_2O_3	Hard paste
Low P_2O_5, High SiO_2, Medium Al_2O_3, High CaO	Silicaceous

B. ***The magnesia MgO protocol:***

High MgO, Low P_2O_5, High SiO_2, Medium Al_2O_3	Soaprock
High MgO, Low P_2O_5, High SiO_2, Low Al_2O_3	Soaprock
Low MgO, High P_2O_5, Medium SiO_2, Low Al_2O_3	Phosphatic
Low MgO, High P_2O_5, Medium SiO_2, Medium Al_2O_3, High CaO	Phosphatic
Low MgO, Medium P_2O_5, Medium SiO_2, High Al_2O_3, Medium CaO	Phosphatic
Low MgO, High PbO, Low P_2O_5, High SiO_2, Low Al_2O_3, Medium CaO	Glassy
Low MgO, High SiO_2, Low P_2O_5, Low Al_2O_3	Soft paste
Low MgO, Low P_2O_5, High SiO_2, High Al_2O_3, High CaO	Silicaceous
Low MgO, LowP_2O_5, High SiO_2, High Al_2O_3	Hard paste

C. ***The lead oxide PbO protocol:***

High PbO, High SiO_2, Low Al_2O_3, Low P_2O_5, Medium CaO	Glassy
Low PbO, High P_2O_5, Medium SiO_2, Low Al_2O_3	Phosphatic
Low PbO, Low P_2O_5, High SiO_2, High Al_2O_3	Hard paste
Low PbO, High P_2O_5, Medium SiO_2, Medium Al_2O_3, High CaO	Phosphatic
Low PbO, Low P_2O_5, High SiO_2, High Al_2O_3, High CaO	Silicaceous
Low PbO, Medium P_2O_5, Medium SiO_2, High Al_2O_3, Medium CaOPhosphatic	
Low PbO, Low P_2O_5, Low Al_2O_3, High SiO_2	Soft paste
Low PbO, Low P_2O_5, High MgO, High SiO_2, Medium Al_2O_3	Soaprock
Low PbO, Low P_2O_5, High SiO_2, Low Al_2O_3, High MgO	Soaprock

The cross-correlation between these three elemental oxide protocols lends confidence to the interpretation of the elemental oxide percentage data and their attribution to specific porcelain types. Of course, it is now possible to further refine these basic protocols even more to include further elemental oxide data such as the percentages of soda ash and potash in the analyses: although perceived to be minor components in the analytical data the relative percentage occurrences of soda and potash can often provide useful information to the analyst. For example, Owen and Sandon (2003) have determined that although most of the analytical data for the phosphatic Coalport and Nantgarw porcelains, with their high phosphate content arising from about 40–45% bone ash in their raw material recipes, the most critical analytical determination in reality is the relative ratios of potash and soda in each factory paste, being 3:1 for Nantgarw and 1:1 for Coalport. Hence, in this instance, the ratios of what are normally seen to be minor components in the body pastes in fact provides the key discriminator for the analytical segregation and differentiation between Nantgarw and Coalport porcelains. This could be a very useful factor in future analytical protocols as it is well known that some unmarked Coalport porcelain has frequently been mis-attributed historically to the Nantgarw factory and presumably *vice-versa* (Edwards, *Nantgarw and Swansea Porcelains: An Analytical Perspective*, 2018).

Previous authors have drawn a rather diffuse distinction between bone china and phosphatic porcelain, claiming that the term bone china could only be applied to a significant content of bone ash of around 50% in the recipe: of course, we can see immediately from Table 4.1 that this is quite an arbitrary line and cannot be thought of as definitive – Nantgarw phosphatic soft paste porcelain contains about 43% bone ash compared with 44% in Spode's bone china, which is hardly to be described as a distinctive discriminator between a phosphatic soft paste porcelain and bone china, and indeed on the strict basis of this definition, Spode china does not even merit the classification of "bone china" at all and Rockingham bone china with an average P_2O_5 percentage of 18% equating with a bone ash content of 46% is rather similarly condemned. The real difference in definition of the two ceramic classifications, of course, is that Josiah Spode operated a one-stage firing process akin to the production of the true Chinese, hard paste porcelain for his bone china whereas William Billingsley operated a two-stage firing process at Nantgarw as befits a phosphatic, soft paste porcelain.

A comment about the lead oxide data needs to be made: although the determinations of lead in the fired body are quite unique and can be unambiguously ascribed to the presence of a lead-containing flint glass frit, the back-calculation of the amount of glass frit added to the paste in the recipe is fraught with complications because of the highly variable compositional nature of the flint glass frit for its lead content. Glass cullet used in the porcelain industry can contain zero lead oxide (from crown glass or soda glass) to approximately 60% for the densest and most highly refractive lead or flint glass: quite commonly a figure of 30% is taken to represent the average percentage of lead oxide in the flint glass cullet used in this context. It is certain that the porcelain factory proprietors in the eighteenth and early nineteenth centuries never analysed their shipment of waste glass frit in the cullet from the glassmaking industry for lead content and they must have relied heavily upon data provided by the glassworks proprietors concerning the amount of lead oxide that was added to their glass-forming recipe supplies. A potential problem here is that glass manufacturers adopted the principle of using their "end-of-day" glass residues for waste glass frit and these could contain essentially mixtures of lead glass with different lead oxide contents and possibly even some lead-free crown glass in admixture. One should also be rather careful in interpreting the presence of a lead oxide analytical signature as definitively indicating that glass cullet has been used as a body paste component as impurities from a glazed specimen or shard could have penetrated the elemental oxide determinations during the grinding, chemical extraction and digestion procedures especially for the earlier wet chemistry determinations, dependent upon the sampling procedures used. For biscuit porcelains, all seems well in this respect and any lead content found therein must surely have been deemed to have come from the porcelain body and not the superficial glaze. Usually, therefore, the absence of lead strongly implies that flint glass cullet was not used as an additive; whereas this still does not exclude or debar the use of a lead-free crown glass cullet, significant additions of this would certainly raise the calcium oxide, soda and potash levels in the body analyses as well as the silica level, which may be detectable in the comparative analytical estimations. An inference about the sourcing of the raw materials can sometimes be gleaned from the amounts of soda and potash given in the analytical determinations: soda ash was derived primarily from the burning of terrestrial plant material, whereas potash was obtained from the burning of marine kelp and seaweed. Hence, although it is not strictly indicative of the sourcing of the alkaline ash used as a flux in the porcelain synthesis because of the contributions made to the total percentages of Na_2O and K_2O in the elemental oxide data from raw materials such as feldspar, china clay and other complex silicates, it nevertheless does offer an intriguing glimpse into the possible alternative sourcing of raw materials by the china works proprietors for these important fluxing agents.

The three protocols based on the P_2O_5, MgO and PbO determinations offered above can act as a first-pass screening device or filter for the analytical interpretation of the experimental data and occasionally other possible conclusions may be drawn; an example of this is provided by the Limehouse porcelains which, according to Bernard Watney (in *Limehouse Ware Revealed,* 1993) towards their last phase of operations in 1747–1748 produced a novel and rather unique high magnesia/high phosphate soft paste porcelain which does not fit into any of the protocols devised here (Owen 2000a, b; Potter 1998; Ramsay et al. 2013, 2015. Other investigators, however, have not found evidence of this high magnesia or phosphatic content in Limehouse porcelain shards (Jay and Cashion, 2013; Freestone, in *Limehouse Ware Revealed,* 1993). The significant presence of both magnesia and phosphate in the analyses would certainly qualify this Limehouse ware to be put into the classification of a "*hybrid*" porcelain but that would strictly depend upon one's definition: if, for example, a *hybrid* porcelain is truly lying between a hard paste and soft paste type then a Limehouse assignment would not be applicable in this context, but in terms of the Eccles and Rackham (*Analysed Specimens of English Porcelain in the Victoria and Albert Museum Collection,* 1922) original category definition in this respect, Limehouse, Coalport and Spode (representing "bone china") certainly qualify for an assignment to this type. It is interesting too that the absence of lime, which is represented by a CaO elemental oxide percentage determination, is in accord with these Limehouse porcelains being recognised as truly *hard paste* in category (Freestone, in *Limehouse Ware Revealed,* 1993), since the Chinese did not use lime as a fluxing agent in their manufacture of hard paste porcelains.

5.1 Analytical Data and Source Attribution of Porcelains

Despite the apparent drawbacks in the interpretation of the wealth of analytical data now appearing in the literature for early porcelains, caused mainly by the experimental nature of the production processes and frequent unrecorded changes being made to the composition in efforts by the china works proprietors to increase the excellence and commercial viability of their product, several points can be made which will form the basis of later discussion: firstly, as enunciated in Panes et al. (2013), who were concerned with the attribution of Isleworth porcelains and their differentiation from those of Bow, a cautionary statement was made regarding the attribution of porcelains on the basis of analytical evidence alone and that an "attribution" may be possible but that the analyst needed to be aware of the potential of an accidental overlap in paste compositions. Taking into account the experimental errors in the analytical determinations as described and evaluated above, this could be a real problem for definitive assignments to be made. Phosphatic wares were regarded as being particularly challenging in this respect on account of the free movement of personnel that was occurring between many factories, and the implication that a transfer of knowledge would have been forthcoming during that process which could result in frequent and perhaps unrecorded changes in composition of a factory's products. To compound this situation, the database of analytical information from shards and perfect pieces of English porcelain which amounts to perhaps several hundred items of information, is actually a very small percentage of the porcelain that was produced, when it is appreciated that a single kiln firing could have perhaps involved some 10,000 pieces of china. Dr William John (*Nantgarw Porcelain,* 1948) claimed that he had personally seen about 5000 pieces of finished Nantgarw china during his lifetime of research on the factory output, yet only some 30 Nantgarw shards have ever been analysed, significantly less than 1% of the finished porcelain produced: Dr John has decreed that thc Nantgarw China Works only ever produced one paste body composition, echoing Herbert Eccles' opinion from his observation of the Glynn Vivian Art Gallery Centenary Exhibition of Nantgarw and Swansea china in 1914, and this is seen to be apparently in direct contradiction with recent analytical results. Modern analytical data strongly suggest otherwise (Edwards, *Porcelain to Silica Bricks: The Extreme Ceramics of William Weston Young, 1776–1847*, 2019) and this will be amplified appropriately in detail later. The distinction between Isleworth and Bow porcelains actually involves the analytically determined presence or absence of minor components, such as lead (from the added flint glass frit or cullet component) and sulfur (from gypsum impurities in the component minerals) – and it seems clear that whereas Isleworth and Bow phosphatic porcelains both contain signatures for sulfur, those from Lowestoft and Liverpool do not, and in their paper Panes et al. (2013) were thereby clearly able to differentiate analytically a plate that had previously been incorrectly attributed to Liverpool and suggest its re-assignment to Isleworth. In their scholarly paper dedicated to Bow porcelains, Ramsay and Ramsay (*The Evolution and Compositional Development of English Porcelains from the 16th Century to Lund's Bristol c. 1750 and Worcester c. 1752- the Golden Chain"*, 2017) propose an analytical protocol for the differentiation between the manufacturing dates of Bow porcelains based on the respective ranges of silica, lead oxide and sulfur content percentages as given below:

1746:	Developmental Period	$SiO_2 < 50\%$, PbO 3–5%, S present.
1747–54:	Early Bow	$SiO_2 < 50\%$, PbO 3–5%, S absent.
1755–69:	Middle Period	$SiO_2 > 50\%$, PbO absent, S present.
1770–74:	Tidswell Period	$SiO_2 > 50\%$, PbO 0.2–3%, S present.

This sort of analytical/temporal protocol can assist greatly in the assignment of known sourced pieces from a factory to particular periods of production and thereby assist in the identification of unknown or questionable pieces of porcelain.

5.1.1 New Discoveries

Further to the case study discussed above regarding the *Burghley House Jars,* where expert opinion is still divided about their origin because a precise match with known early English porcelains is not otherwise realised, the attribution of other hitherto unknown porcelains is also an ongoing exercise. For example, Panes et al. (2013) have analysed a blue and white sauceboat which had previously been assigned to the Bow factory and from their analytical data have suggested its re-assignment to the Bovey Tracey factory. Likewise, Dunster and Panes (2015) have found another hitherto unrecognised Bovey Tracey piece, using non-destructive XRay fluorescence spectroscopy (J.M. Dunster, *Developing a Methodology for the Non-Destructive Analysis of British Soft Paste Porcelain,* 2016) which built upon the earlier XRF studies of Domeney (K. Domeney, *Nondestructive Handheld XRF Analysis of Meissen and Vincennes – Sevres Porcelain: Characterisation, Dating and Attribution,* 2012).

A second theme centres on the existence of historical documentation of porcelain manufactories in the form of advertisements in local newspapers for sales of porcelain and hardware associated with the closing down and cessation of manufacture at local china works. This aspect has proved to be a profitable exercise for ceramic historians who are searching for evidence for the location of long lost, local small porcelain factories and several spectacular successes have emerged in recent years. Panes (2013) reported on his fortuitous discovery of a sale advertisement placed in the *London Morning Chronicle* on 23rd June, 1800 marking the closure of a London china manufactory for which no records otherwise exist! This was located in the

West End of Hammersmith in King Street, near to the *White Hart* coaching inn. The *White Hart* still remains in a now upmarket residential area and Panes theorises that the derelict land resulting from the closure of the factory would have been appropriated for residential property development in Georgian times. Although not situated on the banks of the River Thames as were the other major porcelain manufactories in London, such as Isleworth and Vauxhall, the Hammersmith Creek flowed adjacent to the site and would have provided adequate waterborne transport access opportunities for the incoming shipment of raw materials and outgoing export of the finished china. The advertisement specifically mentions supplies of raw materials that were still available on site, so it seems highly probable that this was a manufacturing facility and not merely a decorating workshop or atelier. Selected snippets of the appropriate advertisement are reproduced below:

> *To China Manufacturers.... Miscellaneous Articles at the West end of Hammersmith, near the White Hart- By Mr Scott. On the above premises on Thursday next at twelve o'clock.... The Remaining Stock of Utensils in Trade and other effects of a China Manufactory; comprising a great quantity of plasters, powders, saggers and tiles for kilns, sundry unfinished china ware, flower pots, cups, saucers, &c ...*

Clearly, this advertisement refers to the closure of an operational china factory at Hammersmith whose existence has thus far escaped attention. It is believed that potential archaeological excavation at the identified site, however, would not be forthcoming in the near future because of the extensive and disruptive urban development that has taken place there since the factory closure.

A second example follows from the discovery by Geoffrey Godden (2013) of a similar advertisement notice in a local newspaper in Lewes, Sussex, in July, 1799, referring to the sale of the products of a local china factory:

> *A large quantity of elegant China consisting of complete tea and coffee services, &c &c. All of the newest patterns of the very best burnished gold, as also various other sets of of all colours, dishes, bowls, basens, mugs, cups, and saucers, coffee cups with and without saucers*

Godden was very surprised that a local china manufactory of potentially some size had again escaped attention for so long, historically. In this case he was unable to shed any light on the potential location of the factory or of its output.

Thirdly, in several cases the sites of well-known porcelain manufactories which closed in the late eighteenth and early nineteenth centuries have been developed and are not compatible with archaeological excavations: one such site which has nevertheless yielded china wasters, but without any associated stratigraphic archaeological context is Caughley in Shropshire, which following its closure in 1799 was turned over to agricultural cultivation following the demolition of the factory buildings. A second exemplar site, described earlier in Chap. 4 and Sect. 4.2., is that of the Pinxton China Manufactory, which was developed for coal mining following its closure in 1813 and after demolition of the buildings upon the closure of the mine workings in 1934 is now a derelict area occupied by a scrap metal yard so the archaeological context is surely now severely compromised.

Another example of how newspaper advertisements for the impending closure of china manufactories and the sale of their residual stock have assisted historians is provided by the Nantgarw China Works site which was closed in 1823 under the tenureship of William Weston Young and Thomas Pardoe: Llewellyn Jewitt (*The Ceramic Art of Great Britain,* 1878) reported that following the final sale of the effects of the Nantgarw China Works in 1823, advertised locally in the *Cambrian* newspaper (Edwards, *Porcelain and Silica Bricks: The Extreme Ceramics of William Weston Young, 1776–1847*, 2019), John Rose of the Coalport China Works purchased the moulds, kilns and china-making equipment which he then transported to Coalport and thereafter no more china was made at Nantgarw. In this premiss Jewitt must have been reinforced by the apparent acquisition by John Rose of the Nantgarw factory as evidenced in his new motif for Coalport, viz., a *CSN* in an ampersand, illustrated earlier in Fig. 4.3, which acknowledged his incorporation of the Salopian, Swansea and Nantagrw china manufactories into the Coalport China Works. All the more surprising, therefore, that when the Nantgarw China Works site was investigated archaeologically first by Isaac Williams of the National Museum of Wales in 1932, he found three intact kilns in various states of disrepair, including a biscuit porcelain kiln, a glost kiln and a pipeclay pottery kiln, various outbuildings for pottery and porcelain manufacture and also supplies of raw materials (Williams, *The Nantgarw Pottery and its Products: An Examination of the Site,* 1932). One of these structures he identified as the kiln adopted by William Billingsley and Samuel Walker for the firing of their porcelain in 1817–1820, and that clearly was still standing. He also discovered the waste pit for the shards (shown earlier in the diagram depicted in Fig. 3.10), which yielded evidence of porcelain and pottery manufacture on six distinct archaeological stratigraphic levels. Until then, all ceramic historians had assumed without question that Llewellyn Jewitt was correct in his assertion that John Rose had effectively bought the Nantgarw operation and had then asset stripped the site, despite the fact that in 1858, William Henry Pardoe, the son of Thomas Pardoe, who took over the derelict site in 1832, advertised locally in the Cardiff *Mercury* the sale of his porcelain manufactory enterprise at Nantgarw including stocks of unfinished and finished china and equipment. It stands to reason that if John Rose had in fact purchased and demolished the Nantgarw china manufactory operation hardware in 1823 there certainly would not have been the wherewithal for William Pardoe to carry on his later efforts at porcelain or pottery manufacture there! The Pardoe family thereafter concentrated upon the manufacture of clay pipes and earthenwares – but the historical surmise that porcelain manufacture ceased at Nantgarw with the effective demolition of the site in 1823 was patently incorrect in substance! So, the information contained in contemporary newspaper advertisements about the fate of closing down china manufactories must be read carefully and more importantly, the interpretation put upon the information by earlier historians must be viewed in hindsight with suitable caution.

5.2 The Sensitivity of Analytical Determinations to the Differentiation Between Body Paste Compositions

The sensitivity of the determination of the analytical percentages of porcelain body paste compositions to the differentiation between porcelains from different factories or to the different periods of production within each factory is at the focal point of the forensic scientific ability of analytical science to accomplish an attribution of a potential source origin to a specific sample or specimen of porcelain. At the root of it all lies the tenet "How reliable is the interpretation of the analytical data for each specific specimen and can an incontrovertible and undeniable, unambiguous conclusion therefrom be made about its origin?".

A good start in our evaluation of the potential use of analytical data for the identification of porcelains and the differentiation that is possible between the products of different factories is an assessment of the data presented in Table 4.1. Fundamentally, the body paste compositions of 20 English and Welsh porcelain factories from the early- eighteenth to the mid-nineteenth century are listed, with 4 separate factories being considered under the umbrella of "Liverpool". The analytical data have also been collected for 8 factories in terms of defined periods of production where these have been definitively identified from the archaeological excavation of shards in a proper stratigraphic context or from matching the shards with style of output of the finished product as exemplified by similar finished and decorated specimens in established collections: these factories are Bow, Caughley, Chelsea, Coalport, Derby, Lowestoft, Rockingham and Worcester, and the production periods are given for these selected factories, comprising from 2 to 7 in number. Classified types of porcelain are also broadly described for each period generically as typically hard paste, soft paste, phosphatic, magnesian, soapstone, and magnesian/plombian. The nine analytically determined elemental oxides listed in the table are silica (SiO_2), alumina (Al_2O_3), lime (CaO), phosphorus pentoxide (P_2O_5), potash (K_2O), soda (Na_2O),magnesia(MgO), iron oxide (Fe_2O_3) and lead oxide (PbO). In the original analytical research papers which furnished the original data further elemental oxide data are occasionally given for the detectable presence of minor components in the fired porcelain bodies, but these are not reproduced here in tabular form for reasons of clarity of comparison and will be referred to only where they have a direct bearing on the major data interpretation: examples of these minor components include titania (TiO_2), sulfate (SO_4^{2-}), and arsenic oxide (As_2O_3). Table 4.1 draws a distinction between "sagged" and "unsagged" porcelain shards from the Nantgarw site (Owen and Morrison 1999) but these data clearly indicate that there has been no significant change occurring in elemental oxide composition between the two types of shard, which supports the conclusion that the sagging of the specimens occurred as a result of the high temperature kiln firing procedure of the biscuit body exceeding the requisite temperature rather than a change in the formulation composition of the original paste recipe. Analytical interpretation of the data presented in Table 4.1 reveals the following conclusions:

- Some 275 specimens were analysed from the 20 selected factories: most of these were shards but some comprised finished porcelain specimens from which samples were taken. If one includes the comparison data from Chinese and Japanese porcelains provided in the footnote then a further 92 analytical shard specimens are noted.
- The data from all of the specimens reported here were derived from SEM/EDAXS measurements for internal consistency. No attempt has been made in this table to correlate this recent data with the historically earliest data which were obtained from wet chemical digestive procedures which have been summarised in earlier chapters – although a comparison made earlier for Nantgarw porcelains demonstrates that there is an excellent agreement between the two types of data set, in that particular case from Eccles and Rackham (Wet chemical digestion analysis:*Analysed Specimens of English Porcelain in the Victoria and Albert Museum Collection*, 1922) and Tite and Bimson (SEM/EDAXS analysis: 1991).
- For factories which supplied the analytical data for multiple samples a data spread of results is noted. If we consider phosphatic porcelains, for example, the Bow factory has an average silica composition of 47.3+/−3.6%, alumina 5.7 +/−3.1%, lime 26.0+/−2.6%, phosphorus pentoxide 18.1+/−2.6%, potash 0.7+/−0.4%, soda 0.5+/−0.3%, magnesia 0.5+/−0.2%, iron oxide 0.3+/−0.2% and 0.2+/−0.2% lead oxide. In comparison, the Coalport phosphatic porcelains have average analytical figures of silica 45.2+/−1.5%, alumina 13.4+/−0.8%, lime 21.4+/−0.5%, phosphorus pentoxide 17.4+/−0.7%, potash 1.8+/−0.2%, soda 1.0+/−0.2%, magnesia 0.5+/−0.0%, iron oxide 0.3+/−0.1% and lead oxide 0.0%. Two things can be deduced from a comparison of these analytical data, firstly, that the standard deviations in the experimentally determined elemental oxide data are much higher for phosphatic Bow porcelains than for the analogous Coalport porcelains and, secondly, that the compositional information is very different for the two factories which, nevertheless, both conform to their categorical classification as phosphatic soft paste porcelain. The former point could be ascribed to multiple changes made to the formulation recipes which were invoked at Bow relative to a more stable formulation recipe being used consistently at Coalport, and the second point relates to the interpretation of porcelain source from the elemental oxide data: the silica percentages of 47.3+/−3.6% and 45.2+/−1.5% overlap between the two factories of Bow and Coalport and are, therefore, not sufficiently discriminatory for a forensic identification or sourcing attribution as a single parameter for silica alone. The alumina results,on the other hand, are significantly different and enable a positive attribution to be made between the two factories, being 5.71+/−3.1% for Bow and 13.4+/−0.8% for Coalport, which mathematically are almost three standard deviations of sigma apart. The soda, magnesia, phosphorus pentoxide and iron oxide levels are also not discriminatory enough for factory attribution to be confirmed using these data alone but the potash and lime percentages are a little better suited for their discriminatory potential in this context. This idea will be considered later in a wider aspect for all porcelains studied here.

- The analytical data for the several different types of porcelain manufactured by the 20 factories summarised in Table 4.1 have been collected and averaged to produce the elemental oxide percentage results listed in Table 5.1. In essence, this describes the average data for hard paste/siliceous porcelains (14 factories), magnesian porcelains (4 factories), glassy porcelains (4 factories), phosphatic porcelains (11 factories) and bone china (1 factory). The production at the Worcester China Works is generally considered to be hybridised between two or more of these types and is therefore included as a separate item in the Table, except for one set of analytical results (Owen 2003), which were derived clearly from a phosphatic body type, and this has therefore been included in the phosphatic porcelain entries appropriately in Table 5.1.
- The porcelain body classifications used in Tables 4.1 and 5.1 are broadly those that have been subsumed from the respective original analytical literature, but they have been modified to preserve a correlation comparison between the individual factories: hence, *hard paste* porcelain is synonymous with siliceous porcelain, and *soft paste* porcelain has been subdivided into phosphatic porcelain and bone china, magnesian porcelain and magnesian plombian porcelain have replaced soapstone or soaprock porcelain, glassy porcelain has been preserved for porcelain containing a high lead oxide content and the term *hybrid* porcelain has been utilised predominantly for the Worcester China Works products which demonstrate a compatibility with more than one classification, such as magnesian porcelain and siliceous or glassy porcelains.
- An interesting numerical fact to emerge from a summary of the content of Table 5.1 is that 14 factories manufactured hard paste/siliceous porcelains out of the twenty being considered here. This makes nonsense of the allegation made by several authors historically that the early English porcelain manufactories were focussed exclusively upon making soft paste porcelains because they did not have the hard paste formulation of the Chinese factories without their access to the *petuntse* or *china stone* component. Several factories are known to have manufactured both hard paste and soft paste porcelains at different periods. Of course, much soft paste porcelain was made, especially the phosphatic and glassy porcelain varieties, encompassing some 15 factories in total on our list. Only one bone china factory, Rockingham, appears on this list of phosphatic porcelain manufactories because of its manufacture of phosphatic porcelain as well as bone china (Cox & Cox, *Rockingham Porcelain*, 2001); bone china is considered to be a peculiarly English invention attributed to the successful marketing venture promulgated by Josiah Spode, who certainly transformed the china scene after about 1815 or so, when most surviving porcelain factories were then engaged predominantly in manufacturing bone china throughout the remainder of the nineteenth century.
- A comparison of the analytical data derived for the English hard paste porcelains and the Chinese hard paste porcelains is also interesting; the Chinese and English data are as follows, respectively – Chinese silica 75.6% versus 73.6% English, Chinese alumina 18.2% versus 19.0% English, Chinese lime 0.4% versus 1.5% English, Chinese phosphorus pentoxide 0.3% versus 0.2% English, Chinese

Table 5.1 Types of porcelain manufactured in early english and welsh factories: average values and ranges of their elemental oxide percentages

Porcelain Type	Factory	SiO_2	Al_2O_3	CaO	P_2O_5	K_2O	Na_2O	MgO	Fe_2O_3	PbO
Hard paste/ Siliceous	Bovey Tracey	63.0	30.6	0.3	0.3	3.8	1.1	0.3	0.5	0.0
	Bow	64.3	21.7	5.6	0.0	2.4	4.1	1.7	0.0	0.0
	Bristol	70.8	24.0	0.5	0.0	3.0	0.7	0.5	0.5	0.0
	Caughley	74.6	19.7	0.6	0.3	2.9	1.4	0.2	0.2	0.0
	Coalport	73.5	18.9	0.2	0.5	2.7	1.3	0.2	0.2	0.0
	Fulham	76.9	18.0	0.3	0.0	2.7	0.4	0.3	0.6	0.0
	Limehouse	75.6	14.6	2.4	0.2	2.0	1.3	0.5	0.6	2.0
	Liverpool (Reid)	82.6	8.3	5.2	0.0	2.0	0.5	0.3	0.4	0.5
	Nantgarw	80.4	9.2	0.6	0.5	5.6	1.6	2.0	0.2	0.0
	Pomona	76.9	11.7	3.7	0.2	2.5	1.0	1.1	1.0	1.2
	Plymouth	69.9	27.0	0.6	0.0	2.4	2.2	0.2	0.3	0.0
	Rockingham	73.3	19.9	0.9	0.2	3.2	1.7	0.3	0.4	0.0
	Swansea	71.6	23.9	0.3	0.2	1.7	0.6	0.2	0.7	0.0
	Worcester	76.9	18.5	0.2	0.0	1.5	0.6	0.3	0.6	0.0
Average %		***73.6***	***19.0***	***1.5***	***0.2***	***2.7***	***1.3***	***0.6***	***0.4***	***0.3***
Range %		***63–83***	***8–31***	***0–6***	***0–1***	***1–6***	***0–4***	***0–2***	***0–1***	***0–2***
Magnesian	Caughley	76.0	4.9	1.6	0.3	2.6	1.6	9.9	0.3	3.4
	Liverpool (Chr/ Penn)	75.6	2.8	1.9	0.2	2.1	1.1	8.4	0.4	7.4
	Liverpool (Chaffers)	70.3	4.3	4.8	1.6	2.9	2.2	11.9	0.7	1.3
	Vauxhall	75.6	4.0	5.6	0.0	3.7	1.2	9.1	0.5	0.4
Average %		***74.4***	***4.0***	***3.5***	***0.5***	***2.8***	***1.3***	***9.8***	***0.5***	***3.1***
Range %		***70–76***	***3–5***	***1.5–6***	***0–2***	***2–4***	***1–2***	***8–12***	***0.3–0.7***	***0–4***
Glassy	Chelsea	71.0	4.2	16.0	0.0	3.9	0.5	0.4	0.3	3.5
	Derby	72.3	3.6	5.2	1.5	2.4	0.2	0.4	0.2	13.8
	Limehouse	76.1	16.0	0.5	0.1	1.3	0.7	0.2	0.5	3.8
	Longton Hall	70.1	3.0	14.1	1.0	2.8	0.4	0.3	0.2	8.1
Average %		***72.4***	***6.7***	***9.0***	***0.7***	***2.6***	***0.5***	***0.3***	***0.3***	***7.3***
Range %		***70–76***	***4–16***	***0–16***	***0–2***	***1–4***	***0.2–0.7***	***0.2–0.4***	***0.2–0.5***	***3–14***

(continued)

Table 5.1 (continued)

Porcelain Type	Factory	SiO_2	Al_2O_3	CaO	P_2O_5	K_2O	Na_2O	MgO	Fe_2O_3	PbO
Phosphatic	Bow	47.3	5.6	25.0	17.7	0.7	0.5	0.5	0.3	0.2
	Chelsea	46.4	11.3	21.8	17.2	1.4	0.6	0.5	0.3	0.0
	Coalport	44.9	13.1	21.3	17.2	1.7	1.0	0.5	0.2	0.0
	Derby	46.7	10.0	21.8	16.5	1.0	0.8	0.6	0.4	1.3
	Isleworth	40.9	7.8	22.5	17.9	2.3	0.9	0.6	0.5	3.5
	Liverpool (Gilbody)	57.2	8.5	17.9	13.7	1.1	0.5	0.5	0.3	0.0
	Liverpool (Pennington)	49.3	11.2	20.5	14.9	1.6	0.8	0.6	0.5	0.0
	Lowestoft	43.5	7.5	25.4	20.1	1.0	0.9	0.7	0.4	0.0
	Nantgarw	43.7	12.8	22.0	17.4	2.2	0.5	0.5	0.2	0.0
	Rockingham	56.4	12.2	15.7	10.7	2.3	1.0	0.5	0.7	0.0
	Swansea	49.0	20.0	14.9	11.4	2.7	1.3	0.5	0.3	0.0
Average %		47.8	10.9	20.8	15.9	1.6	0.8	0.6	0.4	0.5
Range %		***40–57***	***5–20***	***15–25***	***10–20***	***1–3***	***0.5–1.5***	***0.5–0.7***	***0.2–0.7***	***0–4***
Bone China										
Rockingham										
Average %		***36.8***	***15.2***	***24.1***	***18.4***	***1.6***	***0.9***	***0.7***	***0.6***	***1.7***
Range %		***31–44***	***13–16***	***17–29***	***14–22***	***1–3***	***0.8–1.0***	***0.6–0.7***	***0.4–1.0***	***0–5***
Hybrid										
Worcester										
Average%		***72.9***	***4.7***	***1.4***	***2.1***	***3.3***	***1.0***	***9.3***	***0.3***	***6.3***
Range%		***70–74***	***3–7***	***0.5–2***	***0–7***	***3–5***	***0.5–1.5***	***5–11***	***0.2–0.4***	***5–10***

potash 3.1% versus 2.7% English, Chinese soda 1.4% versus 1.3% English, Chinese magnesia 0.2% versus 0.6% English and Chinese iron oxide 0.9% versus 0.4% English. These can be considered to be very close indeed within the ranges that have been calculated for English porcelains (the mathematical overlap ranges in data for the Chinese analogues are not available but these too could be estimated to be rather similar to their English analogues). So, despite much historical discussion about the inability of European porcelains to match the Chinese hard paste porcelains attributed to the non-availability in Europe of the *china stone* or *petuntse* or their equivalent minerals, the analytical data actually indicate evidence to the contrary, namely that early English hard paste porcelains came quite close to achieving a very similar elemental oxide composition to the hard paste Chinese porcelains. The continued success of Chinese porcelains in the European market should then be attributed perhaps rather to their commercial production in very large quantities of porcelain items of a consistently fine quality and this must be further related to the large size of the dragon kilns at

Jingdezhen alone. It is recorded that a Chinese Emperor placed a personal order for 430,000 pieces of the finest quality porcelain from Jingdezhen for his own palace requirements—a phenomenally large quantity that certainly would have exceeded the total output of many English and Welsh small factories even without their associated kiln losses resulting in unusable porcelain items, which, for example, at Swansea was 70% and at Nantgarw 90%! It has been alleged that a single dragon kiln at Jingdezhen could be used for the simultaneous firing of 25,000 items of porcelain in one batch.

- An assessment of the analytical data in Tables 4.1 and 5.1 reveals that compositional differences and recipe formulations contribute to a range of values associated with the elemental oxide determinations, for example, silica percentages ranging between 63 and 83% for hard paste/siliceous porcelains, 70 and 76% for magnesian porcelains, 70 and 76% for glassy porcelains and 40 to 57% for phosphatic porcelains. From this it is clear that one cannot use the analytical data for the silica elemental oxide content alone as an absolute determinant for the confirmation of the type of porcelain being studied. What needs to be done, therefore, is to examine the characteristics of the elemental oxide predominance in each type of porcelain and then to establish the coupling of two or more key data occurrences to establish a suitable assessment regime for source type originality derived from the analytical data. A summary of the type of porcelain correlated with the critical presence or absence of elemental oxides can hereby now be constructed:
- ***Hard paste/siliceous porcelains***: *high silica (63–83%), medium-high alumina (8–31%), low lime (0–6%), zero-low phosphorus pentoxide,, zero -low magnesia (0–2%) and zero-low lead oxide (0–2%).*
- ***Magnesian porcelains:*** *high silica (70–76%), low alumina (3–5%), low lime (1–6%), low phosphorus pentoxide (0–2%), high magnesia (8–12%) and medium-low lead oxide (0–4%).*
- ***Glassy porcelains:*** *high silica(70–76%), low- medium alumina (0–16%), zero-high lime (0–16%), zero-low phosphorus pentoxide (0–2%), zero -low magnesia(0–0.4%) and medium -high lead oxide (3–14%).*
- ***Phosphatic porcelains:*** *low-medium silica (40–57%), low-high alumina(5–20%), high lime (15–25%), high phosphorus pentoxide (10–20%), low magnesia (0.5–0.7%) and low-medium lead oxide (0–4%).*
- ***Bone china:*** *low-medium silica (31–44%), high alumina(13–16%), high lime (17–29%), high phosphorus pentoxide (14–22%), low magnesia (0.6–0.7%) and zero lead oxide (0%).*
- ***Hybrid porcelain:*** *high silica (70–74%), low alumina(3–7%), low lime (0–2%), low-medium phosphorus pentoxide (0–7%), high potash (3–5%), medium-high magnesia(5–11%), medium-high lead oxide(5–10%).*

Using the above data summaries it is now possible to construct a workable protocol using the analytical elemental oxide data to define more closely the type of porcelain specimen under investigation, and this is shown in diagrammatic form in Fig. 5.1.

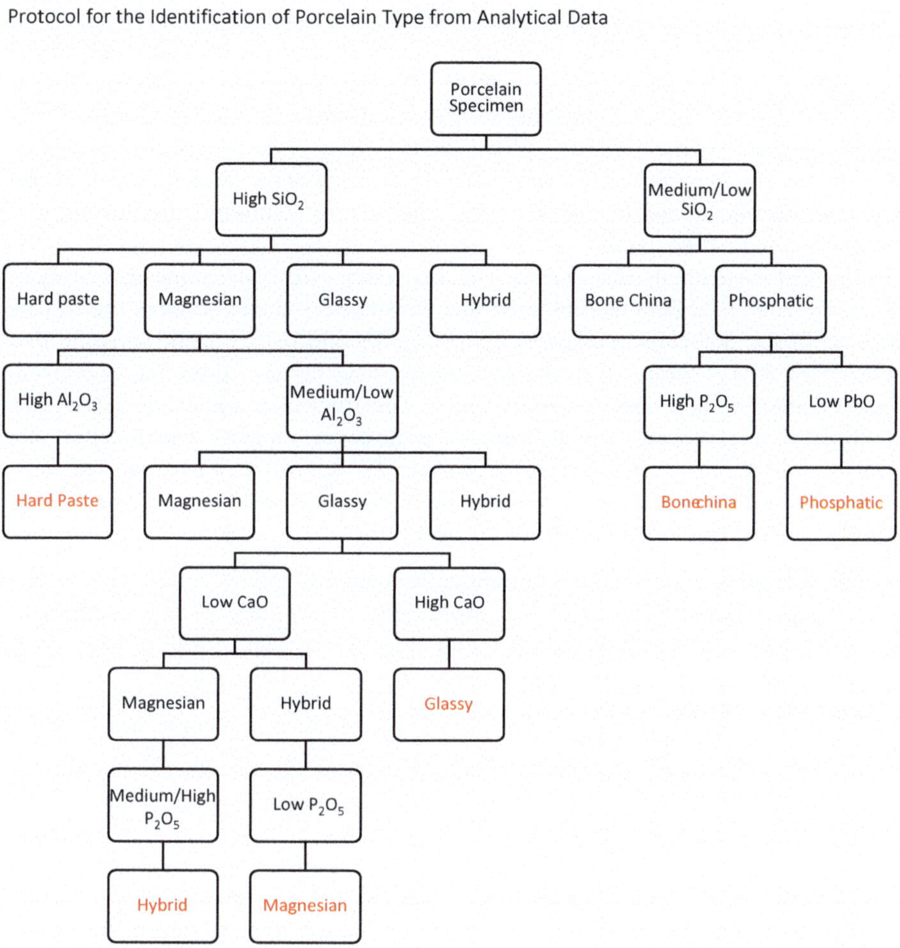

Fig. 5.1 Protocol diagram flowchart for the identification of types of porcelain from the primary analytical data of the elemental oxide percentages

From Fig. 5.1 it should be possible to determine the type of porcelain with which one is dealing analytically as the first step towards defining its factory source and period. In Fig. 5.1, the term "*hybrid*" refers to the Worcester data only and it is accepted that other "hybrids" could well have a different compositional make-up depending upon their cross-type porcelain classifications. In addition, the technical difference between the classification of a bone china and a phosphatic porcelain body is really one of degree from an analytical elemental oxide compositional standpoint – both bodies have a high phosphorus oxide percentage and the crucial decision rests with the amount of phosphate bone ash used and, surprisingly perhaps, the lower lead oxide content of the latter. For example, one of the highest phosphatic porcelain bodies experienced in British porcelains is that of Nantgarw, with a bone ash content in the original formulation approaching 43% compared with Rockingham bone china of 44%, but the zero content of lead oxide in the Nantgarw paste confirms its classification as a straightforward phosphatic porcelain. As discussed earlier, a much more rigorous discriminator between bone china and a phosphatic porcelain rests with the methodology of the synthetic manufacturing process and in particular the single-stage kiln firing of the paste used in the bone china production compared with the two-stage process commonly used for the soft paste phosphatic porcelains.

The next stage in our analytical survey discussion is to evaluate the possibility of using the analytical data in Tables 4.1 and 5.1 to discriminate between the factory sources of the porcelain specimens used. In this context it is useful to recall that most of the data in Table 4.1, from which the averages listed in Table 5.1 are derived, were obtained from shards, where the use of a destructive analytical technique was not forbidden even if only minute samples of the porcelain body were taken for the analyses. In more recent analyses however, the advent and necessity of non-destructive techniques is now becoming more prevalent and this can have several material advantages which can be summarised below.

- The derivation of analytical information from marked and decorated, i.e. finished and perfect, examples of porcelain from particular factories which unimpeachably can be used to characterise the output from specified periods of production and can then be potentially used further to define any changes in formulation that have been introduced therein.
- The assessment of unknown or questionable pieces of porcelain from which samples cannot be excised for destructive testing: it has been noted from a perusal of the literature that in several instances finished porcelain specimens have been sacrificed by museums or collectors in the interests of obtaining the scientific analytical data which may conclusively define their true origin. This of course is entirely dependent upon the choice that has to be made for the incurring of damage, however minimal, in the pursuit of knowledge and information about the specimen. Typically, museum curators are prepared to sacrifice small samples being removed from a damaged piece which maybe a portion of a part service comprising several surviving perfect examples in order to elicit information that will be applicable to the whole service: this was certainly the case for the important but damaged *Biddulph* service platter of Swansea duck-egg porcelain held in the Royal Institution of South Wales in Swansea. Until now, the attribution of a unknown piece of porcelain has been the domain of the expert and connoisseur and often opinions as to its possible identity can be both subjective and variant. This has not been assisted by the contemporary practice of porcelain manufactories to acquire pieces from other rival factories to complete orders, to borrow designs and copy pieces from neighbouring establishments and to use existing stock from newly acquired and recently closed factory premises to incorporate for sale to clients along with their own products. Several authors have already alluded to the common practice in the eighteenth and nineteenth centuries of proprietors hiring personnel from rival factories, as we have outlined

here in Chap. 4, and thereby acquiring first hand information and a detailed knowledge about the paste compositions, along with the processing and kiln operations which applied there.

5.3 Analytical Data Ranges and Values

The data quoted in Table 5.1 may cause some surprise to the reader in that it is quite apparent that it seems impossible to quote a particularly definitive percentage of silica or alumina or any other elemental oxide to a precise figure and that each determination is subject to an experimental error which defines its accuracy and precision – in reality, therefore, the analytical percentages lie within narrow ranges of experimental values. There are several reasons for this situation, namely:

- The analytical determinations themselves are subject to inherent errors which may be estimated reasonably precisely and these may even have been reduced by multiple sampling (replicates) and mathematical averaging processes. These errors can arise from instrument calibration procedures and are then compounded further by cumulative errors in calibration and in instrument response to standardisation. The accuracy of an analytical determination is not the same as its precision: the accuracy reflects the closeness and reproducibility of each single analytical determination to the mathematical average of all the replicate data, whereas the precision of the analytical result reflects the closeness of the average determination to the real value. For example, in a weighing experiment a 100 g sample may be weighed to the nearest 1 g, giving an accuracy of +/−1% in the weighings, but the instrumental balance may have a systematic error in calibration which means that the actual weight of the nominally 100 g specimen is 90–110 g, giving a precision of +/− 10% in the weighing process. Hence, we have a false idea of the actual correctness of the analytical determination of a parameter from the accuracy alone.
- The determinations of the elemental oxide percentages in a bulk sample are critically dependent upon the homogeneity of the specimen and therewith the degree of fineness of mixing of the original components: although we are dealing with high temperature melt phases in porcelain manufacture, the actual kiln temperatures dictated the effectiveness of the reactions in the molten components, which all have different melt characteristics and behaviour, including associated chemical phase transformations and the reactivities of specific components at high temperatures, e.g. silica and calcium hydroxyapatite. The analytical situation is exacerbated by the adoption of microscopic analytical techniques for the determination of quantitative data about elemental oxide compositions; in essence, how close does an analytical determination obtained from a cubic micron-sized specimen sampled "footprint" reflect the bulk composition of the whole specimen? In other words, does the elemental oxide composition of such a small sample, which may weigh only 5 pg (i.e. 5×10^{-12} g, for porcelains with an average

density of 5 g cm^{-3} being interrogated microanalytically with a one cubic micron footprint), when the whole sample supplied for analysis weighs perhaps 50 g? The analytical result obtained from such a microsample is derived from an interrogation of only a tiny part of the specimen, amounting to one ten thousand millionth of the whole: we must then ask the question if this can truly represent the sample and the concept of homogeneous and inhomogeneous (?) speciation, or whole sample homogeneity, now applies. Modern analytical data therefore present determinations made from many replicates studied at the micron level and also averaged determinations scaled up to the bulk level to give a better appreciation of the real bulk sample composition. The latter estimations, i.e. bulk compositional data, have been selected as the source of data used in Tables 4.1 and 5.1 rather than the microdata, which nevertheless give much useful subsidiary information about the mineral compositions in the porcelain individual crystallised melt microdomains.

- The potential errors involved in the making up of the original recipes cannot be ignored, not just perhaps regarding the weighing accuracies highlighted above but also in the storage of the raw materials on site, which can be contaminated with organic volatiles and by ingress of water in damp surroundings.
- The presence of lead oxide is usually regarded as a definitive indicator of the addition of a flint glass cullet being added in the formulation recipe of the biscuit porcelain but a small analytical sample could also be contaminated by inclusion of particles of the glaze used, which invariably for the earlier porcelains consisted of a lead oxide component of rather indefinite composition. At the microsampling level the chance of a rogue determination arising from the detection of an impurity in an inhomogeneous phase is a very real possibility, especially where the sampling involves glazed shards and the microscopic interrogation of the sample near an interface between a glazed and unglazed region of the body due to penetration of the glaze slip into voids in a porous biscuit porcelain base.
- From details provided by analysts using the SEM/EDAXS technique for the determination of the elemental oxide composition of porcelains it appears that the components of silica and alumina can each be estimated to a precision of about 1%, lime to 5% and magnesia and phosphate to 10–20% (figures quoted by Tite et al., 1984). If we now add in the collective estimated errors accrued from the other effects outlined above it can be seen that it would be quite reasonable to propose that the cumulative errors in precision offered in the elemental oxide analytical determinations could easily approach the following values: silica and alumina 3%, lime 8% and magnesia and phosphorus pentoxide 15%. The distinction between analytical accuracy and precision is quite definite and has been exemplified above.
- From Table 4.1 the individual analytical values for each factory are themselves spread over ranges which can easily be seen to be perhaps 5% or more: this could obviously reflect the compositional variations in the paste mixtures which were used within different periods of the factory output. Often, detailed notes from the proprietors about the incorporation of such compositional changes to their biscuit

formulations are not available to modern researchers. Even when these are recorded, such as the excellent handwritten details provided in his notebooks by Lewis Dillwyn for his Swansea porcelain experiments between 1815 and 1817 (Lewis Dillwyn, *Note Books,* cited in Eccles & Rackham, *Analysed Specimens of English Porcelain*, 1922; Edwards, *Porcelain to Silica Bricks: The Extreme Ceramics of William Weston Young, 1776–1847*, 2019), it is impossible to be aware of how much porcelain was actually produced with any particular specified formulation and how easily these data can now be accessed.

The question now remains: how can the elemental oxide analytical data defined by the SEM/EDAXS, XRD or XRF techniques realistically be used to define the origins of a piece of unknown porcelain and how applicable are the protocols that have been established for this verification? We should now also seek to introduce molecular spectroscopic data from similar porcelain specimens in a combined analytical approach, both from shards and from perfect finished pieces, and see if this can corroborate, inform or perhaps better refine, the conclusions made from the elemental oxide results.

References

P. Colomban, H.G.M. Edwards, C. Fountain, Raman Spectroscopic and SEM/EDXS analysis of Nantgarw soft paste porcelain. J. Eur. Ceram Soc (submitted for publication, 2020)

L.W. Dillwyn, Notes on the Experimental Production of Swansea Porcelain Bodies and Glazes. Made by Lewis Weston Dillwyn with Samuel Walker at the Swansea China Works Between 1815 and 1817. Presented to the Library of the Victoria & Albert Museum, South Kensington, London by John Campbell in 1920. Reproduced in Eccles & Rackham, Analysed Specimens of English Porcelain, 1922, see reference

K. Domeney, *Nondestructive Handheld XRF Analysis of Meissen and Vincennes – Sevres Porcelain: Characterisation, Dating and Attribution,* Ph.D. Thesis, Cranfield University, 2012

J.M. Dunster, *Developing a Methodology for the Non-Destructive Analysis of British Soft Paste Porcelain*, Ph.D. Thesis, Cranfield University, 2016

J.M. Dunster, N. Panes, Attribution enhanced: Another Bovey Tracey find? Trans. Engl. Ceram. Circ **26** (2015)

H. Eccles, B. Rackham, *Analysed Specimens of English Porcelain in the Victoria and Albert Museum Collection* (Victorian and Albert Museum, South Kensington, London, 1922)

H.G.M. Edwards, *Swansea and Nantgarw Porcelain: A Scientific Reappraisal* (Springer, Dordrecht, 2017)

H.G.M. Edwards, *Porcelain to Silica Bricks: The Extreme Ceramics of William Weston Young, 1776–1847* (Springer, Dordrecht, 2019)

I.C. Freestone, A technical study of Limehouse, in *Limehouse Ware Revisited*, ed. by D. Drakard (English Ceramic Circle, London, 1993), pp. 68–73

G. Godden, Late Again!, *Northern Ceramic Circle Newsletter*, No. 172, December, 2013

W. Jay, J.D. Cashion, Raman spectroscopy of Limehouse porcelain sherds supported by Mossbauer spectroscopy and scanning electron microscopy. J. Raman Spectrosc. **44**, 1718–1732 (2013)

L.L. Jewitt, *The Ceramic Art of Great Britain: From Prehistoric Times down through Each Successive Period to the Present Day* (Virtue & Co. Ltd., London, 1878)

W.D. John, Nantgarw Porcelain, *The Ceramic Book Company* (Newport, 1948)

J.V. Owen, A preliminary assessment of the geochemistry of porcelain sherds from the Limehouse factory site, in *Limehouse Ware Revisited*, ed. by K. Tyler, R. Stephenson, MoLAS Monographs, No. 6 (Museum of London, London, 2000a), pp. 61–63

J.V. Owen, A preliminary assessment of the geochemistry of porcelain sherds from the Limehouse site, in *The Limehouse Porcelain Manufactory: Excavations at 108–116, Narrow Street, London, 1990*, ed. by K. Tyler, R. Stephenson (Museum of London Archaeology Service, Monograph No. 6, 2000b)

J.V. Owen, The geochemistry of Worcester porcelains form Dr Wall to Royal Worcester: 150 years of innovation. Hist. Archaeol. **37**, 84–96 (2003)

J.V. Owen, M.L. Morrison, Sagged Phosphatic Nantgarw Porcelain (ca. 1813-1820): Casualty of Overfiring or a Fertile Paste? Geoarchaeology **14**, 313–332 (1999)

J.V. Owen, J. Sandon, A rose by another name > A geochemical comparison of Caughley (c. 1772–1799), Coalport (John Rose & Co. c. 1799–1837) and rival porcelains based on sherds from the factory sites. Post-Mediaeval Archaeol. **37**, 79–89 (2003)

N. Panes, A porcelain factory in Hammersmith? Trans. Engl. Ceram. Circl, 135–141 (2013)

N. Panes, R. Howard, J.V. Owen, Attribution enhanced: Isleworth porcelain re-examined. Trans. Engl. Ceram. Circl. **24**, 89–116 (2013)

J. Potter, The Limehouse story, in *Digging for Early Porcelain*, ed. by D. Barker, S. Cole, (City Museum and Art Gallery, Stoke-on Trent, 1998), pp. 40–53

W.R.H. Ramsay, P. Daniels, E.G. Ramsay, The Limehouse Porcelain Factory, Its Output, in *Antecedents and the Influence of the Royal Society of London on the Evolution of English Porcelains based on Composition and Technology* (Invercargill, New Zealand, 2013)

W.R.H. Ramsay, P. Daniels, E.G. Ramsay, *Limehouse Porcelain: Are Limehouse Porcelains in Fact All Limehouse? Evidence from Archaeology, Science, and Historical Documents* (Resurgat Publishers, Oxford, 2015)

W.R.H. Ramsay, E.G. Ramsay, *The Evolution and Compositional Development of English Porcelains from the 16th Century to Lund's Bristol c. 1750 and Worcester c. 1752- the Golden Chain* (Invercargill Press, Invercargill, 2017)

M.S. Tite, I.C. Freestone, M. Bimson, A technological study of Chinese porcelain of the Yuan Dynasty. Archaeometry **26**, 139–154 (1984)

B.M. Watney, Limehouse, its relationship to Newcastle-under-Lyme (Pomona) and other manufactories, in *Limehouse Ware Revealed*, ed. by D. Dakard (English Ceramic Circle, London, 1993)

I.J. Williams, *The Nantgarw Pottery and its Products: An Examination of the Site* (The National Museum of Wales and the Press Board of the University of Wales, Cardiff, 1932)

Chapter 6
The Molecular Spectroscopic Analysis of Porcelains

Abstract The advent of the molecular spectroscopic characterisation, especially Raman spectroscopy, of porcelains. The instrumentation involved and the complementarity of the molecular data when correlated with the elemental oxide data is reviewed. Elemental oxide quantitative percentage compositional data versus qualitative molecular and molecular-ion definition data and how the two can both be used to better define the porcelain type attribution. The potential for the use of molecular spectroscopic instrumentation for the non-destructive interrogation of perfect and finished porcelain specimens without the necessity for invasive sampling is discussed and several case studies are presented to illustrate this concept.

Keywords Raman spectroscopy · Molecular structure of fired porcelain bodies · Correlation with raw materials used · Elemental oxide data versus molecular spectral data · Non-destructive sampling procedures · Case studies

The complementarity of the molecular spectroscopic analytical data and the elemental oxide percentage data for the quantitative and qualitative estimation of the formulation and composition of porcelain bodies and glazes will now be assessed. The major difference between the two analytical approaches is that the elemental oxide determinations obtained by the SEM/EDAXS, XRD and XRF techniques provide the elemental composition in terms of the metals and non-metals in the specimen, in conjunction with information about the physical location of these entities within the porcelain or the glaze; however, as discussed earlier, not all elemental oxides analysed can be uniquely traced back to a their origin in a specific raw material, for which silicon is a prime example. This means that a signature of silicon is determined and its chemical or mineralogical origin in the sample cannot be distinguished from contributions made from silica, sand, flints and china stone used in the recipes and between complex silicates and their polymorphs created through chemical reactions in the kiln firing processes using SEM/EDAXS and XRF, although XRD has additional distinctive signal patterns for crystalline specimens which can be used to characterise the presence of individual minerals and their polymorphic transformations.

H. G. M. Edwards, *18th and 19th Century Porcelain Analysis*,
https://doi.org/10.1007/978-3-030-42192-2_6

In contrast, molecular spectroscopy is dependent upon the chemical bonding between two or more elemental metals and non-metals and therefore has the potential to discriminate analytically between the occurrence of different minerals and compounds in which the same element occurs. For example, the silicon-oxygen bonds in quartz, tridymite, cristobalite, wollastonite and albite can all be detected as different signatures in the molecular laser Raman spectrum of a complex silicate mixture that is found in most porcelains. Hence, the complementary nature of the two types of analytical technique, namely elemental and molecular spectroscopic, is evident and will be demonstrated in selected examples reviewed in this section. One factor that needs to be considered when making an objective comparison between, for example, Raman spectroscopy (RS) and scanning electron microscopy (SEM) as analytical techniques is the "footprint" of the sample being interrogated: the actual sample, in most cases a porcelain shard, may be several g in mass and of several cm^2 in surface area but only a very small fraction of this will be analysed in the analytical experiment, perhaps only a surface circle of diameter two microns in the RS (1 micron is 10^{-6} m) and of a few nm (1 nanometre is 10^{-9} m) in the SEM, a thousand times smaller. The spectral information available therefore exhibits some rather subtle distinctions of a spatial nature and this is reflected in the presentation of local mineral domain compositions in the SEM analyses of porcelain shards, which often have a significantly different analytical composition to the bulk materials. Wherever possible therefore, for comparative purposes, the data presented in the Tables for the specific SEM/EDAXS analyses are those reported for the bulk specimen materials.

Hitherto, we have considered solely the elemental oxide compositions of porcelains and it will now be instructive to include the molecular spectroscopic data for assessment and evaluation. In this coverage the Raman spectroscopic data will be surveyed: the reason for this selection is that Raman spectroscopy has recently been proposed as a viable non-destructive molecular analytical technique for the characterisation of the porcelain bodies, their glazes and enamel pigments and the information obtained has been demonstrated to provide complementary data which is not available uniquely from the elemental oxide determinations. As Raman spectroscopy is a technique which has not seen much application to English and Welsh porcelains hitherto it is appropriate, therefore, to give a brief summary of the technique and the sort of information which can be deduced from the molecular spectroscopic analyses of ceramics.

6.1 Raman Spectroscopy

Raman spectroscopy is a molecular scattering technique, whereby a monochromatic laser beam is imaged onto the surface of a specimen and the back-scattered Raman radiation is collected and spectrally dispersed on a holographic grating and detected on a charge-coupled detector (CCD) device. The Raman spectrum is displayed as a series of wavenumber shifts from the laser-generated Rayleigh line which acts as a wavenumber zero on the wavenumber shift scale (Long 2002). Being a molecular technique, the Raman effect gives a unique spectral signature for materials which have chemical bonds, as found in inorganic and organic molecules and molecular ions, and therefore complements elemental detection techniques such as XRD (X-Ray Diffraction), XRF (X-Ray Fluorescence) and SEM/EDAXS (Scanning Electron Microscopy/Energy Dispersive X-Ray Spectroscopy). It is especially valuable for the detection of molecular and molecular-ion components in mixtures and for the qualitative and quantitative changes in materials which are effected during chemical processing, as a result of procedures such as thermal reactions in ceramic kilns. As far as ceramics and porcelains are concerned, the presence of a transparent glaze has no untoward adverse effect upon the technique since the incident laser beam can penetrate the glazed layer and interrogate the underlying ceramic body without scattering and loss of irradiance (incident power per unit area, usually W $m^{-2)}$; during this experiment, no damage is done to the sample by the imaged beam of light and no chemical or mechanical pre-treatment of the sample is required, such as the physical removal of the glaze. In heavily decorated and gilded samples, such as dinner plates or ornamental pieces, it is usually possible to discover a suitable area devoid of pigment and gilding for the interrogation of the porcelain body underlying the glaze. The presence of individual minerals or materials is recognised from the observed spectral band wavenumber positions, which can be identified from the existing published literature databases on minerals and geological materials, which frequently refer to the spectral "fingerprint" which is characteristic of the material concerned. The intensity of Raman scattering is directly proportional to the individual chemical species concentration and is also proportional to the irradiance (W $m^{-2)}$ of the incident laser beam, i.e. its power per unit area, so the greater the concentration of a particular material present the more intense the Raman band observed and the more intense the laser irradiance used for illumination, the stronger the spectrum. However, not all materials have the same Raman molecular scattering factor cross-sections for laser irradiation, so several species are always relatively more strongly represented and more easily observed in Raman spectra, even when they are occurring in low sample concentrations, such as cinnabar, calcite, anatase and gypsum, which all have large molecular scattering factor cross-sections.

Although the primary investigative instrumental techniques for porcelains and ceramics adopted thus far have been those where the elemental determinations have been paramount, such as X-Ray Diffraction (XRD), X-Ray Fluorescence (XRF) and Scanning Electron Microscopy (SEM), the advantage of a molecular technique such as Raman Spectroscopy (RS) is that it readily associates the combination of the chemical elements and their formulation through chemical bond identification: hence, although SEM, for example, can provide evidence of the presence of mercury, lead, sulfur, oxygen and tin in an artwork or archaeological specimen it is often a matter of conjecture and interpretation as to how these elements are paired together. For example, potentially these elemental data could signify the presence of the pigments cinnabar or vermilion (mercury sulfide), mosaic gold (tin sulfide), lead oxide (massicot, litharge, plattnerite and minium), lead sulfide(galena), calomel (mercury oxide), cassiterite (tin oxide), and Naples yellow (lead tin sulfide) and there are several other possibilities too. The heavier metals and anionic entities that are found in the analytical determination can create several real possibilities for their pairing or association: sometimes, the colour of the particles under investigation will assist in a narrowing down of the possibilities, for example, a red pigment might be indicative of the presence of cinnabar or litharge, whereas a yellow pigment might be indicative of massicot or mosaic gold. In all of these cases, the Raman spectral signatures are different and highly discriminatory and assist in the evaluation of the correct pairing of metals and their anions: even more problems can arise when one considers the elemental detection of calcium, magnesium, sulfur and oxygen: whence, calcite, aragonite, gypsum, anhydrite, bassanite, magnesite, dolomite, epsomite and hydromagnesite are then all real mineral possibilities but, unlike the pigment examples cited above, all of these are white so the particle colour does not help at all with a partial discriminatory identification in this case. Again, the RS signatures here are discriminatory and can be used to identify mixtures of these materials, as in dolomitized calcite and in hydrated lime prepared from dolomite.

For ceramics, the presence of silica, silicates and sand, which all contain silicon-oxygen bonds, and Si=O or Si-O bonding combinations in three dimensional silicate networks, give rise to silicaceous matrices which can be quite complex to describe when they have been created at high temperatures. A description of the types of silicate network that can arise at the temperatures adopted for the kiln firing of soft-paste and hard-paste porcelains has been given in specialist texts (Colomban and Prinsloo 2009) and a summary has been provided by Edwards (2017). As well as the basic orthosilicates containing the discrete SiO_4^{4-} ions with four-valent silicon from the parent *ortho*-silicic acid, H_4SiO_4, condensed ions such as pyrosilicates and metasilicates also occur and network formation through the sharing of Si=O bonds with neighbouring silicon atoms gives rise to -O-Si-O- linkages in extended three-dimensional structures, conferring strength and rigidity upon the fired porcelain articles. All of these will have different Raman signatures but the resultant spectral band response may be rather broadened by the several alternative possible structural conformations of Si=O and Si-O bonding in the three-dimensional structures of clays and glasses in comparison with the discrete, sharp features that are expected for silicates in other more crystalline minerals such as gemstones (Colomban et al. 2020).

A major advantage of RS for porcelain analysis is the ability to interrogate the glazed body of a finished article through the glaze, which is transparent to the wavelength of radiation used by the laser beamin illumination, and this opens up the specimen sampling procedure to include perfect examples of rare and museum pieces without the potential of incurring damage, unless the incident laser irradiance (power per unit area, W m^{-2}) is very large, which is possible when microscopic illumination is being adopted, or if the potential absorption of the laser radiation wavelength by a pigmented specimen causes localised laser heating and hence thermal degradation can occur. Hence, for the assessment of analytical information obtained from perfect finished porcelain specimens, Raman spectroscopy could provide a solution for the requirement of a non-contact, non-destructive analytical procedure and it is with this in mind that the preliminary analyses of some selected porcelain specimens detailed below will have been undertaken – the first on record for much porcelain using this technique. Previous studies of English porcelain using the non-destructive Raman spectroscopic technique have been limited to just two published papers, the first of which (Edwards et al. 2004) reported the spectral interrogation of a suspected Rockingham porcelain table top, shown earlier in Fig. 4.1, inlaid in a mahogany tripod stand: this experiment was ground-breaking in that it successfully confirmed the origin of the porcelain table top as being Rockingham porcelain of the Royal Rockingham period, *ca.* 1835–1840, by comparison with the corresponding spectra of a marked dessert plate of Royal Rockingham porcelain of the same period (Cox and Cox 2001). The second of these experiments resulted in the spectral identification of the source of a rare spill vase dating from *ca.* 1815–1820 as Davenport porcelain (Edwards et al. 2019): in the latter of the two Raman spectral investigations the data were obtained from a perfect specimen without any chemical or mechanical pre-treatment being undertaken whereas the former involved the use of broken fragments of both the table top and the standard dessert plate for spectral comparison (Fig. 6.1) before their professional restoration was

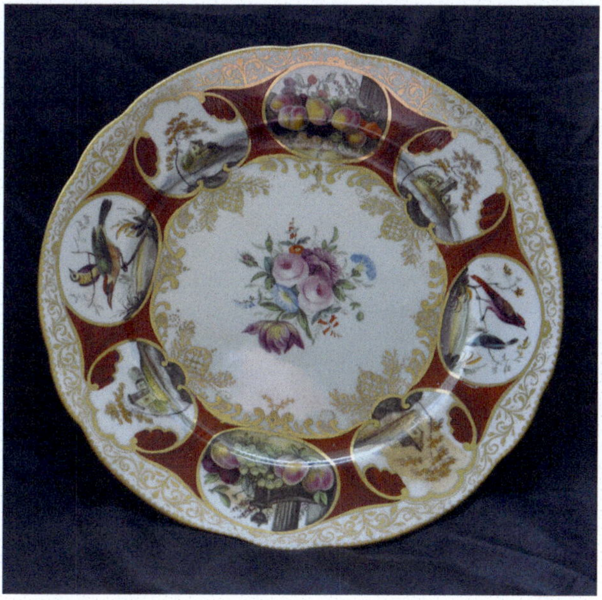

Fig. 6.1 Nantgarw porcelain dinner plate, 1818; London decorated by James Plant at Robins and Randall's atelier for Mortlock's of Oxford Street; commissioned by George, the Prince Regent, as a wedding present for his younger brother, Adolphus, Duke of Cambridge on his marriage in June, 1818, at Buckingham Palace to Princess Augusta Wilhelmina Louise, Landgravine of Hesse-Cassel. Marked NANT-GARW C.W. This specimen plate was selected for exhibition at the Special 200th Anniversary Exhibition of Nantgarw Porcelain, entitled "*Coming Home*" and held at Tyla Gwyn, Nantgarw, in the residence of William Billingsley on the actual Nantgarw China Works Site. *Private collection*

undertaken. Even in this latter case, the broken fragments were not prepared further in any way for analytical interrogation and the experiments were totally non-destructive. A survey of porcelains that have been studied hitherto by Raman spectroscopy is summarised in Table 6.1: much of this work has been reported for European porcelains such as Sevres, Meissen, Buen Retiro, Medici and St Cloud.

Several Raman spectral studies of European and Asian porcelains and celadons have been reported from the group of Professor Philippe Colomban, including early Sceaux, Medici and Sevres porcelains (Colomban 2001, 2004, 2005, 2013; Colomban and Treppoz 2001; Colomban et al. 2001, 2004a, b, c; Mancini et al. 2016; Ricciardi et al. 2006). These have cumulatively demonstrated the viability of Raman spectroscopy as a molecular spectroscopic technique to differentiate between porcelains from different manufactories and has additionally materially assisted in the creation of a database for early European porcelains, which has not hitherto included English or Welsh porcelains: a recent paper from Colomban et al. (2020) provides the first assessment of the composition of Nantgarw porcelain in this respect with the inclusion of Raman spectroscopic data.

Table 6.1 Porcelains studied by Raman spectroscopy

Porcelain Factory	Year of Foundation	Type of Body	Analyst(s) & Year
Tam Thai, Vietnam		Hard paste	Liem et al. (2000, 2003) and Colomban et al. (2004c)
Sevres	1756	Soft paste	Colomban et al. (2004a, b, c) and Maggetti and d'Albis (2017)
St Cloud	1693	Soft pate	Colomban et al. (2004a, b, c)
Mennecy	1738	Soft paste	Colomban et al. (2004a, b, c)
Chantilly	1730	Soft paste	Colomban et al. (2004a, b, c)
Vincennes	1740	Soft paste	Colomban et al. (2004a, b, c)
Medici	1575	Soft paste	Colomban et al. (2004a, b, c)
Ming	1348	Hard paste	De Waal (2004a, b), Kock and de Waal (2007), Widjaja et al. (2011) and Carter et al. (2017)
Rockingham	1826	Bone China	Edwards et al. (2004)
Meissen	1709	Hard paste	Colomban and Milande (2006)
Capodimonte	1743	Soft paste	Ricciardi and Milande (2008), Ricciardi et al. (2006) and Fabbri et al. (2008)

(continued)

Table 6.1 (continued)

Porcelain Factory	Year of Foundation	Type of Body	Analyst(s) & Year
Buon Retiro	1760	Soft & hard paste	Ricciardi and Milande (2008), Ricciardi et al. (2006)
Bow	1746	Steatitic paste	Jay and Orwa (2012)
Limehouse	1745	Soft & hard paste	Jay and Cashion (2013)
Chinese		Hard paste	Van Pevenage et al. (2014)
Sceaux	1748	Soft & hard paste	Mancini et al. (2016)
Chinese Wucai[a]		Hard paste	Colomban et al. (2017a)
Limoges	1771	Hard paste	Colomban et al. (2017b)
Rouen	1673	Soft paste	Colomban et al. (2018)
Lille	1711	Soft paste	Colomban et al. (2018)
Paris[b]	1730	Soft paste	Colomban et al. (2018)
Chinese Huafalang		Soft & hard paste	Colomban et al. (2017)
Davenport	1807	Bone China	Edwards et al. (2019)
Nantgarw	1817	Soft paste	Colomban et al. (2020)

[a]KiangsiKangxi, Yongzheng and Qianlong periods, famille rose, famille verte and fencai
[b]Pavie, Hebert, Chiccaneau and Bellevaux factories

6.2 The Basis for the Raman Spectral Data Interpretation for Ceramics

The Raman spectroscopic analysis of porcelains primarily addresses the identification of the mineral molecular composition of the bodies and glazes but secondary information can also be acquired about the production processes and in particular the temperatures achieved in the kilns in the final stages of porcelain manufacture. Earlier work from Professor Philippe Colomban and his team at the Universite de Pierre et Marie Curie in Paris, now part of the Sorbonne University, has established a sound basis for the interpretation of the intricate and highly complex silicate spectra in terms of both the types of silicon-oxygen bonding present as well as the degrees of polymerisation of the silicate matrices which have been subjected to elevated temperatures in kiln fired porcelains (Colomban 2001, 2004, 2005, 2013; Colomban and Treppoz 2001; Colomban et al. 2001, 2004a, b, c). Glassy silicified ceramics can be classified into five main types, Q^0–Q^4, according to the arrangement of the SiO_4 tetrahedra in their structures, namely:

Q^0: monomeric SiO_4 groups, characterised by SiO stretching vibrations occurring in the wavenumber shift range 800–850 cm^{-1}; an example is forsterite, Mg_2SiO_4, another is fayalite Fe_2SiO_4;

Q^1: dimeric SiO_4 groups occurring as Si_2O_7 moieties with SiO stretching vibrations in the wavenumber range 1050–1150 cm^{-1}; an example is danburyite, $Ca(BO_3)_2Si_2O_2$;

Q^2: chain silicates comprising Si_3O_9 units with SiO stretching vibrations in the wavenumber range 1050–1100 cm^{-1}; an example is hedenbergite, $CaFeSi_2O_6$;
Q^3: sheet silicates comprising Si_4O_{11} groups with SiO stretching vibrations in the wavenumber region near 1100 cm^{-1}; an example is beryl, $Be_3Al_2Si_6O_{18}$;
Q^4: tectosilicates and SiO_2 structures with SiO stretching vibrations in the wavenumber range 1150–1250 cm^{-1}; an example is oligoclase, (Ca,Na)(Al,Si) $AlSi_2O_8$.

Alpha-quartz has a well-defined crystal structure with a characteristic and strong SiO symmetric stretching band assigned to O=Si=O units in the Raman spectrum at 464 cm^{-1}: in fused silica this broadens and is found shifted to higher wavenumbers under mechanical stress – so, for example, in meteorites which have been subjected to high impact stresses the quartz vibrational band shifts to 520 cm^{-1}, characteristic of a shocked quartz, geologically termed coesite, In glasses and glazes found on finished porcelains, an estimate of the degree of polymerisation (DP) of the silicate matrix can be evaluated from the relative band intensities of broad features centred at 500 and 1000 cm^{-1}. Generally, this DP is measured as the A_{500}:A_{1000} band intensity ratio which increases numerically with the temperature to which the glaze has been subjected: hence, for soft paste porcelains fired at low kiln temperatures, which may typically be around 900 °C, the A_{500}:A_{1000} ratio is in the range between 0.8 and 1.1, whereas for hard paste porcelains fired at kiln temperatures of up to 1400 °C, this ratio rises to between 1.5 and 1.7. In the current study the prime intention is not to examine the glazes themselves but to interrogate the porcelain bodies through the glaze coating: also, because the laser wavelength used in our studies is in the near infrared region of the electromagnetic spectrum at 785 nm, this tends to produce luminescence emission bands which appear in the same region as the Raman spectral features. This will be discussed in detail later when the spectral data from the Raman spectroscopic experiments will be collated from a detailed comparative analysis. It is useful at this point to itemise the characteristic Raman spectral bands which are expected to occur for individual minerals and materials which are created in the high temperature kilns during the production of porcelains: the chemical compositions of the raw materials and the formulation of the chemical species into which they are converted during the high temperatures adopted in the kiln firing process for the manufacture of porcelains are presented in Table 6.2 and the characteristic Raman spectral data which can be used to identify these materials and their reaction products are given in Table 6.3.

6.3 Experimental Objectives of the Raman Spectroscopic Analysis of English and Welsh Porcelains

Several objectives are envisaged in these molecular spectroscopic analytical experiments, namely:

Table 6.2 Chemical composition of the raw materials comprising porcelain formulations and the chemical species they are converted into upon firing in the Kiln

Raw material	Chemical formula	Kiln conversion	Final porcelain material
Sand (Quartz, Silica)	SiO_2	SiO_2	Tridymite, Cristobalite
Flint	SiO_2	SiO_2	Tridymite, Cristobalite
Kaolinite	$Al_2Si_2O_5(OH)_4$	$CaSiO_3$	Wollastonite
		$CaNa[AlSi_2O_8]$	Bytownite
		Al_2SiO_5	Mullite
		$NaAlSi_3O_8$	Albite
Potash/Pearl ash	K_2CO_3	–	
Soda ash	Na_2O	–	
Lime	CaO	–	
Bone ash	$Ca_5(OH)(PO4)_3$	$Ca_3(PO_4)_2$	Whitlockite
Feldspar	$KAlSi_3O_8$	$Ca_2Al_2Si_2O_8$	Anorthite
(Soaprock)		$KAlSi_2O_6$	Leucite
		$KNaAlSi_3O_8$	Sanidine
Anatase	TiO_2	TiO_2	Rutile

1. Can one obtain definitive Raman spectra and key diagnostic Raman spectroscopic signatures for English and Welsh porcelain specimens, glazed and decorated which potentially could be used for forensic discrimination purposes and the attribution of origin?
2. Can one assign the observed Raman features to the paste components of the respective recipes after firing in the kilns and is it possible to correlate the bands observed with the absence or presence of specific components that may have been specified in the paste recipes, *viz*., the workbook notes of Lewis Dillwyn for Swansea porcelains (reproduced in Eccles and Rackham, *Analysed Specimens of English Porcelain in the Victoria & Albert Museum Collection*, 1922) and the putative Nantgarw formulation and recipe published by John Shelton (*The Practical Potter*, 1847) which he allegedly obtained from Samuel Walker before he departed for North America?
3. Does the presence of a surface glaze interfere with the interpretation of the Raman bands or seriously mask the minerals and components of the fired paste in the interrogation of a porcelain body lying underneath the glaze?
4. Is it possible to discriminate non-destructively between the porcelains produced by individual factories and possibly differentiate between the products made at different times in the history of the factory – the use of genuine, unambiguous, marked specimens which have perhaps also been identified as belonging to specific periods of production can be adopted as standards for the interrogation of specimens of unknown, dubious origin and the resultant exposure of fakes or mis-attributed porcelains?
5. Was there only a single body composition employed by particular factories as has been maintained by many authors (Cox and Cox 2001) and can one differentiate spectroscopically between porcelain produced at other times during the fac-

Table 6.3 Raman spectral signatures of minerals relevant to porcelain bodies & glazes

Mineral	Chemical formula	Key spectral wavenumbers/cm^{-1}
Whitlockite	$Ca_3(PO_4)_2$	960
a-Wollastonite	$CaSiO_3$	575,988
b-Wollastonite	$CaSiO_3$	636,972
Mullite	Al_2SiO_5	480,600,960,1130
Anorthite	$Ca_2Al_2Si_2O_8$	195,480,510,550,980
Quartz	SiO_2	205,464
Enstatite	$MgSiO_3$	350,405,675,1015,1036,1088
Forsterite	Mg_2SiO_4	820,850
Fayalite	Fe_2SiO_4	835,860
Haematite	Fe_2O_3	229,299,409,640,1320
Carbon	C	1320,1585
Feldspar	$KAlSi_3O_8$	190,275,325,411,488,508,522,645,973
Diopside	$MgSiO_3$	660,1015
o-silicate	SiO_4^-	450–460, ~1000,1155
Borax	$Na_2B_4O_7$	1400
Rutile	TiO_2	144,233,445,610
Anatase	TiO_2	143,200,398,517,638
Cristobalite	SiO_2	230,410,520
Tridymite	SiO_2	204,310,365,420.785
Gypsum	$CaSO_4.2H_2O$	412,481,617,1007,1130
Cerussite	$PbCO_3$	152,1052,1370,1477
Ilmenite	$FeTiO_3$	222,371,680
Serpentine/Chrysotile	$Mg_3Si_2O_5(OH)_4$	230,345,389,620,690,1105
Sanidine	$KNaAlSi_3O_8$	170,280,475,510
Coesite	SiO_2	520
Anhydrite	$CaSO_4$	715,1017,1133
Leucite	$KAlSi_2O_6$	490,515
Albite	$NaAlSi_3O_8$	230,285,320,408, 485,510
Bytownite	$CaNaAl_4Si_4O_8$	226,415,468,623,975,994,1060

tory lifetime, for example, the three recorded and accepted Swansea porcelain bodies, *viz.*, glassy, duck-egg and trident porcelains (Dillwyn 1922)?

6. With the Raman spectral data available is it then possible to deduce whether or not a piece of porcelain of unknown or merely suspected attribution actually originates from a particular factory site? This is of great value for the scientific attribution of rare or unique items manufactured at a particular factory and would materially assist in their discrimination from forgeries and fakes This art forensic aspect is an important input to expert opinion, which would have been derived solely and naturally on the basis of shape, texture and decoration, to provide a holistic scientific assessment of the factory origin.
7. Is there a potential for the detection of "outliers", interlopers or associated pieces in large porcelain services, which have perhaps been procured as replacements

for broken or damaged items after closure of a particular factory or which may have been supplied originally to complete a dinner, dessert or tea and coffee service commission order from the factory?

6.4 Porcelain Specimens for Non-destructive Raman Spectroscopic Analysis

With these objectives and questions to be addressed, the selection of several specimens of early Nineteenth Century English and Welsh porcelains donated for this specific purpose from established private collections was assembled for preliminary analytical studies: these items comprise 4 specimens of Nantgarw porcelain, 5 specimens of Swansea porcelain and also 5 representative shards of Nantgarw porcelain from a batch archaeologically excavated from the Nantgarw China Works site in the 1990s and selected from some two kilos of shards provided for this project from the site archive held at the Nantgarw China Works Museum in Tyla Gwyn, Nantgarw. Also, in this section 2 specimens of porcelain which seem to possess the correct criteria for their proper assignment to a Swansea origin in terms of translucency and appearance are included for preliminary assessment to compare with the genuine specimens – a straightforward test of the analytical protocols established here for the affirmation of a putative Swansea factory origin. Hence, in this first tranche of specimens, some 16 specimens of Nantgarw, Swansea and putative specimens from these factories are taken for non-destructive analysis. In addition, specimens of English porcelains from the Derby, Pinxton, Rockingham, Worcester (Barr, Flight & Barr period, *ca.* 1808–1813) and Coalport factories were analysed conjointly and similarly during eth same exercise.

The 21 porcelain specimens studied for this experiment, with reference to the appropriate figures, are now described as follows:

NG1: Nantgarw phosphatic porcelain dinner plate, *Duke of Cambridge* service, *ca.* 1818, London decorated by James Plant, to the commission of George, The Prince Regent, for the wedding of his younger brother Adolphus, Duke of Cambridge, to the Landgravine of Hesse-Cassel (Fig. 6.1). This specimen plate was shown as part of the "*Coming Home*" Exhibition held at the Nantgarw China Works Museum, July–September, 2019, to celebrate the 200th Anniversary of the foundation of the Nantgarw China Works, at which some 63 pieces of the finest porcelain produced by Nantgarw were displayed publicly, many of these for the first time.

NG2: Nantgarw phosphatic porcelain spill vase, *ca.* 1817–1819, London decorated (Fig. 6.2). This specimen porcelain item was shown as part of the "*Coming Home*" Exhibition held at the Nantgarw China Works Museum, July–September, 2019, to celebrate the 200th Anniversary of the foundation of the Nantgarw China Works, at which some 63 pieces of the finest porcelain produced by Nantgarw were displayed publicly, many of these for the first time.

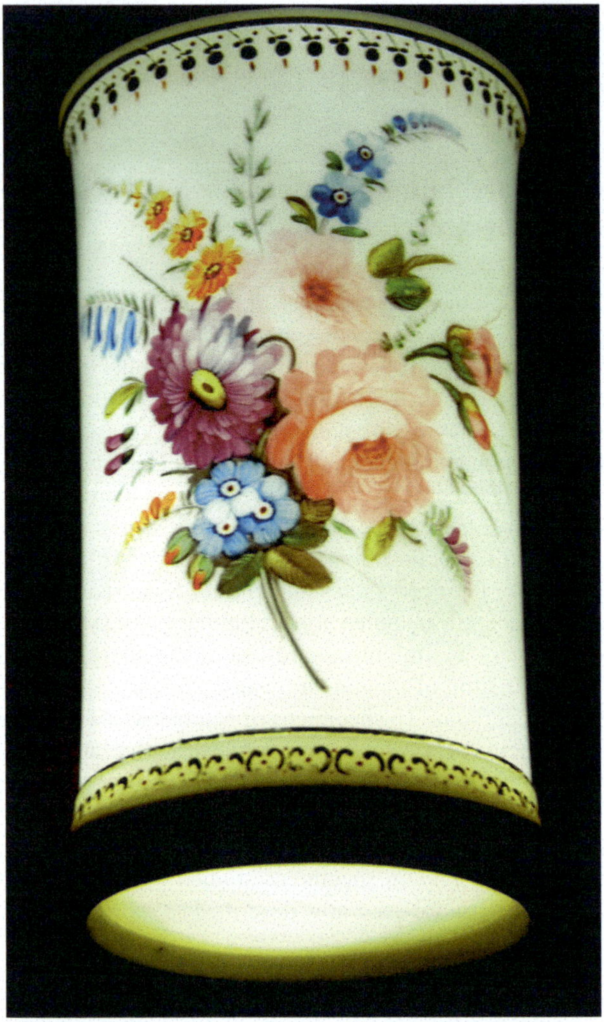

Fig. 6.2 Nantgarw porcelain spill vase, *ca.* 1817–1819; photographed in transmitted light to demonstrate the beautiful and clear white translucency that is so characteristic of the factory output. This specimen spill vase was selected for exhibition at the Special 200th Anniversary Exhibition of Nantgarw Porcelain, entitled "*Coming Home*" and held at Tyla Gwyn, Nantgarw, in the residence of William Billingsley on the actual Nantgarw China Works Site. *Private collection*

NG3: Nantgarw phosphatic porcelain coffee cup, from the *Spence-Thomas* service, *ca.* 1820–1822, locally decorated by Thomas Pardoe (Fig. 6.3).

NG4: Nantgarw phosphatic porcelain coffee cup and saucer, *ca.* 1817–1819, London decorated by Moses Webster in Robins and Randall's atelier for John Mortlock (Fig. 3.7).

NG5: Nantgarw phosphatic porcelain dinner plate, armorial porcelain, *Phippes* service, *ca.* 1817–1819, London decorated (Fig. 6.4).

NG6: Nantgarw phosphatic porcelain dinner plate, *Lady Seaton* service, *ca.* 1817–1819, London decorated (Fig. 3.1).

SW1: Swansea duck-egg porcelain deep dish, *ca.* 1815–1818, decorated locally by William Pollard (Fig. 3.2).

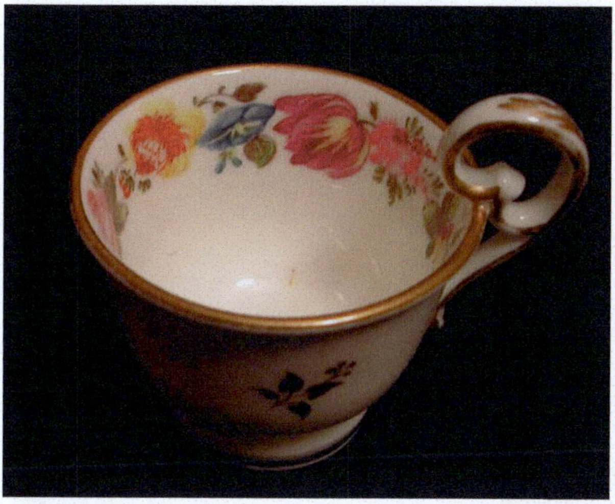

Fig. 6.3 Nantgarw porcelain coffee cup with heart-shaped handle, with Nantgarw glaze No.2. (perfected by Young & Pardoe), locally decorated with a garland of garden flowers by Thomas Pardoe, *ca.* 1820–1823, belonging to the *Spence-Thomas* breakfast service. A muffin dish and cover from the same service is illustrated in W.D. John, G.J. Coombes and K. Coombes, *Nantgarw Porcelain Album,* Illustration 31. Described therein as "*Nantgarw porcelain of exceptional quality*". *Reproduced with permission from the private collection of the Rev. J.B. Dickinson*

SW2: Swansea duck-egg porcelain spill vase, *ca.* 1815–1818, decorated locally by William Billingsley (Fig. 6.5).

SW3: Swansea duck-egg porcelain miniature watering can, *ca.* 1815–1818, decorated locally by William Pollard (Fig. 6.6).

SW4: Swansea duck-egg porcelain violeteer, *ca.* 1815–1818, decorated locally by Henry Morris (Fig. 6.7).

SW5: Swansea "trident" magnesian porcelain dessert plate, *ca.* 1817–1820, decorated locally by David Evans (Fig. 6.8).

SW6: Swansea glassy porcelain teabowl, *ca.* 1814–1815, decorated locally (Fig. 6.9.

D1: Derby phosphatic porcelain dessert comport, *Viscount and Lady Cremorne* service to the commission of Lady Philadelphia Cremorne, 1785 (Fig. 6.10).

D2: Derby phosphatic porcelain dessert plate, *Prince of Wales* service, 1786, decorated by William Billingsley to the commission of George, Prince of Wales, later King George IV (Fig. 6.11).

D3: Derby phosphatic porcelain dinner plate, *Barry-Barry* Service, *ca.* 1790–1810, decorated by William Billingsley to the commission of Pendock Barry (Fig. 6.12).

D4: Derby phosphatic porcelain spill vase, *ca.* 1790–1795, decorated by William Billingsley (Fig. 6.13).

P1: Pinxton phosphatic porcelain tea cup and saucer, *ca.* 1795–1799, decorated by William Billingsley (Fig. 6.14).

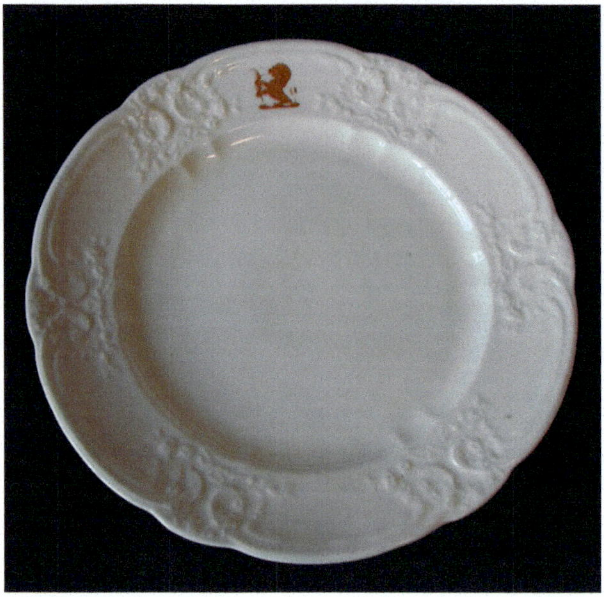

Fig. 6.4 Nantgarw armorial dinner plate, marked NANT-GARW C.W., free of painted enamelled decoration, bearing the crest of the Phippes family – *a demi-lion, or, rampant sinister, holding palm frond* – *ca.* 1817–1819. The Phippes family of London were granted arms by the Commonwealth in 1656: this service was probably ordered in London through Mortlock's on the commission of Henry Phippes, Viscount Normanby, created Baron Mulgrave in the county of York in 1794 and Earl Mulgrave in 1812. *Private Collection*

W1: Worcester bone china dish, Barr, Flight & Barr, *ca.* 1808–1812, decorated by William Billingsley (Fig. 6.15).

W2: Worcester bone china coffee can, Barr, Flight & Barr, *ca.* 1808–1812, decorated by William Billingsley (Fig. 6.16).

C1: Coalport phosphatic porcelain sucrier and lid, *ca.* 1820–1830 (Fig. 6.17).

R1: Rockingham bone china spill vase, *ca.* 1835–1840 (Fig. 6.18).

The Raman spectroscopic data are presented in Table 6.4 for pieces from this comprehensive list have been selected to illustrate the conclusions which may be derived from the analytical information: an essential part of the interpretation of the novel Raman data for these porcelains is the list of standard spectral characteristic wavenumbers produced in Table 6.3 for chemical species which have been identified hitherto in high temperature fired ceramics. It is anticipated, therefore, that a comparison of the spectral data in these two tables will result in a positive identification for the mineral species present in the bodies of the fired porcelains and therefrom lead to a deduction of the raw materials used in the preparation of the body

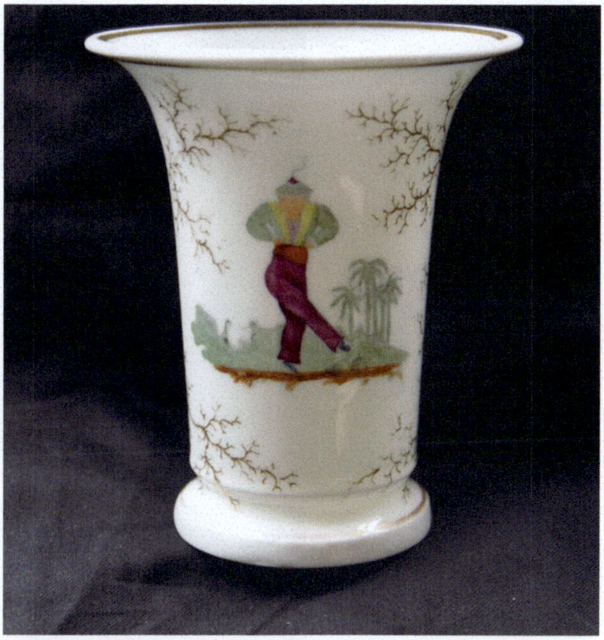

Fig. 6.5 Swansea porcelain trumpet-shaped spill vase, very fine duck-egg translucency, *ca.* 1817–1820; *chinoiserie* and seaweed tendrils decoration attributed to characteristic local decoration personally by William Billingsley. *Private collection*

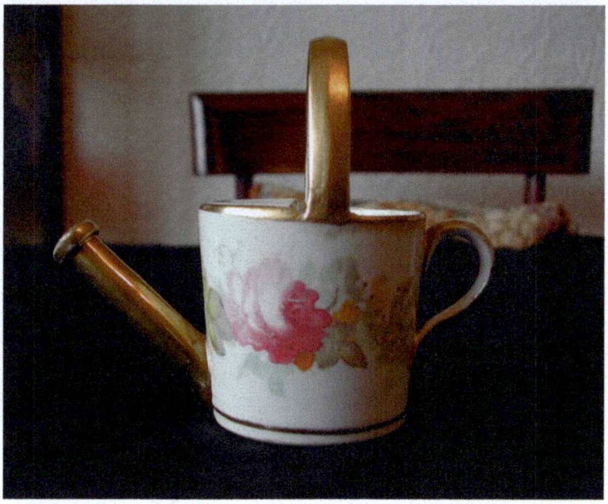

Fig. 6.6 Miniature Swansea watering can, duck-egg body, decorated by William Pollard with a wreath of garden flowers; based on a coffee can with an added gilt spout, part cover to the front rim and a strap handle, *ca.* 1817–1820. Unmarked. *Private collection*

pastes and the processes to which they have been subjected. There now follows a summary listing of the useful information that can be derived from these data and an assessment made of the viability of using such data for porcelain specimen attribution in possibly doubtful cases.

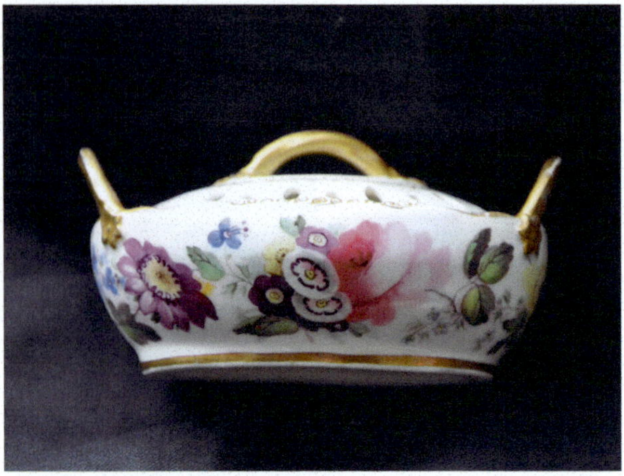

Fig. 6.7 Swansea porcelain violeteer, *ca.* 1817–1820, duck-egg body, with pierced lid, gilt strap handles and Greek scroll gilding border, decorated with sprays of garden flowers by Henry Morris. Unmarked. *Private Collection*

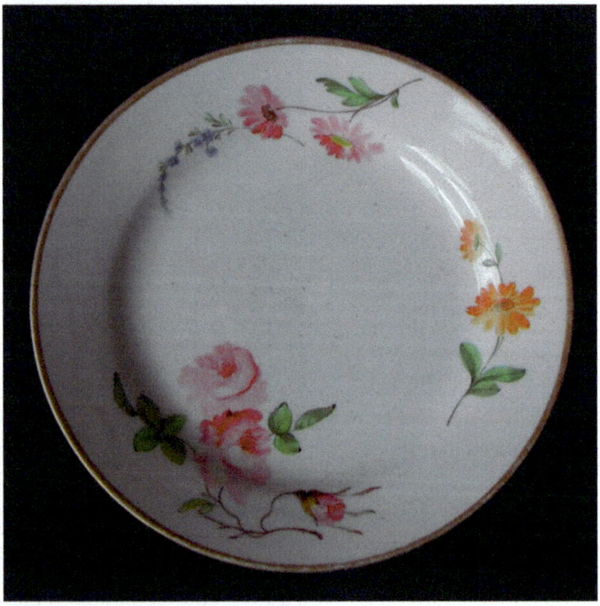

Fig. 6.8 Swansea "trident" porcelain dessert plate, impressed SWANSEA and impressed with trident motif, decorated with garden flowers by David Evans, *ca.* 1818–1820. *Private collection*

6.4.1 Summary of the Raman Spectral Data Interpretation for Porcelains

The excitation of the Raman spectra using visible and near infrared wavelengths, *viz.* 514.6 nm argon ion gas laser (green radiation) and 785 nm diode laser (red/near infrared radiation) produces a series of characteristic bands which can be subdivided into two main wavenumber regions, namely 1800–1000 cm^{-1} and 1000–100 cm^{-1}. The former contains mostly information about the Si=O stretching

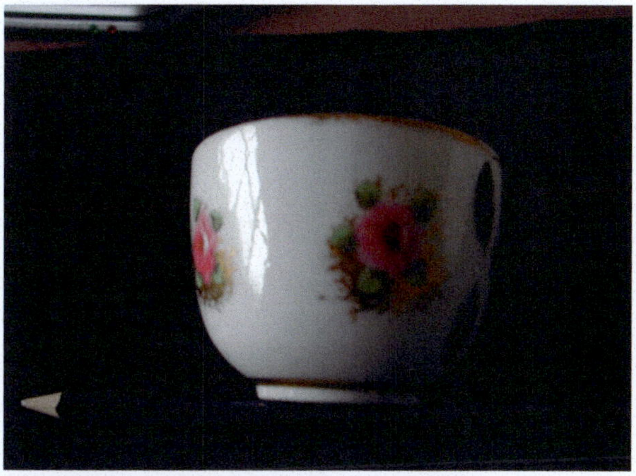

Fig. 6.9 Swansea glassy porcelain teabowl, painted with pink roses on a gilt seaweed ground, *ca.* 1815–1817, unmarked. *Private collection*

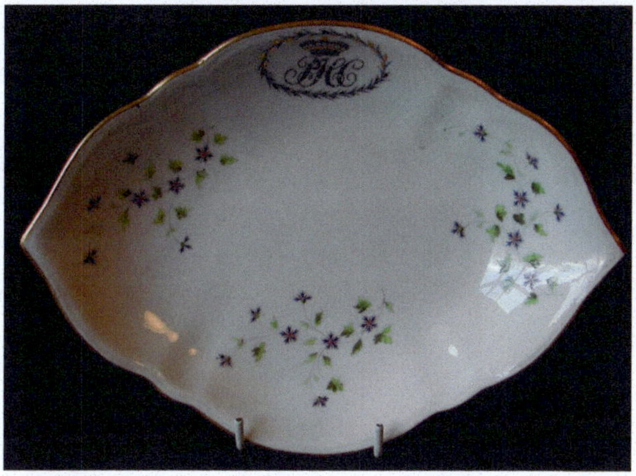

Fig. 6.10 Dessert comport from the *Cremorne* service, Derby porcelain, 1788, marked with puce crossed batons and crown, bearing the coronet and initials "*PHC*" of Lady Philadelphia Harriet Cremorne, wife of Viscount Cremorne, born in Philadelphia in 1741 and grand-daughter of William Penn, founder of the state of Pennsylvania in the USA. Lady–in-Waiting to Queen Charlotte, wife of King George III, Lady Cremorne died in 1828. The initials and coronet facilitate the identification of this as an important, unique historical and documentary service which has been decorated in an otherwise fairly common factory pattern, namely the Bourbon blue cornflower sprig. The scalloped, lozenge shape of the comport in this particular service is also extremely rare in Derby porcelain and was specially commissioned for this service by Lady Cremorne personally. Illustrated and discussed in J. Twitchett, *Derby Porcelain 1748–1848*, 2002, pages 194–195, where it is noted from documentary letters that only two comports of this shape were ordered for this prestigious service. *Private Collection*

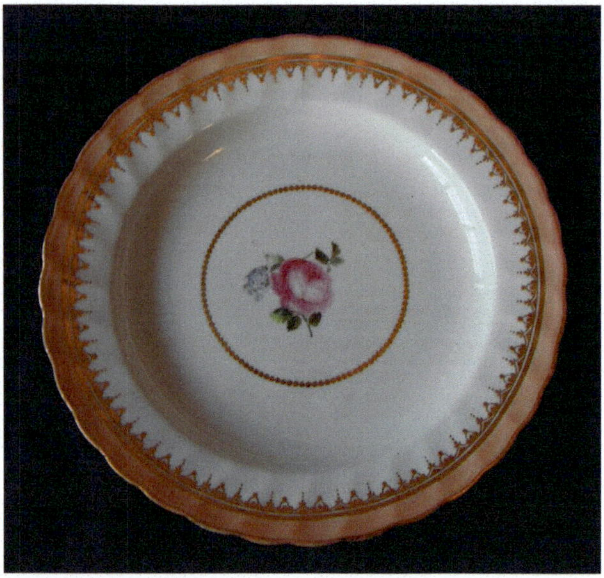

Fig. 6.11 Derby porcelain plate from the *Prince of Wales* service, 1786, pattern 65, decorated by William Billingsley with a single pink rose and forget-me-nots, dawn pink edging and gilded by William Longden, gilder's number "*8*" near edge of the footrim. Puce mark in enamel, depicting a cursive *D* with crown and crossed batons, pattern number. *Private collection*

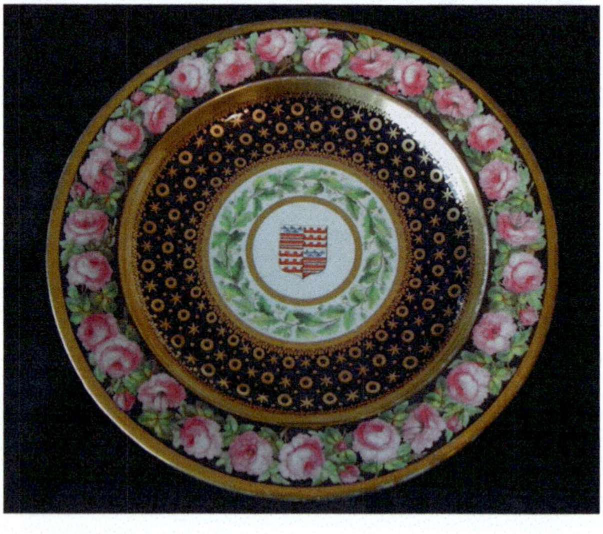

Fig. 6.12 Derby porcelain dinner plate from the *Barry-Barry* dinner service, *ca.* 1800–1820, with sumptuous and exquisitely fine decoration ascribed to William Billingsley after he had left the Derby China Works, and profusely gilded. See Edwards (2017), Edwards and Denyer (*William Billingsley: The Enigmatic Porcelain Artist, Decorator and Manufacturer,* 2017). *Private collection*

vibrations of complex silicates and also for the visible excitation a series of strong, broad electronic transitions from lanthanide Ln^{3+} ion impurities which occur as impurities in Cornish china clays (Edwards, *Nantgarw and Swansea Porcelains: An Analytical Perspective,* 2018). This spectral region therefore has little value *per se* for the forensic analytical differentiation between porcelain ceramics. However, the lower wavenumber region in contrast contains much information about the mineral species molecules present in the fired porcelains, such as whitlockite, wollastonite, forsterite and bytownite. As mentioned earlier, the composition of the raw materials

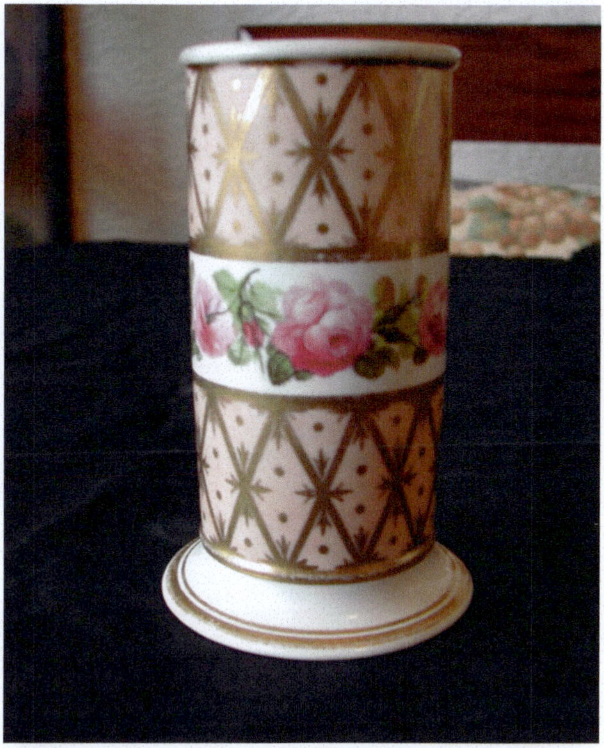

Fig. 6.13 Derby porcelain spill vase, *ca.* 1790–1795, with a salmon beige background and lattice gilding with small stars: a wreath of pink roses around the centre painted by William Billingsley – illustrated in W.D. John, *William Billingsley*, 1968. *Private collection*

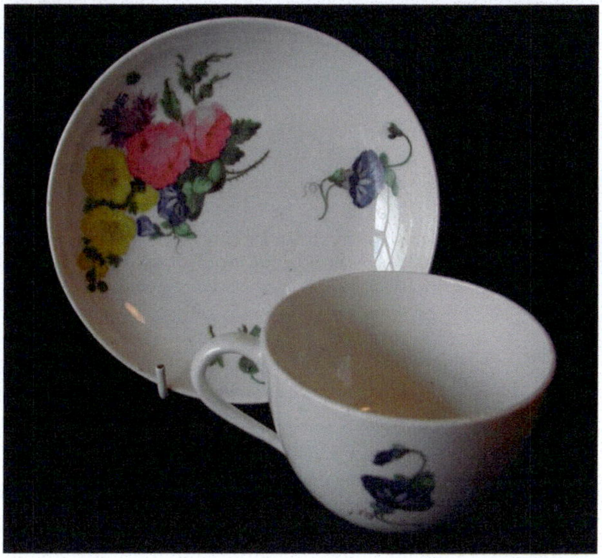

Fig. 6.14 Pinxton porcelain tea cup and saucer, decorated by William Billingsley, *ca.* 1798, extremely rare and ungilded, exhibiting the floral decoration to perfection. *Private collection*

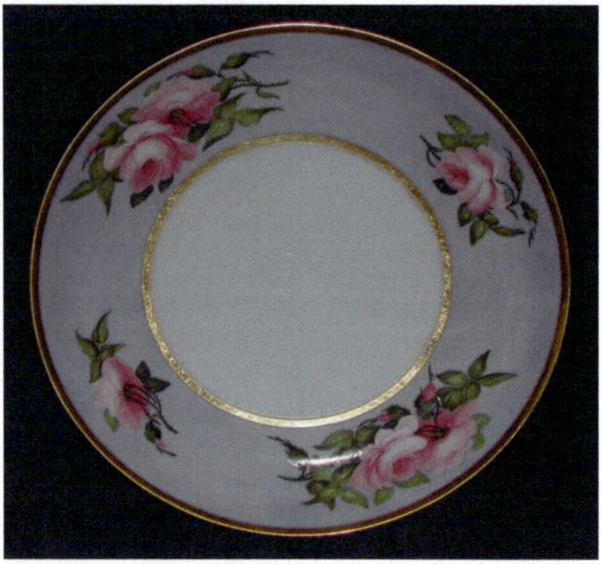

Fig. 6.15 Worcester porcelain deep dish, Barr, Flight & Barr period, *ca.* 1808–1812, with sky blue colour at the verge and decorated with four groups of roses by William Billingsley, two single blooms alternating with two groups of two roses and rosebuds. Billingsley's decoration on Worcester porcelain is extremely rarely encountered. *Private collection*

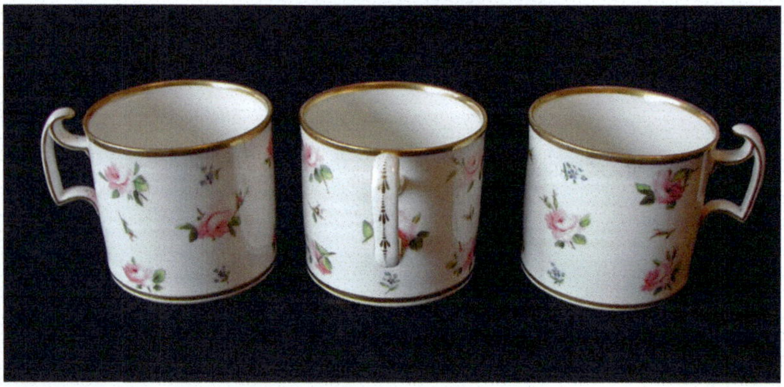

Fig. 6.16 Coffee cans from a Barr, Flight & Barr period Worcester porcelain tea and coffee service ascribed to William Billingsley (W.D. John, *William Billingsley*, 1968), *ca.* 1808–1812. *Private collection*

in the formulated porcelain body paste before firing bears but little resemblance to the final ceramic body in terms of the chemical entities present and whereas the elemental oxide determinations using modern instrumental electronic diffraction and X-Ray fluorescence methods indicate the presence of silicon-oxygen, aluminium-oxygen, magnesium-oxygen, calcium-oxygen and phosphorus-oxygen bonds, these can be regarded as cumulative results and not indicative of specific chemical species and molecules present. This is the reason for the complementarity of the molecular spectroscopic and diffraction/fluorescence techniques as discussed earlier; for example, Si-O bonds occur in the parent quartz, its isomorphs tridymite,

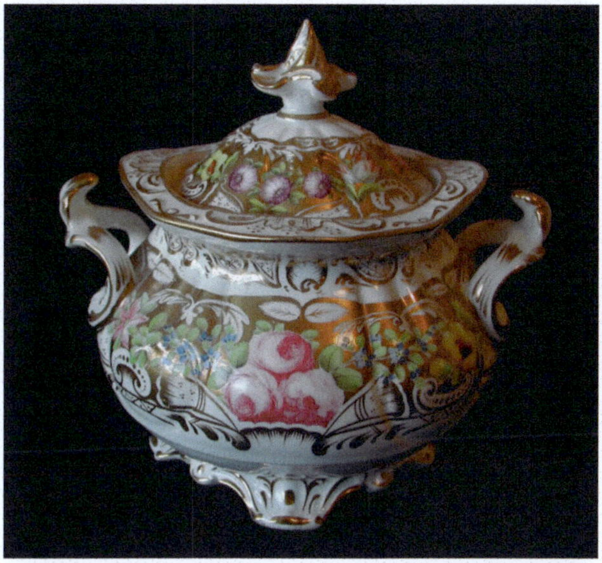

Fig. 6.17 Coalport porcelain sucrier with lid, *ca.* 1815–1825, sumptuously gilded and decorated in baroque style. Unmarked, but pattern number 2/675 in red enamel written to the base places this item in this period. Service illustrated in Godden, *Coalport and Coalbrookdale Porcelains*, 1991, where it is remarked upon for its particular fineness. *Private collection*

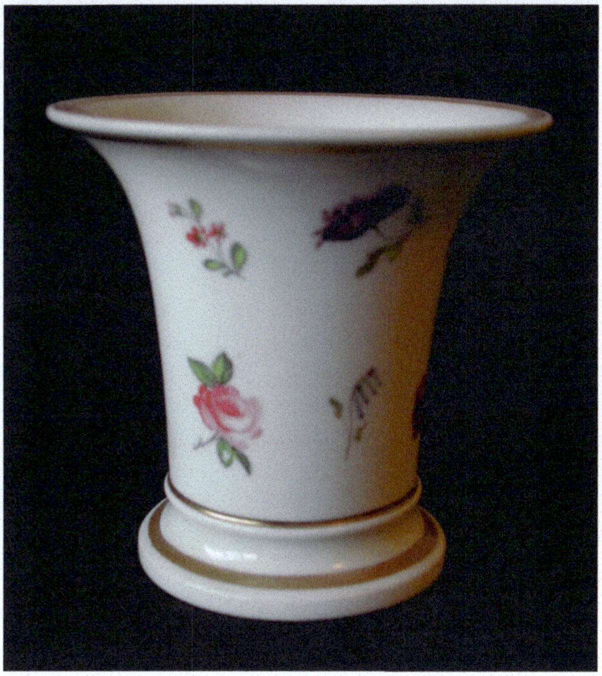

Fig. 6.18 Rockingham porcelain small trumpet-shaped spill vase, delicately decorated with single blooms of garden flowers, red enamel griffin passant mark, *ca.* 1835–1840. *Private collection*

Table 6.4 Raman spectroscopic data for English and Welsh porcelains

Specimen	Descriptor	Date	Raman Bands/cm^{-1}
NG2	Nantgarw spill vase	1817–19	995, 963, 955, 860, 836, 631, 438, 408, 387, 315, 276, 253, 147,120
NG4	Nantgarw saucer	1817–19	995, 963, 955, 860, 836, 631, 438, 408, 387, 315, 276, 253, 147,120
NG5	Nantgarw plate	1817–19	995, 963, 955, 860, 836, 631, 438, 408, 387, 315, 276, 253, 147,120
NG6	Nantgarw plate	1817–19	995, 963, 955, 860, 836, 631, 438, 408, 387, 315, 276, 253, 147,120
SW1	Swansea dish	1815–18	975, 963, 958, 861, 837, 630, 509, 439, 409, 390, 320, 256, 147,117
SW3	Swansea watering can	1815–18	975, 963, 958, 861, 837, 630, 509, 439, 409, 390, 320, 256, 147,117
SW4	Swansea violeteer	1815–18	975, 963, 958, 861, 837, 630, 509, 439, 409, 390, 320, 256, 147,117
SW5	Swansea plate	1817–20	1014, 997, 861, 838, 631, 409, 389, 253
SW6	Swansea teabowl	1814–15	1014, 997, 860, 839, 631, 409, 388, 255
D1	Derby plate	1786	994, 967, 958, 859, 836, 630, 439, 410, 388, 318, 273, 258, 148, 120
D3	Derby plate	1790–1810	994, 967, 958, 859, 836, 630, 439, 410, 388, 318, 273, 258, 148, 120
P1	Pinxton cup	1795–99	995, 966, 958, 859, 837, 630, 408, 389, 317, 253, 147
W1	Worcester BFB dish	1808–12	1771, 1218, 1194, 996, 860, 838, 628, 409, 390, 317, 254, 147
CP1	Coalport sucrier	1820–30	993, 965, 958, 860, 836, 631, 439, 408, 387, 319, 273, 252, 149, 117
R1	Rockingham spill vase	1835–40	997, 931, 861, 838, 630, 409, 390, 320, 254, 147

coesite and cristobalite, in flints and cherts as well as in metal-coordinated species such as feldspars and olivines. A further complication arises in the high temperature kiln firing when the silicate modifications, with ions of calcium, aluminium, magnesium and iron themselves react with phosphate ions from the bone ash utilised in phosphatic porcelains and bone china to form new mineral species.

Therefore, the interpretation of the molecular spectral analytical data on our selected range of porcelains can be interpreted as follows:

- There is no Raman signal observed for quartz, SiO_2, in any of the fired porcelains studied here as evidenced by the characteristic Si=O stretching band at 464 cm^{-1}; this is not unexpected when considering the chemical reactions undergone at the high kiln temperatures even though a major component of the raw materials formulations comprises fine sand, flints and chert, which are predominantly quartz-containing.
- All the spectra contain a strong doublet band near 860 and 836 cm^{-1}, which can be assigned to the iron silicate, fayalite, Fe_2SiO_4. Haematite is an impurity in sili-

caceous rocks, especially sand, and at high temperatures this will form a stable silicate.

- A weak feature is seen at the lower end of the wavenumber shift range at approximately 147 cm^{-1}. This is confidently assigned to the Ti=O titanium oxide stretching band for either anatase or rutile; anatase, TiO_2, is a known impurity in Cornish kaolin. Occurring to the extent of about 1%, and it has additionally a large molecular cross-section for Raman scattering so it can be detected in only small amounts in the spectral data. As most if not all of the English and Welsh porcelains studied here acquired their china clay raw material form Cornwall then it would be expected that these would exhibit a Raman signal for the mineral anatase. The possibility of rutile is also present as at high kiln firing temperatures the anatase is metastable and thermodynamically can convert to its stable high temperature polymorph, rutile, also formulated as TiO_2. Supporting bands for anatase occur at 315 and 387 cm^{-1}.
- The higher wavenumber bands in the lower region contain much information not only about the raw materials used in the preparation of the paste but also about the kiln firing conditions operating at the factory. The highest band recorded in this wavenumber region at 1014 cm^{-1} corresponds to the minerals enstatite and diopside, which both have the formula $MgSiO_3$. These magnesium silicates are formed at high kiln temperatures form the soaprock /soapstone component of the raw materials used in magnesian porcelains such as the Swansea "trident" ware. We would not expect, therefore, to observe these in normal phosphatic porcelains which do not have a soapstone component such as Nantgarw or Swansea duck-egg porcelains.
- At 997–995 cm^{-1} we observe a band of variable intensity and bandwidth which represents a composite signal for several minerals: the most important of these is the *alpha-* form of wollastonite, alpha-$CaSiO_3$, which is quite rarely seen in many porcelains because it is the stable form of wollastonite found only at the highest kiln firing temperatures used. The Nantgarw China Works adopted one of the highest firing temperatures in the two-stage process for British soft paste phosphatic porcelains believed to be around 1400 °C. It is interesting that the later Rockingham China Works specimen (1826–1840) also exhibits this feature but this is to be expected for the bone china single-stage manufacturing process, which also must have therefore have used high firing temperatures of this order, i.e. 1300–1400 °C, in emulation of the Chinese hard paste porcelains. Another occurrence amongst the specimens studied here is provided by the Worcester Barr, Flight & Barr examples (1808–1810) which also have a non-standard type of porcelain body similar to that of bone china. The bandwidth of the 997–995 cm^{-1} feature also contains spectral components of anorthite, $Ca_2Al_2Si_2O_8$, and bytownite, $CaNaAl_4Si_{14}O_8$, two aluminosilicates which are found to form at high temperatures in calcium-rich melts.
- At 975 cm^{-1} we can observe a feature arising from a feldspar raw material component which forms bytownite at the kiln temperatures. However, the most predominant molecular contribution at this wavenumber is undoubtedly from *beta*-wollastonite, beta-$CaSiO_3$, which is generally found for most soft paste por-

celains, especially those fired at the lower end of the accepted firing temperature range.

- At 963 cm^{-1} a rather broad feature is noted which is seen to split into another lower wavenumber component at 955 cm^{-1}: these can be assigned to mullite, Al_2SiO_5, at the higher wavenumber and whitlockite, $Ca_3(PO_4)_2$, respectively. The former is indicative especially of a high firing temperature in excess of about 1300 °C and the latter will only be found in phosphatic porcelains which have had calcined bone ash as the raw material component. Hence hard paste porcelains will show the higher wavenumber signal for mullite but the lower one of whitlockite will be absent from their spectra.
- The strong doublet at 860,836 $cm^{-1}$is easily recognisable as the signature of fayalite, Fe_2SiO_4, which arises as mentioned above from the interaction at high temperatures between haematite, Fe_2O_3, and quartz or other ortho-silicate ions, SiO_4^{2-}, from feldspar. In the presence of magnesium ions we might expect that we should also see evidence for an possible intermediate mineral member of the forsterite/fayalite series, $(Fe, Mg)_2SiO_4$, up to the extreme member forsterite, Mg_2SiO_4, depending upon the relative amounts of iron and magnesium present in the melt.
- The middle wavenumber range of this lower wavenumber region is characterised by a series of common features occurring at 631, 509, 485, 438, 408, 387, 315, 276, 253 and 225 cm^{-1}. These are not always present together in each porcelain specimen and are generally weak in the band spectral intensity: the strongest and most reproducible bands for the specimens selected here are those at 631 and 408 cm^{-1}, which can be assigned to bytownite, anatase, and pseudo-wollastonite (*beta*-wollastonite), respectively. The latter band has a spectral coincidence with features arising from cristobalite (a high temperature SiO_2 polymorph), albite, bytownite and enstatite. The band at 387 cm^{-1} is assignable to anatase, a recorded impurity in Cornish china clays, whereas that at 315 cm^{-1} is also attributable to anatase, tridymite (another high temperature polymorph of SiO_2) and albite. The features at 276 and 253 cm^{-1} are consistent with the original feldspar additive and cristobalite. Finally, the weak feature at 509 cm^{-1} is consistent with an assignment to anorthite and also possibly anatase.
- In conclusion, it appears that the non-destructive analysis of porcelains by molecular spectroscopy as exemplified here reveals complementary information about the composition of the fired bodies. At first, from these results it could be inferred that many porcelains have a similar composition and this is to be anticipated since the raw materials at the high kiln firing temperatures required in the manufacture of porcelain are reduced to a range of molecular species such as cristobalite, wollastonite, fayalite and whitlockite but the following criteria can now be established which will assist in the correlation of the molecular spectroscopic information with the elemental oxide data. Further details of the molecular assignments for Nantgarw porcelain shards can be obtained from the paper by Colomban et al. (2020).

6.5 Correlation Between the Elemental Oxide and Molecular Spectroscopic Data

Firstly, it can be seen that many of the key molecular species itemised in Table 6.3 have appeared in the collection of porcelains analysed here: examples include those discussed above and it is now important to match these data with those obtained from the elemental oxide determinations derived from diffraction and fluorescence spectroscopic experiments.

Phosphatic porcelains: in the molecular analyses are indicated by the presence of whitlockite, $Ca_3(PO_4)_2$, a calcium orthophosphate produced from the calcined hydroxyapatite added as a raw material at the high kiln firing temperature.

Magnesian porcelains: these are indicated by the presence of enstatite and diopside, both formulaically represented as the magnesium silicate $MgSiO_3$.

Glassy porcelains: these show evidence of free silicate matrices derived from the flint or crown glass added as raw materials with broad spectral features around 1150–1160 cm^{-1}.

Bone china: shows evidence of feldspar addition as a raw material and of the single-stage firing process adopted.

The presence of molecular titania in the form of anatase indicates the Cornish clay source for kaolin adopted by almost all the English and Welsh porcelain manufacturers, in which it occurs as an impurity to the extent of about 1%. The presence of anatase in the high temperature fired body is interesting because most textbooks confirm that it is the metastable polymorph of rutile and that chemical conversion is accomplished at a temperature in excess of 850 °C. However, recent experiments have demonstrated that this is not the whole story as anatase can exist at temperatures significantly in excess of 1200 °C for a long time without changing its structure into rutile. This may be surprising but it is attributed to the medium in which the anatase is held and clearly molten silicates protect it kinetically from its thermodynamic favourable conversion to the apparently more stable rutile. Hitherto, the detection of anatase in fired ceramic bodies has led to the conclusion that the ceramic was fired at lower temperatures, somewhat less than 850 °C, but this didactic inference must now be seriously questioned, especially since several specimens of Chinese hard paste porcelains have been detected to contain molecular anatase and these have undoubtedly been fired at a much higher temperature than the official conversion temperature of anatase to rutile. In the elemental oxide determinations from fired ceramic bodies, the distinction is not usually made between anatase and rutile for a TiO_2 evaluation.

The differentiation between the two polymorphs of calcium silicate, namely *alpha*- and *beta*- wollastonite, both represented formulaically as $CaSiO_3$, is also an important one for the molecular spectroscopic analytical method as the "normal" signature of the *beta*-form is found in most soft paste porcelains fired at temperatures approaching 1200 °C but at a higher kiln firing temperature the conversion to the polymorphic *alpha*-form is made and this is a relatively rare occurrence.

However, the presence of *alpha* – wollastonite is clearly seen here in several porcelain specimens which undoubtedly indicates that they have been fired at much higher temperatures – the case of Nantgarw porcelain is exemplary in that it is known that a kiln operating temperature approaching 1420 °C was achieved and indeed maintained at this level within about +/– 20 °C for several days, so affording time for the conversion of the metastable *beta* to *alpha* polymorph to take place (Edwards, *Porcelain to Silica Bricks; the Extreme Ceramics of William Weston Young, 1776–1847*, 2019). In several cases we can see that both polymorphs are present which must indicate that the *beta* form is equilibrating with its more stable *alpha* analogue but has not yet completely made the full transition.

The component bands of the strong doublet centred at 860 and 836 cm^{-1} are both assignable to the Si=O silicate stretching of the orthosilicate ion species in fayalite, Fe_2SiO_4, the end member of the olivine series, $(Fe,Mg)_2SiO_4$, the other end member being forsterite, Mg_2SiO_4. Generally, in mixed iron (II) and magnesium (II) silicates there is a doublet in this region whose component wavenumbers reflect the mixed ionic nature of the lattice and the ratio of the iron to magnesium can be estimated from a quantitative estimation of the band intensities of the two components of the doublet. Here, the individual wavenumbers of the two band components at 860 and 836 cm^{-1} indicate that we are identifying fayalite, Fe_2SiO_4. In forsterite, Mg_2SiO_4, these bands shift to 837 and 814 cm^{-1}, respectively, and in the olivine solid solution series there are progressive wavenumber shifts observed over th ranges 860–837 cm^{-1} for the higher wavenumber band and 814–828 cm^{-1} for the lower wavenumber band of the doublet. Where we have a magnesian porcelain, therefore, we should expect to observe the higher wavenumber doublet band down-shifted in wavenumber value and the lower wavenumber band up-shifted in wavenumber value (Breitenfeld et al. 2017). This is confirmed analytically in the specimen of Swansea trident porcelain studied here and this can be correlated with the recipe for which Lewis Dillwyn recorded his substitution of a significant amount of soaprock for calcined bone ash in the production of his "trident" body (Dillwyn 1922; Edwards, *Nantgarw and Swansea Porcelain: A Scientific Reappraisal,* Appendix II, 2018).

References

L.B. Breitenfeld, M.D. Dyar, C. Carey, T.J. Tague Jr., P. Wang, Predicting olivine composition using Raman spectroscopy through band shift and multivariate analysis. Lunar Planet. Sci. Abstr. **XLVIII**, Article 1898 (2017)

E.A. Carter, M.L. Wood, D. de Waal, H.G.M. Edwards, Porcelain shards from Portuguese wrecks: Raman spectroscopic analysis of marine archaeological ceramics. Herit. Sci. **5**, 1–8 (2017)

P. Colomban, Differentiation of antique porcelains, celadons and faiences from the Raman spectra of their bodies, glazes and paintings. Asian Chem. Lett. **5**, 133 (2001)

P. Colomban, Raman spectrometry: A unique tool to analyze and classify ancient ceramics and glasses. Appl. Phys. A **79**, 167–170 (2004)

P. Colomban, A case study: Glasses, glazes and ceramics, in *Raman Spectroscopy in Archaeology and Art History*, ed. by H. G. M. Edwards, J. M. Chalmers, (Royal Society of Chemistry Publishing, Cambridge, 2005)

P. Colomban, The destructive/non-destructive identification of Enameled pottery, glass Artifacts and associated pigments – A brief overview. Arts **2**, 77–110 (2013)

P. Colomban, V. Milande, On-site analysis of one of the earliest known Meissen porcelains and stoneware. J. Raman Spectrosc. **37**, 606–613 (2006)

P. Colomban, L.C. Prinsloo, Optical spectroscopy of silicates and glasses, in *Spectroscopic Properties of Inorganic and Organometallic Compounds*, ed. by J. Yarwood, R. Douthwaite, S. B. Duckett, vol. 40, (Royal Society of Chemistry Publishing, Cambridge, 2009), pp. 128–149

P. Colomban, F. Treppoz, Identification and differentiation of ancient and modern European porcelains by Raman macro- and micro-spectroscopy. J. Raman Spectrosc. **32**, 93–102 (2001)

P. Colomban, G. Segon, X. Faurel, Differentiation of antique ceramics from the Raman spectra of their coloured glazes and paintings. J. Raman Spectrosc. **32**, 351–360 (2001)

P. Colomban, V. Milande, H. Lucas, On-site Raman analysis of Medici porcelains. J. Raman Spectrosc. **35**, 68–72 (2004a)

P. Colomban, I. Robert, C. Roche, G. Sagon, V. Milande, Identification des porcelains tendres du 18eme siècle par spectroscopie Raman: Saint Cloud, Chantilly, Mennecy et Vincennes/Sevres. Revue d'Archaeometrie **27**, 153–167 (2004b)

P. Colomban, G. Sagon, L.Q. Huy, N.Q. Liem, L. Mazerolles, Vietnamese 15th century blue-and-white tam tai and lustre porcelains and stonewares: Glaze, composition and decoration techniques. Archaeometry **46**, 125–136 (2004c)

P. Colomban, F. Ambrosi, A.-T. Ngo, T.-A. Lui, X.-L. Feng, S. Chen, C.-L. Choi, Comparative analysis of Wucai Chinese porcelains using mobile and fixed Raman microspectrometers. Ceram. Int. **43**, 14244–14256 (2017)

P. Colomban, L. Arberet, B. Kirmizi, On-site Raman analysis of 17th and 18th century Limoges enamels: Implications on the European cobalt sources and the technological relationship between Limoges and Chinese enamels. Ceram. Int. **43**, 10158–10165 (2017a)

P. Colomban, Y. Zhang, B. Zhao, Non-invasive Raman analysis of Chinese huafalang and related porcelain wares: Searching for evidence of innovative pigment technologies. Ceram. Int. **43**, 12079–12088 (2017b)

P. Colomban, V. Milande, T.-A. Lu, Non-invasive on-site Raman study of blue decorated early soft paste porcelain. The use of arsenic – rich (European) ores compared with huafalang Chinese porcelains. J. Eur. Ceram. Soc. **38**, 5228–5233 (2018)

P. Colomban, H.G.M. Edwards, C. Fountain, Raman spectroscopic and SEM/EDXS analysis of Nantgarw soft paste porcelain. J. Eur. Ceram. Soc. (2020)

A. Cox, A. Cox, *Rockingham Porcelain* (Antique Collectors Club Publishing, Woodbridge, Suffolk, 2001)

D. de Waal, Raman investigation of ceramics from 16th and 17th century Portuguese shipwrecks. J. Raman Spectrosc. **35**, 646–649 (2004a)

D. de Waal, Raman identification of the pigment in blue and white Ming porcelain shards. Asian Chem. Lett. **8**, 57–65 (2004b)

L.W. Dillwyn, *Notes on the Experimental Production of Swansea Porcelain Bodies and Glazes*. Made by Lewis Weston Dillwyn with Samuel Walker at the Swansea China Works Between 1815 and 1817. Presented to the Library of the Victoria & Albert Museum, South Kensington, London by John Campbell in 1920. Reproduced in Eccles and Rackham, Analysed Specimens of English Porcelain, 1922, see reference

H. Eccles, B. Rackham, *Analysed Specimens of English Porcelain in the V&A Museum Collection* (Victoria & Albert Museum, South Kensington, London, 1922)

H.G.M. Edwards, *Swansea and Nantgarw Porcelain: A Scientific Reappraisal* (Springer Publishing, Dordrecht, The Netherlands, 2017)

H.G.M. Edwards, *Nantgarw and Swansea Porcelains: An Analytical Perspective* (Springer, Dordrecht, The Netherlands, 2018)

H.G.M. Edwards, *From Porcelain to Silica Bricks: The Extreme Ceramics of William Weston Young* (Springer, Dordrecht, The Netherlands, 2019), pp. 1774–1847

H.G.M. Edwards, M.C.T. Denyer, *William Billingsley: The Enigmatic Porcelain Artist, Decorator and Manufacturer* (Perose Antiques Ltd, Thornton, Bradford, UK, 2017)
H.G.M. Edwards, P. Colomban, B. Bowden, Raman spectroscopic analysis of an English soft paste porcelain plaque mounted table. J. Raman Spectrosc. **35**, 656–661 (2004)
H.G.M. Edwards, A.P.H. Surtees, R. Telford, Chapter 4: Dancing on eggshells: A holistic analytical study of a ballet dancer on Regency porcelain, in *Raman Spectroscopy in Archaeology and Art History*, Royal Society of Chemistry Analytical Chemistry Spectroscopy Monographs Series, ed. by H. G. M. Edwards, P. Vandenabeele, vol. II, (Royal Society of Chemistry Publishing, Cambridge, UK, 2019)
B. Fabbri, P. Ricciardi, P. Colomban, *The study of the Capodimonte production: An occasion for the proposal of a Raman database for ancient porcelains*, in Proceedings of the ART08 conference, Jerusalem, 2008. http://www.ndt.net/article.art2008/papers/104Fabbri.pdf
G. Godden, *Encyclopaedia of British pottery and Porcelain Marks* (Herbert Jenkins, London, revised and expanded edition, Barrie & Jenkins, London, 1991)
W.H. Jay, J.D. Cashion, Raman spectroscopy of Limehouse sherds supported by Mossbauer spectroscopy and scanning electron microscopy. J. Raman Spectrosc. **44**, 1718–1732 (2013)
W.H. Jay, J.O. Orwa, Raman spectroscopic application to early (ca. 1746-1764) English steatitic porcelains : A tentative study for comparison. J. Raman Spectrosc. **43**, 307–316 (2012)
W.D. John, *William Billingsley* (The Ceramic Book Company, Newport, 1968)
D. Kock, D. de Waal, Raman studies of the underglaze blue pigment on ceramic artefacts of the Ming Dynasty and of unknown origon. J. Raman Spectrosc. **38**, 480–487 (2007)
N.Q. Liem, G. Sagon, V.X. Quang, H.V. Tan, P. Colomban, Raman study of the microstructures, composition and processing of ancient Vietnamese (proto) porcelains and celadons. J. Raman Spectrosc. **31**, 933–942 (2000)
N.Q. Liem, P. Colomban, G. Sagon, H.X. Tinh, T.B. Hoanh, Microstructure, composition and processing of 15th Century Vietnamese porcelains and celadons. J. Cult. Herit. **4**, 187–197 (2003)
D.A. Long, *The Raman Effect: A Unified Treatment of the Theory of Raman Scattering by Molecules* (Wiley, Chichester, 2002)
M. Maggetti, A. d'Albis, Phase and compositional analysis of a Sevres soft paste porcelain plate from 18781, with a review of early porcelain techniques. Eur. J. Mineral. **29**, 347–367 (2017)
D. Mancini, C. Dupont-Logie, P. Colomban, On-site identification of Sceaux porcelain and faience using a portable Raman instrument. Ceram. Int. **42**, 14918–14927 (2016)
P. Ricciardi, V. Milande, Non-destructive Raman characterisation of Capodimonte and Buon Retiro porcelain. J. Raman Spectrosc. **39**, 1113–1119 (2008)
P. Ricciardi, P. Colomban, B. Fabbri, V. Milande, Towards the establishment of a Raman database of early European porcelain. e-Prev. Sci. **6**, 22–26 (2006)
J. Taylor, *The Complete Practical Potter* (Shelton, Stoke-upon-Trent, 1847)
J. van Pevenage, D. Lauwers, D. Herremans, E. Verhaeven, B. Vekemans, W. De Clercq, L. Vincze, L. Moens, P. Vandenabeele, A combined spectroscopic study on Chinese porcelains containing Ruan-cai colours. Anal. Methods **6**, 387–394 (2014)
E. Widjaja, G.H. Lim, Q. Lim, A.B. Mashadi, M. Garland, Pure component Raman spectral reconstruction from glazed and unglazed Yuan, Ming and Qing shards: A combined Raman microscopy and BTEM study. J. Raman Spectrosc. **42**, 377–382 (2011)

Chapter 7
The Earliest Porcelain in Europe … Meissen?

Abstract Following on from the detailed considerations of the early English and Welsh porcelain manufactories it is appropriate to revisit the idea of who actually produced the first porcelain in Europe. There is no doubt that historically this honour has been accorded to Meissen, who under the tutelage of Ehrenfriede von Tschirnhaus and Johann Bottger, did undoubtedly make a viable porcelain body in 1709 with the financial support and encouragement of Augustus, Elector of Saxony. However, this has denied the material contributions made by the early French and English factories and these will now be evaluated here, with the benefit of analytical chemical evidence relating to their wares.

Keywords Meissen · Rouen · St Cloud · Fulham · Soft paste porcelains

7.1 The Early French Porcelain Factories

We have already discussed the earliest English porcelain manufactories and their foundations, many of which were generated from established earthenware and faience businesses, and we have considered in some detail the role played by the Royal Society of London through the personages of Robert Hooke and Robert Boyle in John Dwight's experiments at his Fulham Pottery in the early 1670s, which seemingly led to thc synthesis of unique artefacts which possessed a translucent porcelain-like body, called "porcellaneous" by several historians and observers. A relevant piece of information to emerge from these accounts was the evidence of the personal interaction between key players in England and in Continental Europe, particularly France, who would have been in a position to exchange factual details about recipes, formulations and processing conditions for what was essentially a highly empirical and experimental enterprise. It will be highly apposite, therefore, to now consider briefly the growth of French porcelain manufactories in the late seventeenth and early eighteenth centuries and then to make a comparison with their

H. G. M. Edwards, *18th and 19th Century Porcelain Analysis*,
https://doi.org/10.1007/978-3-030-42192-2_7

Table 7.1 Early porcelain factories of the seventeenth and eighteenth centuries in France

Factory	Period	Founder and sponsor
Rouen	1673–1696	Edme & Louis Poterat; King Louis XIV
St. Cloud	1693–1766	Pierre Chicaneau; Philippe, Duc de Orleans, King Louis XIV
Ville l'Eveque	1711–1766	Marie Moreau & Dominique Chicaneau
Lille	1711–1730	Barthelmy Dorez
Chantilly	1725–1800	Ciquari Cirou; Louis-Henri de Bourbon, Prince of Conde
Mennecy	1734–1773	Francois Barbin; Duc de Villeroy
Vincennes	1741–1756	Claude-Humbert Gerin; Marquis Orry de Fulvy, King Louis XV & Mme de Pompadour
Sceaux	1748–1810	Jacques Chapelle, LF de Bey; Richard Glot
Orleans	1753–1768	Gerault Daraubert; King Louis XV
Sevres	1756–1804	JC Chambellan Duplessis: King Louis XV
Crepy-en-Valois	1762–1770	Louis-Francois de Gagnepain, Philippe, Duc de Valois
Etiolles	1766–?	Messrs. Monnier & Pelleve
Arras	1770–1790	Messrs. Boussemaert & Delhaye
Bourg-la-Reine	1773–1804	Comte d'Eu

English analogues. Table 7.1 provides a listing of the key French porcelain manufactories which existed in the late 17th through to the end of the eighteenth century.

Perusal of the contents of Table 7.1 reveals several key features which require some further elaboration:

- The fourteen porcelain manufactories listed here, commencing in their date of foundation with Rouen in 1673, received some critically important Royal and aristocratic patronage which was not apparent with their English analogues. It is for this precise reason that the French porcelain manufacture has been described by several authors and historians hitherto as being a regal and aristocratic affair, unlike the equivalent English business enterprise, which had no aristocratic patronage (Munger 2000). This is also the reason that the onset of the French Revolution and the Reign of Terror which followed it in 1794 created such a major effect upon the French porcelain industry with its decimation of the aristocracy, which was in part rectified by Napoleon's appointment of the esteemed Alexander Brogniart as Director at the Sevres Porcelain Manufactory in 1800, a position that he then retained for 47 years.
- Royal patronage and support for porcelain manufacture in France started with King Louis XIV at Rouen in 1673 and was followed later at St Cloud, Vincennes, Orleans and Sevres.
- As found in the history of the foundation of the English factories, the movement of key personnel between factories was also prevalent in France in the eighteenth century, which could have encouraged the verbal dissemination of practical knowledge concerning the processes and raw materials used for porcelain manufacture. This was perceived as unacceptable to King Louis XV, who decreed

that none of his workforce at the Royal Porcelain Manufactories, and in particular at Vincennes, were allowed to leave their employment there without his express permission and agreement, and if they did so against his wishes they would be prosecuted as criminals. As in the English businesses, the ownership of the French factories tended to pass down within families and upon the closure of one factory, the staff relocated frequently to alternative employment elsewhere at neighbouring and often rival sites.

- The first manufactory of soft paste porcelain in France at Rouen opened on the 1st October, 1673, where it made *porcelain francaise,* claimed as "*la veritable porcelain de Chinie*", the "true porcelain of China", believed to be encouraged by the *Compagnie Francaise des Indees Orientales* (The French East India Company) which was founded in 1664 to counter the activities of the Dutch VOC, who had the monopoly at that time of importing Chinese porcelains into mainland Europe (Dillon, *Porcelain*, 1904). The establishment of a porcelain manufactory at Rouen was the idea of King Louis XIV, who commissioned Edme and Louis Poterat to manufacture porcelain "*similar to the one brought from China*", which they commenced to undertake in their faience workshops (Lacombe 2006; Solon 1905; Barber, *Artificial Soft Paste Porcelain in France, Italy, Spain and England*, 1907). Rouen faience had already achieved an established reputation a few years before by their provision of 100 m^2 of faience tiles to the Trianon de Porcelaine near Versailles. Edme died in 1684 and Louis then started a new factory exclusively for porcelain manufacture, shrouded in secrecy "*fearing that his assistants would steal the secret*". Louis Poterat died in 1696 and porcelain production at Rouen then ceased, but in the meantime the creation of a delicate soft paste blue and white porcelain of a high translucency was successfully achieved, only surpassed much later by Sevres, even if the Rouen production was limited and it was not commercially viable because of the high firing losses (Froyssart 2007). The original composition comprised limestone, sandstone, ashes, salt and black soap. Black soap, also known as *Marseilles soap*, was a mixture of olive oil, soda and salt, and this would presumably have provided an alkaline fusion flux for the porcelain frit: the main production site of this component was the Camargue in Provence, utilising local raw materials. Its alternative name of "potassium soap" also implies that perhaps it also contained an infusion of potash from calcined Camargue kelp. Savary des Brulons, cited in (Froyssart 2007), wrote in 1722 in his *Dictionnaire Universal de Commerce*, that "*There are for 20 years made in France an attempt to copy the Chinese porcelain: the first attempts were successful in Rouen,, these objects in faience of the new types are not to be classified as French faience but are genuine porcelain invented by the French in recent years and have been successfully produced first in Rouen, Passy near Paris and then in St Cloud*". He also suggested that the Poterat family could have obtained the secret of successful porcelain manufacture at Rouen from elsewhere, pointing out that the recipe for the first European attempt at porcelain synthesis at the Court of the Medicis in Florence a hundred years earlier was still unknown, but in contrast citing that the porcelain production by John Dwight at the Fulham Pottery in England between 1671–1673 was actually well known to them! At this time, the detailed information concerning Chinese porcelain manufacture in Jingdezhen had not yet percolated through to Europe from the Jesuit priest father Francois Xavier d'Entrecolles, so the information for "elsewhere" could not sensibly have been emanating from China. The reference to John Dwight's factory at Fulham in England is therefore doubly interesting especially since earlier we have mentioned that Ehrenfriede von Tschirnhaus had been recorded as being a visitor at the Fulham manufactory around the same time.

- The porcelain manufactory at St Cloud started up in 1693 under the direction of Pierre Chicaneau and is credited with the production of the first commercially viable soft paste porcelain in France under the official patronage of Philippe, Duc de Orleans, who was awarded its patent in 1702, although by 1700 it is recorded that over 3200 pieces of St Cloud porcelain were already on its factory account books, so by then it was already a successful enterprise (Solon 1906a, b). St Cloud porcelain was said to have been "*as perfect as the Chinie*" with a pleasant, warm amber to ivory translucency (Honey, *European Ceramic Art*, 1952). The composition of the paste is reported as comprising sand, chalk, soda, alum, gypsum, a calcareous marl and an alkaline flux, very different from that of the Rouen manufactory which it displaced. Again, difficulties in kiln control restricted the production at St Cloud to smaller items, including snuff boxes, beakers, table wares and cutlery handles.
- At Chantilly, the factory sponsored by the Prince of Conde, Louis-Henrie de Bourbon, and founded in 1725, produced a high-quality porcelain said to be as good as that obtained from Meissen. Unlike other early factories which modelled themselves upon blue and white chinoiserie scenes, it produced copies of the polychrome Japanese Kakiemon wares which had been imported by the Dutch VOC during the turbulent times experienced in China at the end of the Ming Dynasty and before the stabilisation of the Qing Dynasty (Barber, *Artificial Soft Paste Porcelain in France, Italy, Spain and England*, 1907).
- Sevres, perhaps for many historians the epitome of French porcelain production, was under the direct patronage of King Louis XV, whose mistress, Mme de Pompadour, was instrumental in the absorption of the Vincennes factory at its closure and rehousing the synthetic china operation in a wing of her chateau under the astute direction of Chambellan Duplessis. Starting off with the manufacture of soft paste porcelain, the Sevres manufactory benefitted from the discovery of a supply of kaolin at St Yrieux-la-Perche, near Limoges, in 1764 by a Bordeaux chemist, Jean-Baptiste Dariet, resulting in a change-over to a hard paste porcelain paste in 1770. Soft paste porcelain was still made in small quantities at Sevres until 1804, when it thereafter ceased. The fortunes of the Royal Porcelain Manufactory at Sevres, unlike its contemporaries, actually prospered after the French Revolution when Napoleon appointed Alexander Brogniart to its Directorship in 1800, a position he held until 1847. An interesting record of the firing process at Sevres from these times is that the kilns were wood-fuelled, using specifically birch wood, and that a high temperature of 1300 °C was achievable in this process: the biscuit porcelain was fired at this temperature for 16 h, and a further 12 h was required for the glazing process, but the critical part

of the manufacturing process was the cooling stage which was controlled gradually and carefully over a period of 15–20 days. Only 100 pieces were stocked in the kiln for each firing batch, which might explain the very high cost of Sevres porcelain commercially even then: the comparison numerically and economically with the dragon kilns at Jingdezhen, which could fire approximately 25,000 items of hard paste Chinese porcelain at 1400 °C in a single batch process is illustrative.

7.2 Paste Compositions of the Early European Porcelain Factories

It is instructive to compare the recipes and ingredients utilised by the earliest European porcelain manufactories: namely, those of Rouen, St Cloud, Meissen, Fulham, and with the major Chinese porcelain factory at Jingdezhen included for comparison, are listed in Table 7.2. It can be seen immediately that the proposed idea by some historians that there must have been an exchange of knowledge, voluntarily given or desperately sought as a result of strategic industrial espionage, is clearly not sustainable here. For a start, both French factories, Rouen and St Cloud, used a sand and sulfatic mineral composition with the addition of calcite in the form of chalk, marl (a mixture of clay, mud and limestone) or limestone. In contrast, those of Fulham and Meissen are both silicate-based and sulfate–free, but these again are sensibly different in their individual composition. Finally, the Chinese

Table 7.2 Composition of porcelain bodies in the earliest European factories (and Chinese for comparison)

Factory	Country	Foundation	Composition	Formulation
Rouen	France	1673	Limestone, sandstone, ashes, salt,	$CaCO_3$,SiO_2
			Black soap	Na_2CO_3,K_2CO_3
				NaCl, organics
St Cloud	France	1693	Sand, soda, alum, gypsum, chalk, marl	SiO_2,Na_2CO_3,
				$CaSO_4$,K_2SO_4
				$CaCO_3$,$Al_2(SO4)_3$
Fulham	England	1671	Sand, clay, lime, potash	SiO_2,$Al_2(SiO_4)_3$,
				Na_2SiO_4,CaO,
				K_2CO_3
Meissen	Germany	1709	Clay, gypsum, alabaster, sand	SiO_2,$CaSO_4$,
				Na_2SiO_4,
				$Al_2(SiO_4)_3$
Jingdezhen	China	1400	Kaolin, petuntse	Na_2SiO_4,$Al_2(SiO_4)_3$
				Al_2O_3,K_2SiO_4,
				SiO_2

recipe from Jingdezhen as revealed by Father Xavier d'Entrecolles is totally different again in its use of the china stone or feldspar-like *petuntse.* It is realised, of course, that the revelation of the Chinese recipe in the first quarter of the eighteenth century could not temporally have had any impact upon the foundation recipes of these European factories so their porcelains cannot be strictly compared like for like. However, the broad distinction that all four European porcelains are *soft paste* and the Chinese analogue a *hard paste* must still apply. This, of course, now brings us to the next question which has created somewhat of an impasse historically: which European manufactory should be credited with the invention of porcelain? Most, if not all, historical accounts have defined the Meissen factory as the originator of porcelain synthesis by Ehrenfriede von Tschirnhaus and Johann Bottger in Europe in or around 1709 (Honey, *Dresden China: An Introduction to the Study of Meissen Porcelain,* 1934). However, the reasoning behind this allegation is now obscure and perhaps, therefore, a re-examination of the historical documentation is now required to clarify the issue.

In essence, in 1671, John Dwight at the Fulham Pottery in England first manufactured a "porcelain" in Europe, followed by a more successful operation at Rouen in France in 1673, both of which were not viable commercially for the production of porcelains at the scale required by a demanding clientele – that was left initially to the manufactory at St Cloud in France and then to Meissen in Saxony, which clearly provided the best commercial basis for the European manufacture. However, all were soft paste porcelains in contrast to the Chinese hard paste porcelain analogue imported from Jingdezhen and because of this they were rather deprecatingly perhaps assigned the term "*artificial porcelains*" (E.A. Barber, *Artificial Soft Paste Porcelain – France, Italy, Spain & England,* 1907). This in no way diminishes the standing and viability of Meissen porcelain in its successful objective to provide a European alternative to the Chinese imports, but to simultaneously downplay the importance of the contemporary early French and English factories is surely no longer acceptable historically. It is equally incorrect to assume that the French porcelain industry required the input of Xavier d'Entrecolles' information about the Chinese recipe before they could go into full scale production and it is equally incorrect to ascribe the perceived time-lag in the English operation to its late start up and wake-up call occasioned by William Cookworthy's discovery of a local source of kaolin in Cornwall in 1747. The very obvious result of Xavier d'Entrecolles' revelation of the secret ingredients of Chinese porcelain was the polarisation of porcelain production towards hard paste bodies, which set Meissen and Sevres along their successful paths, particularly with the discovery of the local sourcing of kaolins in France and in Saxony. Curiously, this did not happen in England, where the deviation towards the manufacture of glassy porcelains occurred, which included a significant proportion of cullet in their composition as seen at Chelsea and Derby, and then calcined bone ash, which eventually created the "English standard" bone china of the early nineteenth century, such as Spode and Rockingham. It seems quite incongruous, therefore, to maintain that there was a free-flow of information sharing between early English and Continental porcelain manufactories when in fact their porcelain paste body compositions were so different and varied. Rather, it is better to realise that there were several operations which did in fact produce acceptable porcelains or porcellaneous items between about 1670 and 1710: that Meissen still holds the accolade of the "first European porcelain" is equally debatable historically, even though its commercial enterprise was perhaps the front runner before the others commercially. All porcelains are synthetic materials of varying compositions and to describe the European versions as "artificial" in comparison with the Chinese "true" porcelains is not accountable either as this implies that the Chinese porcelains are somehow "natural" and that perhaps the European examples are fake?: On the basis of the historical evidence seen thus far an alternative regime for porcelain production in Europe may be the descending series: Fulham > Rouen >St Cloud>Meissen, with Meissen still being accorded the premier position of the temporally first successful European commercial production facility, although records at St Cloud definitively indicate that several thousand pieces were made and sold there some two decades before Meissen – so how "commercial" does one need to be to score?

7.3 Summary and Conclusion

In summary, therefore, this digression from our entitled theme of English and Welsh porcleains into relevant examples of early European porcelains has been instructive in that it is now apparent from historical documentation that not only did the synthesis of a porcelain-like or porcellaneous material commence much earlier than was first thought in England but that the interchange of recipes and ingredients between the rival sites of manufacture in England and France seemed unlikely to have occurred, and that this is supported from the analytical and compositional evidence. It is interesting that the intense secrecy accorded by the Chinese towards their porcelain production was mirrored also in Europe, as the activities of Kings Louis XIV and XV towards the preservation of their industrial knowledge in the porcelain factories that were fortunate to be accorded their Royal patronage demonstrate. Nevertheless, this author can find no evidence from the early literature that a deliberate obfuscation was practised by porcelain manufactory proprietors in this respect, as was evidently undertaken by their compatriots in the fine art field of oil painting pigments, where the advice and recommendations given by prestige painters on the selection of pigments and their composition for art usage was perhaps deliberately and frequently proposed to be misleading. Such a case is demonstrated by the recipe given by the renowned limnologist to the Tudor Court, Nicholas Hilliard, who divulged his recommendation for the pigments used in his execution of the famed "*Armada Jewel*" (also known as the *Heneage Jewel*), which depicted Queen Elizabeth I reigning over the destruction of the Spanish Armada invasion fleet in 1588: Hilliard in contemporary documentation advocated the use of several pigments of the highest quality which he claimed that he utilised for this work (N. Hilliard, *The Arte of Limning*, repr. 1992), but spectroscopic analysis has revealed that these were never used for this purpose and it has now been suggested that his publication of these details could have been designed possibly as a subterfuge to mislead his rivals and competitors (Derbyshire and Withnall 1999; Edwards 2004)!

References

E.A. Barber, *Artificial Soft Paste Porcelain – France, Italy, Spain & England* (Hodder & Stoughton, London, 1907)

A. Derbyshire, R. Withnall, Pigment analysis of portrait miniatures using Raman microscopy. J. Raman Spectrosc. **30**, 185–188 (1999)

E. Dillon, *Porcelain* (Methuen Books, London, 1904)

H.G.M. Edwards, Probing history with Raman spectroscopy. Analyst **129**, 870–879 (2004)

C. Froyssart, Porcelaine de Rouen, 1673–1696, in *Porcelaine Tendre de Rouen Provenant de la Collection Wenz,* Article 1791, 2007

N. Hilliard, *The Arte of Limning*, reproduced by eds. R.K.R. Thornton and T.G.S. Cain (Carcanet Press, Manchester, 1992)

W.B. Honey, *Dresden China: An Introduction to the Study of Meissen Porcelain* (Faber & Faber, London, 1934)

W.B. Honey, *European Ceramic Art* (Faber & Faber, London, 1952)

C.S. Lacombe, L'apparition de la porcelaine tendré a Rouen chez le Poterat, l'hypothese protestante? Revue de la Socitete des Amis du Musee Nationale de Ceramique Sevres **15**, 29–35 (2006)

J. Munger, French porcelain in the 18th Century, in *The Heilbrunn Timeline of Art History*, (Metropolitan Museum of Art, New York, 2000)

M.L. Solon, The Rouen Porcelain. The Burlington Magazine for Connoisseurs **7**, 116–124 (1905)

M.L. Solon, The St Cloud Porcelain, Part I. The Burlington Magazine for Connoisseurs **10**, 24–28 (1906a)

M.L. Solon, The St Cloud Porcelain, Part II. The Burlington Magazine for Connoisseurs **10**, 89–96 (1906b)

Chapter 8
The Role of Analytical Data in the Holistic Interpretation of Porcelains

Abstract The incorporation of analytical data into a holistic assessment of porcelain identification in a forensic sourcing and provenancing exercise is not without difficulty because of two avenues: firstly, the potential uncertainty that can be generated by the analytical data interpretation relating to overlapping measurements of compositional percentages and, secondly, the possible conflict that it engenders with an already established expert opinion regarding the origin of the unknown piece being investigated. In the ceramics field, cases where the analytical evidence goes against the established findings and conclusions of art expertise and connoisseurs relating to ceramics sometimes brings into question the relevance of the analytical data. This situation is in direct contrast with the high regard in which the analytical interrogation of old master oil paintings is now held: in fact, it appears that potentially valuable oil paintings intended for auction in the fine art world should now have the credentials of scientific analysis in support of their expert opinion for authentication. The reason for this dichotomy is not straightforward but it could be explained by the fact that significantly fewer pieces of porcelain have actually been analysed compared with the analogous oil paintings, where it is accepted that both destructive and non-destructive analytical techniques are necessary to provide the wherewithal for their identification. Indeed, during the restoration and conservation of an oil painting for public display it is quite normal that small samples are excised from the artwork for the determination of the pigment and binder composition, so affording the possibility for microanalysis, a situation which is clearly not strictly comparable with modern porcelain analyses as recounted here. Several case studies are presented here where analyses have raised some questions about the sourcing of early porcelains and in some case confirm, and in others differ, from expert opinions expressed by connoisseurs. The advent of improved methods of non-destructive qualitative and quantitative analytical testing of porcelains is awaited which could therefore herald a new era of acceptance of scientific analytical testing procedures in ceramics.

Keywords Analytical methods · Destructive and non-destructive testing · Controversial results · Exposure of fakes · Reliability of marks · Use of sourcing protocols

H. G. M. Edwards, *18th and 19th Century Porcelain Analysis*,
https://doi.org/10.1007/978-3-030-42192-2_8

8.1 The Knowledge Base and the Compatibility of Analytical Data and Expert Opinion

The major theme of this book has been the evaluation of not only the analytical data which have been derived from a wide range of porcelain specimens from English and Welsh factories which originated in the eighteenth and early nineteenth centuries, but also their interpretation within the knowledge base of information which is currently available from available historical records and comparison with the associated documentation about the operations of the manufactories: this can be described as the *forensic holistic analytical approach*. The analytical data can be subdivided into the individual elemental and molecular spectroscopic from combined SEM/EDAXS, XRF and Raman techniques, which therefore provides a complementary assessment of both the elements and the molecules and molecular ions present in the specimens studied. The interpretation of the analytical results can be undertaken from first principles without resorting to the parallel historical documentary information existing about each factory, which may encompass the raw materials used at various stages in the production operations and their sourcing, the presence of impurities, any strategic modifications introduced by the proprietors in their attempts to improve the quality and robustness of their fired porcelain bodies and details of the pre-processing stages that were undertaken to ensure the adequate fineness of preparation of their proto-porcelain frit. Whilst it would be tempting to suggest that these two avenues are actually, therefore, derivatives of two independent *hard* and *soft* science approaches, in fact, of course, one should technically and realistically feed into and inform the other and, surely, a definitive recipe detailing the raw materials used and their quantities specified for the porcelain production undertaken during a specific factory period is as *hard* a piece of scientific information as one requires to establish the characteristic formulation pursued by a particular factory. In fact, this documentary evidence could be deemed to be as *hard* as the analogous analytical data which are derived from the quantitative elemental oxide analyses of the completed fired porcelain items. This statement presupposes that there is an observed agreement between the two sets of analytical and historical recipe formulation data – and, indeed, then no problem should be envisaged – but what happens when the analytical elemental oxide data are found to be in obvious disagreement with some aspect of the formulation recipes or raw materials used for the particular factory concerned according to the historical documentation? The reality of such a situation is that an assessment must then made realistically about the likelihood that a potentially unrecorded change in raw material composition was instituted in the manufactory which gave rise to the significant changes in elemental oxide percentages that were later detected in the analytical experiments. Whereas this would seem to address the area of disagreement between the recorded recipe and the actual formulation and the analytical data, however, in severe cases of mismatching between the analytical data and the historical record some questions may then naturally arise as to the perceived genuineness, or otherwise, of the specimen being analysed (the analyte) and its consequent attribution to the factory concerned based upon the derived analytical information. In other words, has the analytical data exposed the presence of a *fake* piece of porcelain, or alternatively, one which has been mis-attributed to that particular factory but which properly really belongs elsewhere: as the recorded attribution has normally been achieved in the first place by an expert opinion, this then brings into sharp focus the potential disagreement between analytical science and expert opinion or connoisseurship? The interpretation of the analytical data hence constitutes the focus and area under dispute between an analytical approach and that afforded by an expert connoisseur who may have already have used some available historical information in their judgemental conclusion, which may have also been open to a mis-interpretation, to arrive at an objective assessment.

8.1.1 The Analytical Dilemma: A Contretemps Between Analytical Attribution and Expert Opinion

This theme highlights the situation which can arise when analytical data presented in evidence for the attribution of a piece of porcelain are in disagreement with the expert opinion expressed by ceramics connoisseurs. The analytical data are derived from experiments carried out on intact pieces of porcelain, fragments thereof, broken or damaged items or from shards stored in museum archives and excavated from waste tips located on the sites of porcelain manufactories. In contrast, expert opinion is usually based upon many years of study and an accompanying expertise acquired through the handling and research into the specific products of a particular factory, encompassing the shape, body texture, glaze, potting characteristics and style of decoration that characterise a particular factory's output. A third category, which transcends both the scientific analysis from the acquired experimental data and the expressed opinion derived from expert connoisseurship, is provided by the existence of relevant historical documentation: the collation of historical letters and correspondence, the diaries and life histories of factory proprietors, surviving factory workbooks and pattern books is a mine of information but only if it has been properly accessed and interpreted correctly. To summarise, these three categories of evidential information and their reliability in a forensic scientific context can be described as follows:

1. *Scientific evidence*: analytical data and their interpretation – *hard* evidence.
2. *Connoisseurship*: an *opinion* based upon the experience of a particular factory's product by an established expert in the particular products of a factory – *soft* evidence.
3. *Historical documentation*: writings which detail what porcelain was made, and when, during the operations of the factory, which can be open to potential misinterpretation by researchers, especially in accounts which rely upon the opinions and statements of earlier historians, and those from possible eye-witnesses and statements made by contemporaries which do not necessarily conform with the

original documentation from other sources. This must really be considered as neither *hard* nor *soft* evidence as it is open to different historical interpretations which may have been expressed for the same factual statements or data. Certainly, the existence of raw materials composition recipes and details of manufacturing and processing procedures noted by factory proprietors for their particular periods of production must also be considered as *hard* evidential information, but modern analysts need to exercise caution in relying upon historical accounts of factory operations which are based upon hearsay or the casual interpretation of work diary entries, which may have been expressed in code (such as those written in the workbooks of Lewis Dillwyn at Swansea), or in a form of shorthand, which can be further compromised by use of the incorrect nomenclature for different raw materials as highlighted earlier.

For a truly *holistic* approach to an attribution exercise for factory porcelains in all three categories enumerated above should be employed but not necessarily on an equal evidential footing, as this would imply that the information provided from each one individually was precise, infallible and unchallenging and independent of the others in terms of the interpretation of the facts that generate the attribution of the item of porcelain under consideration to a selected Factory X, or not. The key to the eventual acceptance of any set of analytical data invariably presumes that the shard or fragment concerned can be unequivocally assigned to a particular factory, i.e. that the elemental oxide data fall into the range of values expected for products expected for Factory X and, if this is not the case, then the clear conclusion is that the piece analysed then represents a new type of porcelain body for that factory, or alternatively that it is an interloper acquired from another factory, perhaps even as "grog" or as a competitor's item purchased for study at that factory. Expert connoisseurship is based upon the experiential evidence of pieces handled hitherto, which may not of course comprise an example of the novel type exposed by the scientific analysis – in which case the expert opinion may be a negative one and the particular piece under consideration is then most likely rejected as a true *bona fide* product of Factory X. Therefore, in cases where the analytical data and the expert opinions are divided regarding the attribution of the source origin of a piece of porcelain, there will be an impasse created: invariably, a definitive judgement cannot then be forthcoming about the porcelain artefact, which is then usually left in abeyance, awaiting further relevant information relating perhaps to an undiscovered provenance or perhaps even the fortuitous discovery of a hitherto unknown marked service of a similar pattern pertaining to that factory. On several occasions where this situation has occurred, the resultant judgement is naturally taken in favour of the expert connoisseurship, and the specimen is thereby accorded an "unknown" or doubtful status. However, as can be related here, not always is this judgement fully substantiated by all expert opinion, where an individual disagreement between several experts upon the attribution of a rare porcelain piece of perhaps an unusual construction can then be equally doubtful for its full acceptance into the ceramics genre. This certainly occurs in the field of fine art and oil paintings where disagreement between expert opinions as to the artist are quite frequently encountered, especially when it comes down to an attribution to the master or one of his pupils working in his studio under his direction.

In most cases hitherto, the porcelain factory products have been classified and identified solely on the basis of texture, decoration, shape, translucency and overall appearance, by ceramics experts and connoisseurs, who nevertheless may not always agree with each other's opinions regarding the attribution of a particular specimen to a nominated factory. This difference in opinion usually then results naturally in a doubtful assignment of attribution, especially where little other provenance exists which supports the controversial attribution one way or the other. Such cases should clearly be resolvable, at least in theory, by the input of analytical data such as those presented in this current study. However, in general, the situation could be rendered even more complex by the non-acceptance by connoisseurs and the ceramics community of an unusual or non-standard set of analytical data as belonging to a porcelain specimen from a particular factory, and then the piece would still be accorded its "doubtful" or "possible" label: in such a case, of course, the interpretation of the controversial analytical data needs to be reviewed scientifically, objectively and verified independently, if possible by input from several techniques. This is quite commonly found to occur in the attribution of porcelain specimens which are finished, rare and not therefore favoured for the unique or multiple sampling required by some analytical techniques: in contrast, there is usually no problem in accessing a relatively large number of porcelain shards from factory site excavations for destructive analysis, but it is a very different matter when it comes to solitary museum exhibits and collectors' items, which may themselves be rather rare and perhaps also very fragile, which would negate their removal from museum storage for transfer to an analytical laboratory. Hence, although we have seen that for the porcelain factories discussed here several hundred shards have been subjected to elemental oxide analyses, only a relatively few and accordingly very precious, finished porcelain pieces have actually been analysed similarly, perhaps only just ten or twenty in number, representing perhaps then a mere one percent at most of the total number of porcelain specimens studied analytically and reported in the literature.

In several well-known cases, such as the recent studies published on the early Bow porcelains (Ramsay and Ramsay 2008), the elemental oxide analytical data does not correlate at all well with the established expert attribution of a porcelain artefact given by inspection of the body paste texture, shape, pattern and decoration alone as highlighted above. It is fair to say that this generally results in a perhaps a rejection of the analytical data interpretation by the ceramics community in favour of the accepted and established expert opinion and perceived wisdom, and the debate which follows can result in an impasse regarding the confirmation of the attribution by the scientific analyses. This is particularly prevalent with the assignation of very early porcelains which by their very nature were experimental in concept and for which the analytical data may not conform to acceptable standards. Another problem arises when it is appreciated that many rare specimens of early English and Welsh porcelains were not marked at all at the factory: at the Nantgarw China Works for example only the flatware was generally marked with the factory impressed mark, namely NANT-GARW C. W., and most of the remainder of the same service was unmarked (Edwards, *Swansea and Nantgarw Porcelains: A Scientific Reappraisal*, 2017b). Much, if not all, of the ornamental porcelain output of many factories, including that of Nantgarw cited above, was unmarked and is therefore open to a potential mis-attribution. This inspired Dr William John to urge collectors never to destroy even badly damaged but marked Nantgarw and Swansea items from a porcelain service in favour of the retention of their perfect but unmarked analogues in their collections, as posterity would welcome the ability to correlate and match the patterns and structural types from even badly damaged specimens at a later time (John, *Nantgarw Porcelain*, 1948; *Swansea Porcelain*, 1958). In the same vein, in the absence of surviving factory work books and pattern books from many sites, the accumulation of data regarding existing patterns from a factory output is a valuable exercise for the future assignment of unknown pieces to that source. An example of this approach is the classic and highly important scholastic work of Jones and Joseph (*Swansea Porcelain: Shapes and Decoration*, 1988), which is dedicated to the illustration of all known examples of the Swansea China Works set pieces known at that time, with pattern numbers where available, including detailed measurements and cross-sectional drawings of items and the depiction of the various mouldings for the cups and plates used. Yet, it must be said that from the pattern numbers in the numerical range 1–705 recorded for the Swansea factory from a large number of examples of porcelain dinner, dessert, coffee, tea and breakfast services studied, only 67 pattern numbers can now be listed and matched with a surviving pattern (Edwards, *Nantgarw and Swansea Porcelains: An Analytical Perspective*, 2018). less than 10% of the recorded numerical output. This surely affords an opportunity for many genuine Swansea porcelain pieces to have escaped a definitive identification or to have been mis-ascribed in the past and it could be sensibly proposed that the role of an analytical protocol to be adopted for their supportive identification is perhaps now long overdue. In other factories, such as Rockingham (Cox and Cox, *Rockingham Porcelain 1745–1842,* 2001) and Derby (Twitchett, *Derby Porcelain: An Illustrated Guide, 1748–1848*, 2002), concise accounts are provided of known patterns derived from the existing pattern books and, in the case of the Derby factory, comprehensive listings of assigned pattern numbers have been provided for dessert/dinner services, tea services and even ornamental figures, along with useful associated information about the painters and gilders assigned to these patterns, where these were available. Generally, however, even these factory pattern number listings have large gaps (Edwards, *Derby Porcelain: The Golden Years 1780–1830*, 2017a). For example, the Derby plate book has pattern numbers ranging from 1 to 408 with 123 pattern numbers missing (30%) and the Derby tea book has pattern numbers ranging from 1 to 770 with 89 pattern numbers missing (12%) (Edwards, *Derby Porcelain: The Golden Years 1780–1830*, 2017a). Hence, there must be a considerable number of Derby porcelain artefacts in circulation which have a factory mark but not accompanied by a pattern number or, alternatively, unmarked pieces could exist with neither the identifying factory mark nor a pattern number, which then formally would generate the appellation of

Derby(?) from stylistic considerations only and these may well be copies made at other factories.

However, even for a factory such as Derby, where the original pattern books still exist in museum curation the list is still incomplete significantly, comprehensive though it appears to be superficially: for the Derby China Works figures, which were much esteemed for their modelling in biscuit porcelain and in glazed and decorated wares during the late 18th and early nineteenth centuries, the work book listings are also comprehensive but, alas, are also incomplete. The Derby China Works figures listing commences at No.1, *Group of the Virtues*, and is complete at No 390, *Group of Gaultherus and Griselda*: however, these books also list some 45 unnumbered figures which were copied from analogous Bow and Chelsea models, 48 more unnumbered figures modelled by Edward Keys, and a further 18 others that were published and advertised in the records but are not included in the numbered listing, this it is reasoned possibly occurring after a date of 1830 (Twitchett, *Derby Porcelain*: *An Illustrated Guide, 1748–1848*, 2002). Also, a hidden difficulty with such a numbered listing arises from the descriptor not precisely matching the corresponding number in the list. For example, we find in this category the rather fine and unique biscuit, unglazed figure (Bradshaw, *English Porcelain Figures of the 18th Century*, 1981) illustrated in Fig. 1.1 depicting "*Flora*" in a classical pose of "*An Opera Girl in Paris*" wearing a late Georgian period dress resting against a ruined Greek pillar and carrying a garland of flowers which is beautifully crafted by Joseph "Jockey" Hill, whose rebus comprising a triangle is impressed on the base along with the crown and crossed batons of Derby and the number *No. 390* in cursive script. This very fine porcelain figure is genuine Derby biscuit porcelain but certainly does not match the listed descriptor of *"Gaultherus and Griselda"* given in John Twitchett's list obtained from the Derby work books ... hence, the reliability of such a list is immediately drawn into question. A further problem is that two existing separate and apparently authoritative lists of figures appearing in the Derby literature actually have serious discrepancies in this respect: an earlier listing of figures produced by John Haslem in 1876 (Haslem, *The Old Derby China Factory*, 1876) has several distinct discrepancies with the list produced by John Twitchett (Twitchett, *Derby Porcelain*: *An Illustrated Guide, 1748–1848*, 2002). An example of this mysterious situation is provided by a similar Derby biscuit figure (Fig. 8.1) from the same period, *ca.* 1790–1795, which has been described alternatively as *"The Gardener's Companion"* (Twitchett, *Derby Porcelain*: *An Illustrated Guide, 1748–1848*, 2002) and as *"Spring"* (Haslem, *The Old Derby China Factory*, 1876) depicting a girl in Georgian dress with a bonnet carrying cut flowers. The base of the figure is incised with the cursive Derby *D*, with crossed batons and crown, and the pointed star rebus of Isaac Farnsworth and *No.361* incised on the base. Finally, to complicate matters even further, John Twitchett (*Derby Porcelain*: *An Illustrated Guide, 1748–1848*, 2002) lists a further 500 + unnumbered ornamental and figurine china moulds which formed part of the estate of William Duesbury II, proprietor of the Derby China Works, in 1795 and documented afterwards by his employees, Messrs. Soar, Longdon, Farnsworth and Hardenberg, which do not appear in any other list of Derby figures at all.

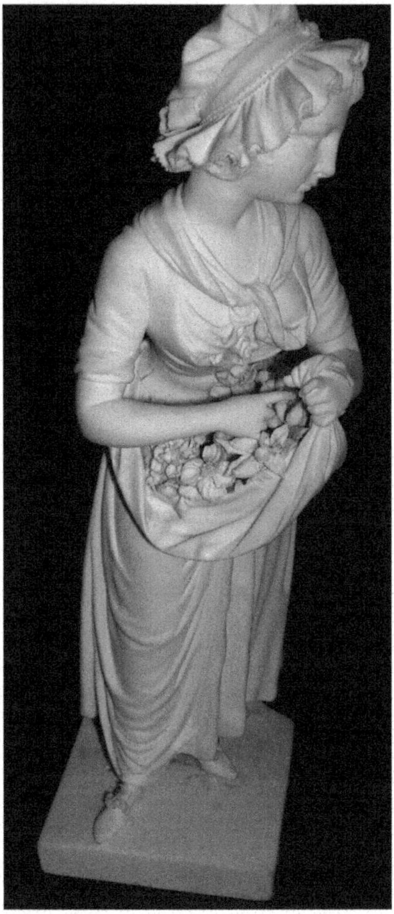

Fig. 8.1 Derby porcelain, *ca.* 1790–1795, unglazed biscuit porcelain figurine depicting a girl in Georgian dress and a bonnet carrying cut flowers in her apron. Height 12″ (30 cm). This figure is beautifully crafted by Isaac Farnsworth, whose rebus comprising a five-pointed star is impressed on the base along with the crown and crossed batons of Derby and the number 361 in cursive script. This figure appears in the Derby figure record listing (Twitchett, *Derby Porcelain, 1748–1848: An Illustrated Guide,* 2002) as *Spring* or alternatively as the *Gardener's Companion* in Haslem's figure listing (Haslem, *The Old Derby China Factory*, 1876). It was priced originally in biscuit porcelain at £2. 2s. 0d. and stated by John Twitchett to be extremely rare. (*Private Collection*)

What this means, therefore, is that even when authoritative and apparently unimpeachable documentary factory records exist about their products, the reality is that during its manufacturing history on occasions some departure could have been made from an assigned factory venture or commission: we should always be aware of this analytically and historically and ideally we should reinforce the historical statement or assertion with a definitive analytical scientific comparator, i.e. the holistic approach should be employed *de rigueur* and the reliance of an attribution which has been made upon one sole aspect to the detriment of the others should not be over-compensated in forming an apparently reliable assessment which could otherwise be deemed bogus.

So, it appears that the potential solution to doubtful attributions could lie in the carrying out of destructive or non-destructive analyses on shards and, of course, if achievable, on the non-destructive analyses of perfect specimens of finished porcelain, especially those which have been accepted as being unambiguously representative through time of a particular factory for the establishment of an irrefutable database of analytical signatures. In this respect, the analyst and connoisseur both need to exercise caution as forensically the presence of an applied painted factory mark does not necessarily indicate that the piece is totally authentic: the presence of an impressed or an incised mark is rather different as this cannot be easily forged at a later stage into a fired and completed, glazed piece of porcelain from another factory. Case studies have appeared in the literature where genuine painted factory marks have been duplicated on porcelain interlopers (so creating *"fakes"*, thereby) and several examples could be cited to indicate the extent of this particular problem. Even today, pieces are offered at auction which bear a *Nantgarw* enamelled cursive painted script mark, which was apparently never used at the factory: a discussion of some of these fake marks is provided in Edwards (*Nantgarw and Swansea Porcelain: An Analytical Perspective,* 2018). This aspect again accentuates the need for a holistic approach in the attribution of the source of a piece of porcelain, whether this involves the use of analytical data or not, and several previous authors have cautioned potential purchasers of rare and unusual porcelain pieces from particular factories against the acceptance of a factory mark alone on an otherwise rare piece as dictating the sole guarantee of authenticity. In the mid- to late-1800s, the firm of *Samson et Cie* in Paris manufactured hard paste porcelains in simulation of some of the earlier then defunct English and Welsh factory shapes and designs which were already commanding premium prices from a clientele at auction sales, and these sometimes bore an appropriate red- or puce- enamelled stencil mark from *Samson et Cie* identifying them as such. However, unscrupulous traders have sought to remove the *Samson et Cie* marks and substitute those purporting to be from the original factories for financial gain at the point of sale, so causing much confusion and suspicion amongst collectors, museum curators and ceramics historians. The Samson products were generally more flamboyant in style and arguably quite out of period in character when compared with the genuine articles from 50 to 60 or more years previously, and experts and connoisseurs have generally been able to identify these appropriately on design, style and body texture, but occasionally one or two have slipped through the net and these have then been hailed as very rare examples of the original factory genre. On occasion, a *pastiche* has even been created, whereby one genuine piece from a factory, perhaps even marked appropriately and correctly, has been amalgamated with another from a different source factory and refired to create a new piece of a unique shape or design, which is then claimed to have originated from the original source factory and even bearing an original factory mark in support of its originality! Clearly, the input of novel analytical information could determine the issue of whether or not a particular piece of porcelain in these circumstances was a genuine example, or perhaps a *pastiche* or a *fake,* especially exposing the instances of where two genuinely antique pieces of porcelain have been fused together in a "marriage", as has been found to occur with antique furniture, where the offending piece can then be termed properly as being "associated". The eminent ceramic historian Ernest Morton Nance, who spent 40 years of meticulous ceramics research on his seminal volume *The Pottery and Porcelain of Swansea and Nantgarw* (Batsford 1942) amassed a small collection of fake Nantgarw porcelain he had acquired along the way, including several pieces which had fooled even his expert eye originally and which were exposed as such only as his own experience increased: these pieces now reside in the National Museum of Wales, Cardiff, including items with a script mark *Nantgarw* enamelled in red, several of which were later re-classified properly as probably Staffordshire, Coalport and Samson products (Edwards, *Swansea and Nantgarw Porcelain: A Scientific Reappraisal*, 2017b).

Some case studies will now be cited where analytical data have successfully provided some new information about eighteenth and nineteenth century porcelains which are sometimes found to be at deviance with the existing established expert opinion, and then, on the other hand, we shall consider the contrary position where there is a supportive agreement found between the information obtained from the analytical data, from expert opinion and connoisseurs and from established historical opinion in confirmation of the porcelain factory source.

8.2 Case Studies in Porcelain Identification Where the Analytical Information Is at Variance with Established Expert Opinion

The case studies presented here are derived from analytical evidence which is quite unimpeachable in its interpretation but which raises some serious potential controversies in conflict and deviance with established thought and existing expert opinion.

8.2.1 The Curious Case of Hard Paste Nantgarw China

Hitherto, it has been accepted without question historically that the Nantgarw China Works under the proprietorship of William Billingsley, Samuel Walker and William Weston Young between 1817 and 1820 made a soft paste phosphatic porcelain of the highest quality, much appreciated for its clear translucency, whiteness of surface glaze and beautiful enamelling decoration. As recounted earlier, it is now believed that because of poor kiln temperature control around 1400 °C at the biscuit firing stage, the kiln firing losses were a staggeringly high 90% and this eventually rendered the manufacturing operation nonviable commercially. Several statements made historically have fuelled the idea that Nantgarw china manufactured during the lifetime of the factory had only one formulaic composition and recipe: for many years, during the operation of the Nantgarw China Works, and subsequently following its closure, this recipe was never revealed publicly, until in 1847 John Taylor (*The Complete Practical Potter*, 1847) published details of the Nantgarw body paste recipe apparently provided by Samuel Walker nearly 20 years after the death of William Billingsley (Edwards, *Nantgarw and Swansea Porcelains: An Analytical Perspective,* 2018). This was believed to reflect the one and only body composition for the Nantgarw China Works and generated the following remarks which were made at the time with varying degrees of confidence, but which have all been assumed to be completely correct hitherto:

1. Billingsley, Walker and Young had only about 18 months active production ongoing at the Nantgarw China Works and they would not have had the time to create another porcelain body experimentally (Turner, *The Ceramics of Swansea and Nantgarw,* 1897).
2. At the Centenary Exhibition of Nantgarw and Swansea china held at the Glynn Vivian Art Gallery in Swansea in 1914, Herbert Eccles pronounced that all the Nantgarw porcelain exhibited had the same and identical porcelain body compositionally, without exception, unlike Swansea china which had three identifiable porcelain bodies, namely glassy, duck-egg and "trident" porcelain. This statement was made on the basis of visual observation and inspection only.
3. Llewellyn Jewitt (*The Ceramic Art of Great Britain from Prehistoric Times to the Present Day*, 1878) concluded that at the premature closure of the Nantgarw China Works in 1820 upon the departure of Billingsley and Walker for Coalport, all materials and hardware were sold at auction and the kilns and pottery proving sheds were dismantled, the site then being "abandoned". The inference was then made quite naturally that there could be no further porcelain manufacture thereafter undertaken at Nantgarw, which in hindsight was an incorrect assumption even if it proved to have been the case (see Appendix III for a more detailed discussion surrounding this event and following the departure from Nantgarw for Coalport of William Billingsley and Samuel Walker). All is therefore not as clear as it purports to be historically.
4. Dr William John (*Nantgarw Porcelain,* 1948) stated that he had personally examined some 5000 pieces of surviving Nantgarw porcelain over many years research and there was in his opinion just one body composition.
5. Ernest Morton Nance (*The Pottery and Porcelain of Swansea and Nantgarw,* 1942) concluded that there was only one porcelain body used at Nantgarw, echoing the opinion of Herbert Eccles which was made some thirty years previously, namely, a soft paste phosphatic porcelain that was very rich in bone ash.

With the weight of this written documentary evidence from respected researchers over a period of 70 years or more it is quite clear how this belief in "one Nantgarw body" arose and that the demolition of the defunct Nantgarw China Works followed very quickly upon its closure and impending auction sale in 1820. However, the first seeds of doubt historically appeared when Isaac Williams (*Excavations at the Nantgarw Site*, 1932) undertook an archaeological excavation at the site of the Nantgarw China Works and found that the potting buildings and three kilns were, in fact, still standing and he discovered and identified the waste pit some thirty feet from the east wall of the still-standing potting shed which upon excavation contained china shards and porcelain fragments still buried within their stratigraphic and undisturbed archaeological context. Moreover, the kilns which had been definitively stated by Llewellyn Jewitt to have been demolished and purportedly removed to the Coalport China Works by John Rose in 1820 were still obviously standing! Recent elemental and molecular spectroscopic combined analysis of several shards

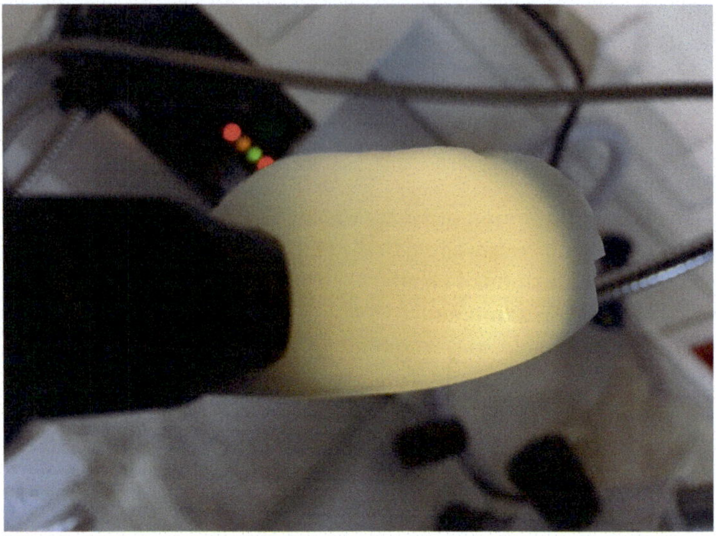

Fig. 8.2 Nantgarw shard NG6 in transmitted light showing its excellent translucency. (Courtesy of the Nantgarw China Works Museum, Tyla Gwyn, Nantgarw. Analysis reported in Colomban, Edwards and Fountain, 2020)

(Colomban et al. 2020) from the Nantgarw China Works site corresponding to the period 1817–1820 have revealed the presence of a large glazed shard, NG 6, of exceptionally good translucency (Fig. 8.2), which exhibits clearly the characteristics of a hard paste non-phosphatic porcelain, intermingled with many more shards which analyse as "standard" soft paste phosphatic porcelain, as would have been expected from the Nantgarw recipe provided in 1847 by John Taylor (*The Complete Practical Potter*, 1847). The SEM/EDAXS analytical details for this unusual Nantgarw shard (numbered NG6) as reported by Colomban, Edwards and Fountain (2020) are as follows:

NG6: 80% SiO_2, 13% Al_2O_3, 0.3% CaO, 0% P_2O_5, 1.2% K_2O, 2.6% Na_2O, 3.1% MgO and 0.1% Fe_2O_3. Total: 100.3%

The analytical data interpretation is therefore quite conclusive: namely, *a hard paste, non-phosphatic, highly siliceous porcelain.* This is by no means the only Nantgarw shard that appears to demonstrate an unusual composition as NG9 from the same batch of shards also exhibits a non-standard composition with a much reduced bone ash content: the only conclusion that can be drawn from these analytical data is that some experimental work was carried out at Nantgarw on porcelain composition but this itself generates the question "… when was this experimental production of hard paste porcelain at Nantgarw carried out" and, of course, "who carried out this work—Billingsley and Walker, William Weston Young, or someone else" ?

It transpires that William Weston Young, the remaining partner from the Nantgarw China Works business set up with William Billingsley and Samuel Walker in 1817 (the so-called Phase II operation), took over the business at its bankruptcy in 1820 upon the departure of Billingsley and Walker for Coalport, and thereby attempted to recoup some funds to keep the Nantgarw China Works going: in this, he engaged the services of Thomas Pardoe from Bristol for the decoration of the residual china left at the works "in the white", mainly in the form of fired biscuit porcelain which probably comprised a mixture of glazed and unglazed porcelains, which would then be sold off after further glazing (if required) and enamelling and also the decoration of some already glazed specimens which had not been sent to the London agents, John Mortlock of Oxford Street. Considered historical opinion affirms that no more china was manufactured at Nantgarw after the departure of Billingsley and Walker in 1819/20 – and indeed it has been suggested, as revealed above, that there would have been no opportunity to do this because of the assumed demolition of the site following the auction of hardware and already completed services of porcelain – which we now know to be factually debatable and most probably incorrect! Documentation exists in the form of work books and diaries that strongly indicates that William Weston Young, conversely, did in fact attempt to manufacture porcelain at Nantgarw after 1820 and used the high temperature biscuit porcelain kiln to effect this, but because he was not aware of the original formulation and recipe used there by his partners William Billingsley and Samuel Walker, he made a hard paste non-phosphatic porcelain in the mistaken belief that this was the "Nantgarw porcelain" body. This product he firstly attempted with a mixed response to sell as completed decorated china to local clients, then as a "dry mix powder", which he actually called "*Nantgarw Dry Mix*", and this too was unsuccessful, although it was briefly tried and evaluated by other china works proprietors in Staffordshire (Edwards, *Porcelain to Silica Bricks: The Extreme Ceramics of William Weston Young, 1776–1847,* 2019, and extracts from William Weston Young's Diaries therein). The major problem was that Young did not realise that the original Nantgarw formulation recipe adopted by his partners William Billingsley and Samuel Walker included a significant proportion of bone ash, Billingsley's "top secret" ingredient, which he mixed himself alone in a secure locked room in the basement of his house at Tyla Gwyn, Nantgarw (which is now the location of the Nantgarw China Works Museum)! It seems quite inconceivable now that William Weston Young as a third partner in the Nantgarw China Works enterprise with William Billingsley and Samuel Walker was not fully aware of the compositional recipe for the porcelain his factory was manufacturing but, it is believed that his being only an occasional visitor to the Nantgarw site even during its heaviest production period, it would have been relatively straightforward to conceal the formulation from him if one so desired.

Clearly, NG6 is a shard that could possibly be assignable to William Weston Young's unsuccessful attempt to create Nantgarw porcelain – it is on record (Morton Nance, *The Pottery and Porcelain of Swansea and Nantgarw,* 1942) that he did sell his own completed, finished Nantgarw No.2 glazed porcelain locally and that he was able to produce several examples of his finished, decorated china as specimens in support of his marketing of the "*Nantgarw Dry Mix*" powder to potential china manufactory proprietors. It is believed that these "hard paste", highly siliceous, Nantgarw specimens were also marked with an impressed NANTGARW C.W., without the normal hyphen between NANT and GARW, in slightly smaller letters and enclosed in a rectangular box stamp. Undoubtedly, such an item of apparent Nantgarw flatware if found and offered for sale today would immediately be rejected by ceramics experts quite properly as a fake on the grounds of its incorrect marking and possession of the incorrect body paste characteristics. A contrasting but still objective view would maintain that this mysterious William Weston Young experimental porcelain, of which no complete specimens are recognised currently, is still *Nantgarw* porcelain, although admittedly of an alternative formulation, as it was made at the Nantgarw China Works site by the proprietor and decorated by either himself personally or by Thomas Pardoe, both of whom were esteemed enamellers in their own right (John, *Nantgarw Porcelain*, 1948; John et al. *Nantgarw Porcelain Album*, 1975). In the latter volume are illustrated some fine examples of the decorative work of William Weston Young and Thomas Pardoe on true Nantgarw porcelain remnants from the Billingsley/Walker era, presumably genuine phosphatic paste porcelain which were part of the residual stock left over at Nantgarw and which were decorated after the arrival there of Thomas Pardoe in 1821, which would grace any modern porcelain collection. The author has learned that on at least one occasion such a potentially Young – manufactured "*Nantgarw*" plate has been presented for sale at an auction house but has been summarily rejected by expert porcelain scrutineers as a fake, which is understandable and correctly appreciated under the circumstances, so possibly some examples may still be in circulation and may yet surface.

In conclusion, the analytical investigation of shards excavated at the Nantgarw China Works site has revealed that a highly siliceous, *hard paste* type, porcelain was made there and that examples could well still be in circulation that would be unrecognised for what they really are on merely textural and incorrect marking grounds. In an earlier study of Nantgarw shards by Professor Victor Owen, several shards of a non-standard Nantgarw composition were identified, including one of a highly siliceous porcelain which can only be assigned as *hard paste,* in support of the above interpretation (Owen et al. 1998; Owen and Morrison 1999): however, Professor Owen does also mention that many of these shards had lost their archaeological context which, therefore did not facilitate their temporal assignment.

It remains to now propose several interpretations which could be drawn from the holistic appraisal of the analytical data and documentary evidence regarding the manufacture of Nantgarw porcelain between 1817 and 1823 and following the closure of the China Works thereafter. Some possibilities are:

- William Billingsley and Samuel Walker did, contrary to all avowed opinion and existing evidence, manufacture a completely novel type of Nantgarw china between 1817 and 1819, before they left to engage with John Rose at his Coalport factory. Previously, authors have maintained that they would not have embarked upon this course to produce an inferior bodied china considering that they already held the market lead for the finest quality phosphatic porcelain, even if the presumptive new composition practically addressed their appallingly high kiln losses of 90%. Lewis Dillwyn did embark upon such a course of experiments carried out between 1815 and 1817 which resulted in the manufacture of his novel Swansea "trident" body, and he thereby suffered a disastrous consequence commercially which led to the closure of the Swansea China Works in 1820. Billingsley and Walker may well have been aware of this development and of the result, especially since Walker was actively involved with Dillwyn in the experimental formulations at Swansea which resulted in the firing of the trident body before he joined Billingsley at Nantgarw for their Phase II operation in 1817.

- William Weston Young did manufacture a hard paste, highly siliceous porcelain body at the Nantgarw China Works during his tenure there between 1820 and 1823; his Diaries do record his efforts in this direction and his attempts at marketing, firstly, the completed, decorated pieces, and secondly, a ready-to-fire, pre-processed *Nantgarw Dry Mix* upon which he devoted much time and effort in selling to china works proprietors for many years afterwards, even into the late 1830s. There is documentary record that William Weston Young and his associate John Wright as late as 1838 were still working hard at Bristol to market this novel porcelain mixture, approaching potential purchasers such as Apsley Pellatt in London and Mintons in Staffordshire, among others. Do shards NG6 and NG9, and possibly also the shards from Professor Owen's study, therefore, truly represent examples of Young's efforts in this direction? Aside from his diary entries relating to his activities in re-creating Nantgarw porcelain, William Weston Young is noted for his invention of the *Dinas* refractory silica brick in 1821/22, which was a major factor in the growth of the iron ore and copper smelting industries in a heavily industrialised Great Britain in the early nineteenth century and his trial experiments were initially undertaken using the high-temperature kiln at the Nantgarw China Works (Edwards, *Porcelain to Silica Bricks: The Extreme Ceramics of William Weston Young, 1776–1847,* 2019). It is, therefore, not unreasonable to propose that Young's experimentation, whilst in charge at Nantgarw, with his highly siliceous refractory silica bricks (which contain approximately 97% silica) would have been compatible with his experiments on an analogous, high-silica *hard paste* porcelain as well, containing some 80% silica, and which would also require a high firing temperature!
- Following William Weston Young's departure from Nantgarw in early 1823, Thomas Pardoe carried on his enamelling operations at the site for only a few months until his untimely death in June, 1823, when the China Works was then left abandoned until his son, William Henry Pardoe, started up earthenware production there in 1833. The idea of porcelain production at Nantgarw was rekindled by William Henry Pardoe in the 1850s, and for a short time it appeared that he did produce porcelain of an indefinite type there, which he advertised in a closing sale for his porcelain venture in 1858. Earthenware production and especially the manufacturing of clay pipes was continued for some decades afterwards into the 1880s. The archaeological excavations of Isaac Williams at the derelict Nantgarw China Works in 1931 (Williams, *Excavations at the Nantgarw*

Site, 1932) unearthed evidence of William Henry Pardoe's manufacturing operations from the 1830s, in the form of earthenware fragments and shards, including the remains of clay pipes. The porcelain shards, however, were all found and excavated from the lowest level, number 6, and were buried beneath large quantities of earthenwares, hence the potential stratigraphic location of the unusual porcelain shards exemplified by NG6 and NG9 dictates that these should probably be excluded from the later William Henry Pardoe operations and that they properly belong to the much earlier William Billingsley/Samuel Walker or William Weston Young period, i.e. 1817–1822. What the current research in body composition is unable to focus upon is the differentiation between these two possibilities: however, later in this text, we shall explore the conclusions from the data obtained using some very recent analytical studies carried out on the same batch of shards in which it appears that the glaze on shards NG6 and NG9, which possess the high silica body compositions, can be assigned to the *No.2 Nantgarw* glaze perfected by Young and Pardoe in stark contrast to the so-called *Nantgarw No 1* glaze used on the other shards which conforms to that of Billingsley and Walker. Shards NG6 and NG9 (Colomban et al. 2020) cannot therefore be routinely ascribed to a Billingsley and Walker creation and the conclusion, therefore, must inevitably be that William Weston Young did manufacture a hard paste porcelain at Nantgarw, as he claimed in his documentary notes and diary, unless contrary evidence appears from other as yet undiscovered sources. This conclusion will naturally come as a surprise to many historians who have relied upon the historical tenet that "only one porcelain body was ever manufactured at Nantgarw" – the codicil to this should now be applied, "True, but only in the Billingsley and Walker period, viz. 1817–1819", and we now really should explore the several avenues which opened up following the departure from Nantgarw of William Billingsley and Samuel Walker.

8.2.2 The Strange Case of Bow Porcelain – Or Perhaps Not?

One of the earliest English porcelain manufactories was Bow, a truly pioneering venture (Ramsay and Ramsay 2007a, b) which has been formally listed historically as starting up in 1743 and for which recent research has uncovered the true meaning behind an early patent which for a long time historically it has been claimed was misunderstood and misinterpreted with regard to its far-reaching implications. Perhaps one of the greatest and longest-running mis-interpretations in this respect is the patent of Edward Heylyn and Thomas Frye taken out in 1744 for the production of Bow porcelain; several authors have rejected this important early English patent document outright as "unworkable" especially with regard to its stated use of their imported Cherokee clay or "*unaker*" (for an account of these controversial allegations which have been put into a cogent scientific perspective, see Ramsay et al. 2001; Ramsay and Ramsay 2007a, b), whereas others have simply downgraded its importance to the growth and birth pangs of the English porcelain industry, or have even ignored its existence completely. In an important analytical and synthetic paper, Ramsay et al. (2004) have described successful scientific experiments in their re-creation of the early Bow porcelain body using *unaker* clay. Ramsay (2006, 2008) have thoroughly researched this particular situation and have identified a common malaise whereby many of the earlier writings, concepts and interpretations based upon an assumed mis-reading of this patent are justifiably unsupportable now and are, therefore, rather misleading for future researchers unless presented for further consolidated argument and debate in support for against the earlier conclusions. Ramsay and Ramsay (2006) have identified the historical origin of these misinterpretations primarily to Chaffers (*Marks and Monograms on Pottery and Porcelain, with Short Historical Notices of Each Manufactory and an Introductory Essay on the Vassa Victilia of England,* 1863), followed by Church (1881), Burton (*A History and Description of English Porcelain*, 1902), Tait (1959) and Watney (*English Blue and White Porcelain of the Eighteenth Century*, 1963) from which other authors have assumed incorrect and apparently factual information without making some verification historically. On the basis of their own extensive analytical studies of Bow porcelains, Ramsay and Ramsay (2006, 2008) can clearly identify the manufacture of *hard paste* porcelain at Bow before 1743, but this conclusion seems to be unacceptable to several connoisseurs who feel that Bow was an exclusively and uniquely a *soft paste* manufacturing facility. The situation gets even more convolved when an attempt is made to compare the analytical similarities (Ramsay et al. 2003; Ramsay and Ramsay 2008) between the so-called "Factory A" porcelains and Bow porcelain production, and the riposte to this assertion has been that a serious misinterpretation of the analytical data has been made and that it should be rejected. Yet, the most recent analytical work of Professor Victor Owen (Owen and Panes 2012) clearly demonstrates unequivocally that the body paste compositions of the Factory A-marked and Bow porcelains are experimentally identical, so providing, as they claim, the first tangible archaeological and analytical evidence of the porcelain body similarity between the two factories. The analytical data of Professor Victor Owen is also in agreement with the analytical evidence published some years before by Ramsay and Ramsay (2005) on a Factory A-marked porcelain polychrome tea canister. This naturally has fuelled a re-opening of the debate once again in favour of the Bow and Factory A-marked porcelains being analytically treatable as originating from the same source, namely the Bow factory, in Stratford, East London. An alternative explanation forthcoming that needs to be considered historically is that sufficient interchange of detailed manufacturing and recipe formulation knowledge existed between Factory A and Bow which resulted in the identical manufacturing processes for the same recipe being undertaken at both sites – which is perhaps not an unrealistic concept, but documentation relating to the location of Factory A which might support such an allegation is not evident?

8.2.3 A Very Rare Swansea Mug?

In this particular case study, a rather fine quality porcelain mug was presented for analysis with the proviso that removal or excision of even minute samples was expressly forbidden: hence, the porcelain specimen could only be analysed non-destructively using offset Raman spectroscopy to determine its body paste characteristics from a molecular spectral standpoint. The mug (Fig. 8.3) comprises a fine duck-egg translucent porcelain body and an associated white glaze with beautiful floral decoration which is rather reminiscent of the decoration on the *Marquess of Anglesey* combined tea and dessert service composed of both Swansea and Nantgarw porcelains. The flower painting is finely executed in the manner of William Billingsley and texturally and stylistically the mug was labelled as "Swansea" and was purchased as such by an esteemed collector and porcelain expert. However, the handle is not typical of the Swansea China Works, as documented by Jones and Joseph (*Swansea Porcelain: Shapes and Decoration,* 1988): further research into the shapes used for the handles resulted in a most interesting discovery relating to the handle form and shape on this particular mug being a replica of an analogue in the Cambrian Pottery, which had been subsumed into the Swansea China Works operation under the joint proprietorship of Lewis Dillwyn (Edwards, *Nantgarw and Swansea Porcleain: An Analytical Perspective,* 2018) and this particular handle features as such in several pictures of the Cambrian pottery mugs illustrated in Morton Nance' classic text (*The Pottery and Porcelain of Swansea and Nantgarw,* 1942). Hence, if indeed the mug is attributed to a Swansea China Works porcelain origin

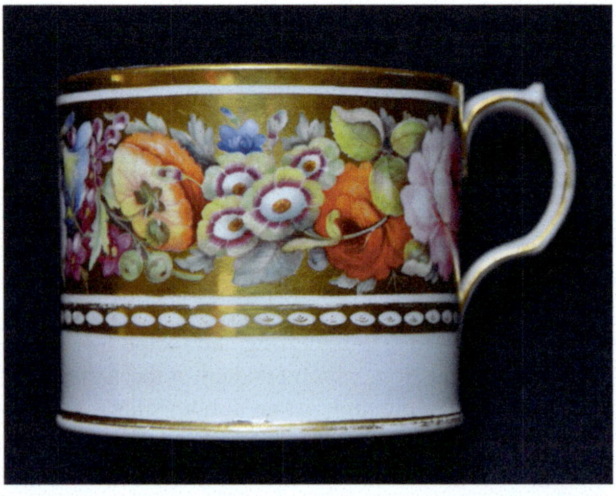

Fig. 8.3 A fine duck-egg porcelain mug, beautifully painted with a garland of garden flowers in the identifiable manner of William Billingsley and finely gilded, unmarked: believed to originate from the Swansea China Works in *ca.* 1817–1819, proprietor Lewis Dillwyn. However, the very unusual handle, which is found in wares of the Cambrian Pottery (also owned by Lewis Dillwyn), has not been recorded on Swansea porcelain hitherto. (*Private Collection*)

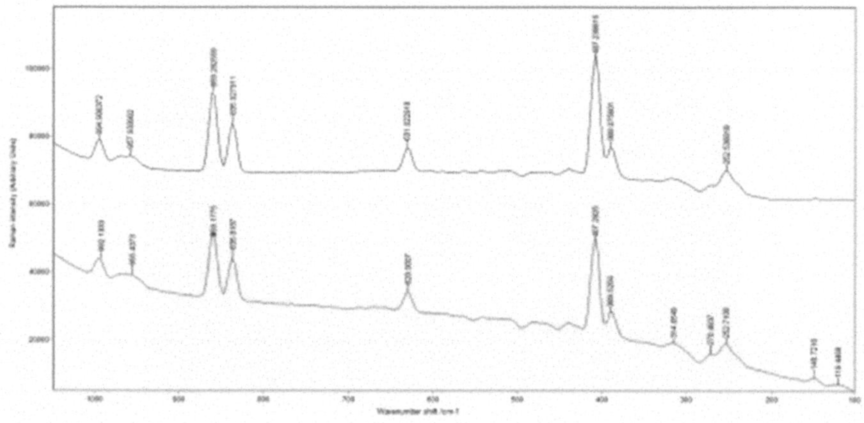

Fig. 8.4 Raman spectra of the putative Swansea mug shown in Fig. 8.3 (upper spectrum) and the deep dish, marked SWANSEA (lower spectrum), shown in Fig. 3.2, demonstrating the similarity between the spectral components of the two bodies analytically and confirming the potential attribution of the mug to a Swansea origin. Wavenumber range 100–1050 cm^{-1}, spectral resolution 10 cm^{-1}, excitation wavelength 785 nm

then it is an extremely rare, if not a unique example on Swansea porcelain, which borrows a handle design in porcelain from the associated pottery: the scarcity means that there are no other examples known and, it being also unmarked, expert opinion has categorised the mug as being of an "*unknown*" origin despite its stylistic and textural status, its characteristic duck-egg translucency and putative typical Swansea enamelling by a highly esteemed and recognised Swansea artist, namely William Billingsley. Raman spectral studies of the body paste of the mug and comparison with an equally fine and marked duck-egg Swansea dish painted by another Swansea artist William Pollard (Fig. 3.2) shows an identical pattern of spectral bands (Fig. 8.4) which, therefore, confirms its potential assignment as being consistent with its attribution to the Swansea China Works factory. However, the unusual handle, which has never before been seen on Swansea porcelain, but is known on Swansea Cambrian Pottery earthenwares, dictates that expert opinion does not fully accept the analytical evidence of the potential factory origin and the mug is hence classified as being "unknown", despite all other factors being correct for a Swansea China Works attribution.

8.2.4 A Unique Rockingham Porcelain Table

Approximately 15 years ago an opportunity was afforded to undertake the first molecular spectroscopic analysis of an English porcelain table-top, non-destructively, with a view to confirming its factory of production. The analysis was accomplished of the porcelain plaques comprising the top of the table. The table

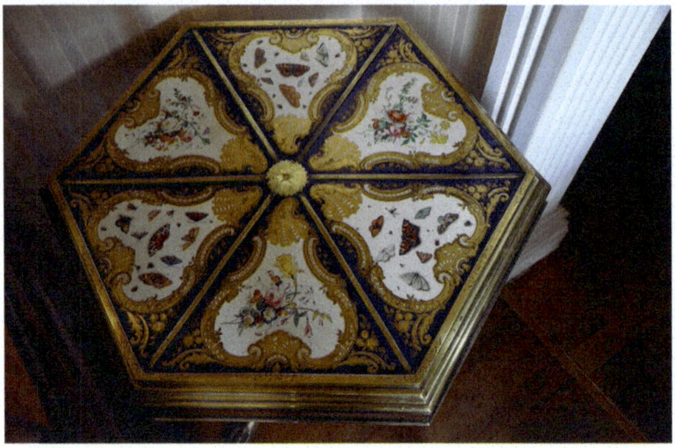

Fig. 8.5 Photograph of the six triangular porcelain inserts, identified analytically as Rockingham porcelain, *ca.* 1835–1840, in the hexagonal padouk wood/mahogany table-top of the tripod table shown in Fig. 4.1. The very high quality decoration with beautiful gilding, comprising three panels of free-flowing flower groups and assorted insects and butterflies with cobalt blue, "Bleu de Roi", ground colour are typical of the hand of John Wager Brameld. (Courtesy of Bryan Bowden Esq)

base was of an early Georgian origin, a fine tripod padauk example with brass inlay in the manner of Channon (Fig. 4.1). The top was of nineteenth century construction with six triangular porcelain plaques laid in fine chased brass mounts on a hexagonal mahogany support. The plaques were decorated alternately with exquisite groups of butterflies and insects and equally fine flower groups (Fig. 8.5). The provenance of the table is that it came from the sale of effects from Wentworth Castle, South Yorkshire.

8.2.4.1 Recent History of the Table

Eaglestone & Lockett (*The Rockingham Pottery*, Municipal Museum and Art Gallery, Clifton Park, Rotherham, 1964) record an interview with a Mr W Mason, a well-known antiques dealer with a speciality in Rockingham porcelain, in which he accurately describes the table, stating that at one time it was in his possession. It was 2ft 3 inches in height and attractively painted with butterflies and flowers – and that very few such pierces can have been made. The next appearance of the table was in Tennant's Auctioneers Autumn Catalogue sale of the 21st November 1979, lot number 126. The provenance of the table as being Wentworth Castle is stated, with a cautionary attribution of the plaques as "possibly Rockingham manufacture". This led to the Raman spectroscopic analysis being undertaken and agreed to by the present owner. Extensive documentary research has now shown that the table top was constructed *ca.* 1839 under Dale's patent. The beautiful painting of the plaques is attributed to John Wager Brameld. At this time most of the artists had left the Rockingham factory due to the critical financial situation and John Wager Brameld would certainly have had the will and ability to undertake this assignment as he no longer had managerial duties at the factory. An earlier example of his work appears on the important table snuff box in Rotherham Museum, Clifton Park, Rotherham.

8.2.4.2 Background History to Wentworth Castle

The linked history of Wentworth Castle at Stainborough and of Wentworth Woodhouse, Rotherham, its neighbour located just six miles away, is a complex one and can be briefly recounted here to set the scene. Thomas Wentworth -Watson, a colonel of dragoons in the late seventeenth century under King William III and ambassador to Prussia under Queen Anne, was created Viscount Wentworth and Earl of Strafford in 1712. He purchased Stainborough Hall in 1708 and immediately set about creating a grand house which Walpole described as "*the most perfect taste in architecture, where grace softens dignity and lightness attempers magnificence*" (Clewes-Salmon 2003). During this process he renamed his estate Wentworth Castle in 1731. Upstaging his cousin, the second Earl Fitzwilliam, who was elevated to the peerage in 1716, died in 1728, and based at Wentworth Woodhouse nearby, became an ongoing obsession with the Earl of Strafford and an intense rivalry between them was paramount. In 1782, he later inherited the titles and associated estates, making him one of the greatest landowners in the country. In 1979, the tenth Earl Fitzwilliam died and the titles then became extinct, so the contents of Wentworth Woodhouse were sold. A similar fate was visited upon Wentworth Castle following the death of the third Earl of Strafford in 1799 when the inheritance and titles passed to the neighbouring Fitzwilliams, but Wentworth Castle itself passed to Frederick Vernon-Wentworth in 1816. In 1902 the estate was inherited by Captain Bruce Vernon-Wentworth, who put the contents of Wentworth Castle up for sale, first in 1919, and then the remaining contents at a six-day closure sale in 1948 (Lancaster and Sons, Barnsley, June 7th–12th 1948). The table under investigation here does not feature in the auction catalogue and was believed to have been sold by private treaty beforehand.

8.2.4.3 Raman Spectroscopic Analysis

The table under study here is shown in Fig. 4.1. The six triangular section porcelain plaques comprising the table top (Fig. 8.5) could be detached from their brass mountings and this facilitated the interrogation of the unglazed porcelain body beneath the enamelled decoration using a transect across the enamels, glaze and porcelain substrate. The table-top porcelain composition was analysed using non-destructive Raman spectroscopy, which provided an exact match spectroscopically for the porcelain body with an enamelled puce griffin marked Rockingham plate from *ca.* 1835–1840, and with matching spectra also acquired for the similar enamelled pigments, therefore placing firmly analytically the plaques as of Rockingham porcelain (Edwards et al. 2004). In *Tennant's Auction Catalogue of 21st November, 1979)* and this particular table is included with a full description as Lot 126, a

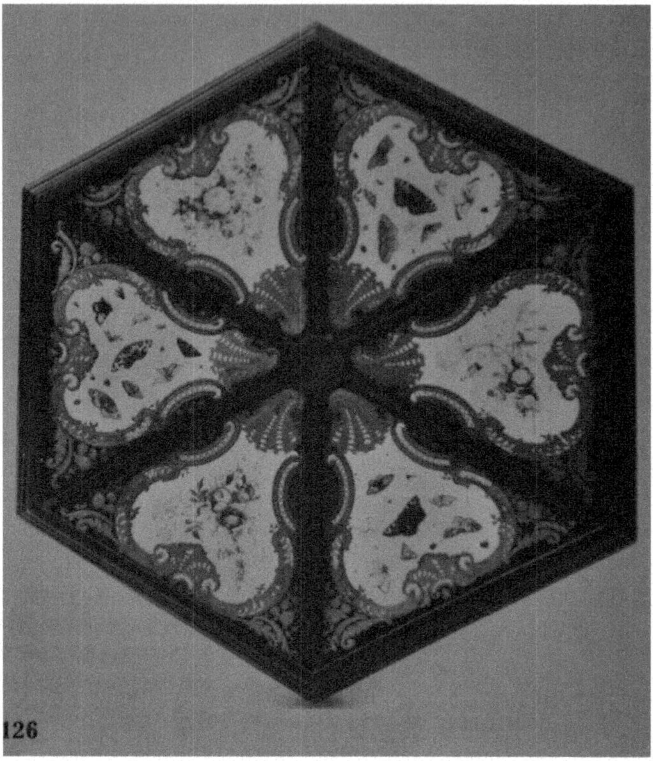

Fig. 8.6 Photograph of the table-top in black and white monochrome as depicted in the Sale Catalogue of Tennant's in Richmond, November 21st, 1979, describing Lot 126 from the estate of Wentworth Castle. This is identical to the colour photograph shown in Fig. 8.5. (Courtesy of Bryan Bowden Esq)

black-and-white photograph (Fig. 8.6), which confirms its placement as part of an original acquisition to Wentworth Castle, so supporting the analytical data interpretation. The table was apparently sold by private treaty prior to the sale itself. The historical documentation is quite explicit and demonstrates its justifiable inclusion in a holistic approach to the scientific attribution of important porcelain items. Llewellyn Jewitt (*The Ceramic Art of Great Britain from Prehistoric Times to the Present Day*, 1878, Volume I, pp. 501 and ff.) mentions some very interesting details relevant to this particular item as follows:

> *In 1838 the manufacture of china and earthenware bedposts, cornices, etc., a somewhat novel feature in the art was added to the other products of the Rockingham Works. In that year a patent was taken out in the name of William Dale for certain improvements in constructing columns pillars, bedposts and other suchlike articles of furniture ... consisting of several ornamental pieces or compound parts of china or earthenware.*

A memorandum of agreement was signed between the Bramelds of the Rockingham China Works and William Dale on the 28th February, 1838, to

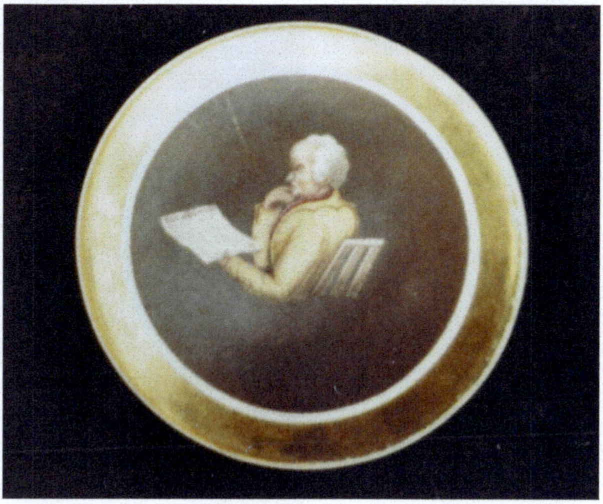

Fig. 8.7 The highly important snuff box described as *"The Politician"* which is the only known example of porcelain decoration signed by John Wager Brameld and therefore provides a comparator for the identification of other potential works by this artist on Rockingham porcelain. (Courtesy of Bryan Bowden Esq. Now in Clifton Park Museum, Rotherham)

incorporate Rockingham china into articles of furniture prior at their manufactory. Llewellyn Jewitt goes on to say that: "*these are now of very great rarity and are exquisitely decorated with small groups and sprigs of flowers, of highly efficient designs and elegant workmanship*". Eaglestone and Lockett (*The Rockingham Pottery,* 1964) cite Llewellyn Jewitt as recording such a china table made at Swinton.

In the sales catalogue of Tennant's Auctioneers of Richmond, North Yorkshire of the 21st November, 1979, this particular porcelain-topped table is Lot 126. Comparison by the present owner has indicated that the butterflies and insects painting on the table-top panels is similar to that on a porcelain snuff box painted by John Wager Brameld, who according to Llewellyn Jewitt (*Ceramic Art of Great Britain from Prehistoric Times to the Present Day*, 1878) was a "*clever painter of flowers , figures and landscapes*", citing a highly important and documentary Rockingham snuff box with an exquisite painting of "*The Politician*" which had been personally signed by him (Figs. 8.7 and 8.8). Also, a part service of Rockingham porcelain with a pattern number "1475" in a private collection has a very similar stylistic flower groups to that of the table, and this pattern can be dated firmly to the period, 1838–1842. It is probable therefore that Brameld personally painted the triangular porcelain sections of the table-top. This conjecture is supported by the fact that most of the artists had by this time left the factory and that John Wager Brameld had both the opportunity and will to undertake the decoration. John Wager Brameld died in 1851, having seen the esteemed Rockingham China Works close in 1842, due to the withdrawal of financial support from Earl Fitzwilliam, and having bankrupted itself in preparing a very high quality and prestigious Royal service commission for King William IV, which was used eventually at the Coronation banquet at Windsor Castle for Queen Victoria. Llewellyn Jewitt (Jewitt, *The Ceramic Art of Great Britain from Prehistoric Times to the Present Day*, 1878) makes a very astute statement about the Rockingham China Works specifically and the Bramelds' quest for artistic excellence there:

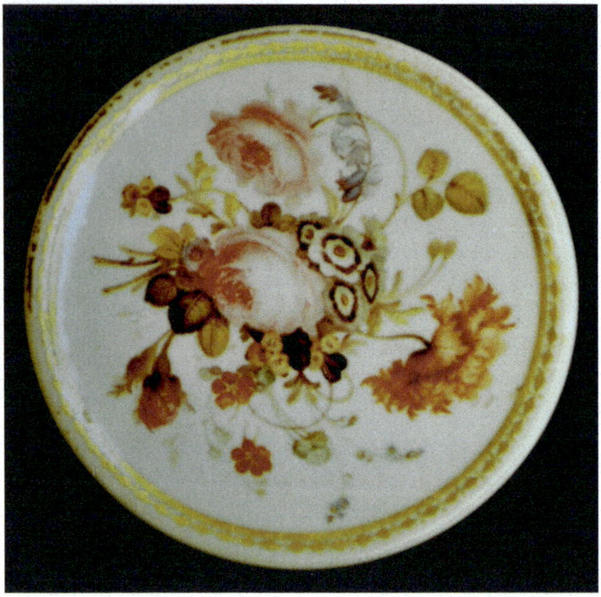

Fig. 8.8 Interior of the snuff box base shown in Fig. 8.7: here, the insects and flower groups identified as painted by John Wager Brameld can be seen and the close similarity evident between these and the same items rendered on the hexagonal table-top in Fig. 8.5. (Courtesy of Bryan Bowden Esq. Now in Clifton Park Museum, Rotherham)

In making art advances in his manufactory, the firm became, as is too frequently the case with those who study the beautiful, instead of being strictly concerned with the management of their works, slightly embarrassed financially.

This, of course, equally applies to several key china works proprietors in our survey in this text: William Billingsley, Lewis Dillwyn and William Weston Young, at Nantgarw and Swansea are classic examples upon whom a similar financial disaster had befallen in their search for artistic perfection in their creation of porcelains of the highest quality. It should be noted that this particular case study exemplifies the positive collaborative support provided by analytical scientific data, historical documentation and expert opinion.

8.3 Case Studies in Porcelain Identification Where the Analytical Information Is Supportive of Established Expert Opinion

The case studies described here are selected to illustrate examples of where analytical science and the data obtained therefrom have provided novel information about several of the English and Welsh factories discussed above which supports expert opinion, even if in some cases this is itself not in complete agreement.

8.3.1 A Very Rare Swansea Watering Can

The miniature porcelain watering can shown in Fig. 6.6 was analysed spectroscopically to determine its factory of origin: it is a very rare item of porcelain, commonly unmarked, which has been attributed convincingly to the Swansea factory on the basis of several similar known examples, of which only some five or so specimens are known in museum and private collections, and is clearly constructed from a modified coffee can in duck-egg porcelain with an added gilded strap handle and perforated spout, and simply decorated with a wreath of garden flowers by William Pollard, a local and esteemed Swansea artist. An identical example is illustrated in Dr William John's book on Swansea porcelain (W.D. John, *Swansea Porcelain,* 1958, Illustration 51A). The Raman spectral data for this item are identical with those of the marked Swansea deep dish in finest duck-egg porcelain, which was also painted by William Pollard, and this has been illustrated in Fig. 3.2. Hence, its factory attribution to Swansea by established expert opinion is correct and is found to be consistent analytically with this attribution; the analytical and historical information are mutually supportive with the expert opinion and the porcelain artefact is unequivocally accepted, therefore, as a genuine, rare piece of Swansea porcelain.

8.3.2 A Nantgarw Trumpet-Shaped Spill Vase

The trumpet-shaped spill vase shown in Fig. 8.9 is of a rather unusual shape for the Nantgarw factory and has given rise to some debate and difference in expert opinion about its factory of origin. The usual type of Nantgarw spill vase is of a cylindrical shape as demonstrated in Fig. 6.2, which is photographed in transmitted light to illustrate its superb and characteristic translucency. Because of this little known style of spill vase its authenticity of attribution to the Nantgarw factory has been questioned by some connoisseurs, whereas others have indicated that they concur with its Nantgarw factory attribution as a rare piece of porcelain. The Raman spectrum of the trumpet-shaped spill vase is identical with that of its analogous Nantgarw cylindrical version and is therefore consistent scientifically with its Nantgarw factory assignment in agreement with some expert opinion. Figure 8.10, shows the spill vase *in situ* being interrogated using a stand-off, non-contact probe head and portable Raman instrument which effectively renders the acquisition of Raman spectral data rapidly and without danger of damaging what appears to be a rare porcelain piece. Since the analytical experiment was carried out, a second trumpet – shaped spill vase (Fig. 8.11) has been identified in a country mansion, although in damaged condition and there, it seems, it has been assigned erroneously to either a Swansea or a Davenport factory product. However, in this case the decorative pattern is identical to that of the Nantgarw tea cup and saucer shown in Fig. 8.12, which has been verified unanimously as a genuine product of the Nantgarw China Works by expert opinion. Again, here, the rarity of a porcelain piece has generated some diverse

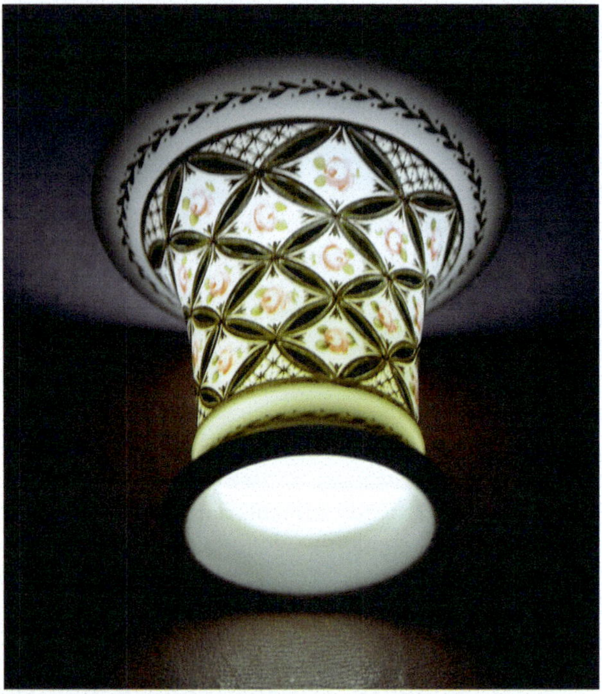

Fig. 8.9 Trumpet-shaped spill vase of beautiful translucency, with local decoration in a chrome green ground and individual pink roses contained in lozenges between the gilded edges of a lattice work pattern. (*Private Collection*)

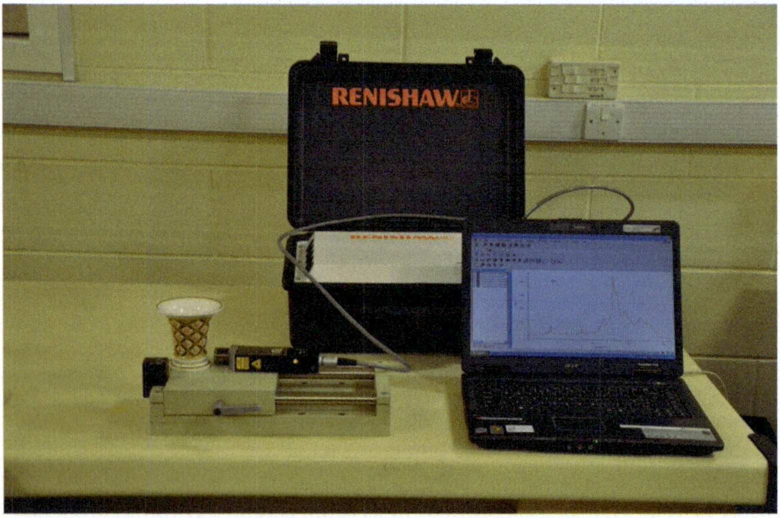

Fig. 8.10 The experimental arrangement for the non-destructive analytical Raman spectroscopic interrogation of porcelain using a one-metre 5:1 probe and stand-off probe head attached to a Renishaw RIAS portable instrument operating with 785 nm laser excitation in the near infrared region of the electromagnetic spectrum. Note that there is no contact between the probe head lens and the item under investigation and the laser focus can be made into eth porcelain body under the clear glaze with a spectral footprint of about 100 microns diameter. Here the Nantgarw spill vase shown in Fig. 8.9 is shown actually studied and the spectrum displayed on the small computer screen attached to the instrument

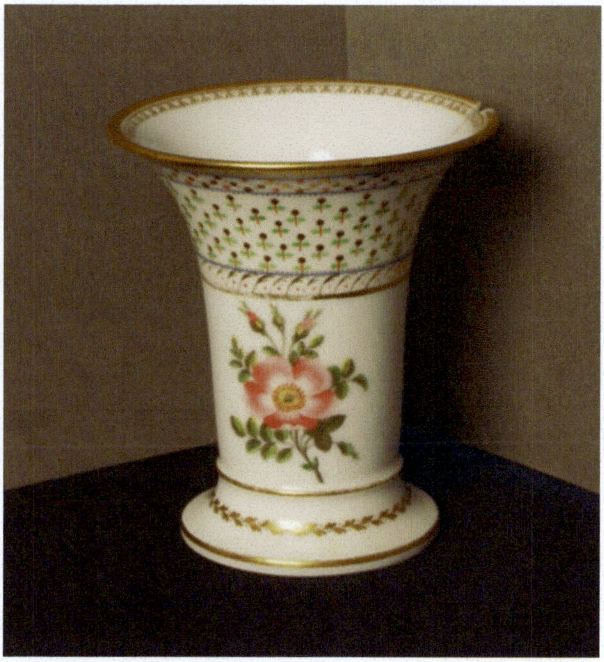

Fig. 8.11 Trumpet shaped spill vase decorated in a Nantgarw set pattern, but incorrectly identified as belonging to either the Davenport or Swansea factories

expert opinions, but the input of analytical science has suggested an attribution which is consistent with some expert opinion, although not all connoisseurs are in agreement with this result.

8.3.3 Analysis of a Rare Barry-Barry Derby Porcelain Plate

The *Pendock Barry-Barry* service in Derby porcelain has been discussed at length by several authors on porcelain (Twitchett, *Derby Porcelain 1747–1847*, 2002; John, *William Billingsley*, 1968; Edwards, *Swansea and Nantgarw Porcelains: A Scientific Reappraisal,* 2017b; Edwards, *Derby Porcelain: The Golden Years, 1780–1830*, 2017a) and the consensus of opinion from the superb rose decoration on this plate (Fig. 6.12) is that it is most likely to have been painted by William Billingsley, the accredited master of rose-painting at the Derby China Works. The only serious objection to this long-held view has been that of John Twitchett, who reasoned that it could not have been possibly painted by Billingsley since in his opinion from textural considerations the body post-dates Billingsley's departure from Derby in 1795. Prior to the analytical study reported here, there were no analytical data currently available for *Barry-Barry* service porcelains. It is well documented that William Billingsley left his position at Derby with William Duesbury II quite amicably in 1795 and for many years pursued the decoration of porcelains

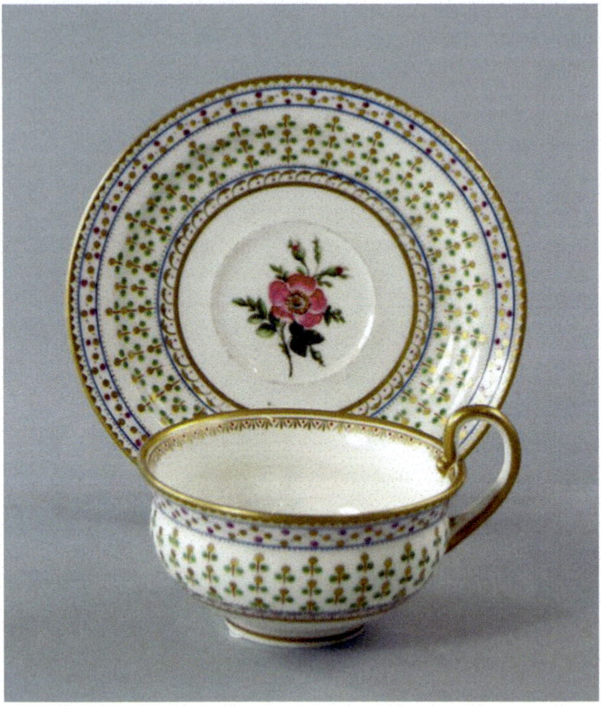

Fig. 8.12 Nantgarw porcelain tea cup and saucer in the same set pattern as the putative Nantgarw spill vase shown in Fig. 8.11 which confirms the attribution of the spill vase

from Derby and possibly several other factories under contract as a freelance artist, especially after he had parted company with John Coke at Pinxton in 1799 to set up on his own at Mansfield and Brampton-in-Torksey before moving on to the Worcester Royal China Works in 1808. Hence, the idea that the *Barry-Barry* service could not have been decorated by Billingsley because *per se* he had already left Derby in 1795 does not preclude his attribution as artist for this important and beautifully decorated service which is so typical of his style of execution. John Haslem (*The Old Derby China Works,* 1876) makes reference to this particular *Barry-Barry* service in his account of the activities at the Derby China Works at the end of the eighteenth century and ascribes it to Billingsley, but no other reference is found to it afterwards and strangely it does not appear in the Derby factory pattern books, which is rather unusual for the importance and avowed excellence of this porcelain service and its superior quality of art work. Another classic example of such missing information in factory records relates to an exquisite service commissioned also from the Derby factory, namely that of *Lord Ongley*, which again is mentioned by John Haslem (*The Old Derby China Works,* 1876), but which does not appear anywhere in the factory record or pattern books – and again this was apparently painted by an artist who had already departed from the Derby China Works to work freelance, but who seemingly carried out commissions for the Derby China Works, namely William Corden, in the period 1819/1820 (Twitchett, *Derby Porcelain,*

1747–1847, 2002)! These two parallel cases are quite revealing in that they lend credibility to the idea of the Derby China Works commissioning external artists, who were nevertheless former employees, to decorate some of their most expensive services.

The decoration on the *Pendock Barry-Barry* service is a classic example of William Billingsley's work and it is in the style of his early nineteenth century accomplishment. Genealogical research into the Pendock Barry-Barry family by Rachel Denyer (Edwards et al. *The Pendock-Barry Derby Porcelain Service: A Holistic Reappraisal,* 2019) has recently uncovered some fascinating details of the complexity of the family fortunes in the early 1800s. It has suggested strongly that the commissioning of the Barry-Barry service from the Derby China Works with its central emblazonment of the Barry coat of arms is consistent with a date of around 1807, when Pendock Neale, eager to possess and acquire the arms of Barry-Barry (shown at the centre of the plate in Fig. 6.12), short-circuited the official College of Arms and Royal grant of the same arms some years earlier than he strictly should have and this was recorded in several legal tracts dated around this time which highlighted and admonished his rashness in doing so. So, it is quite feasible from the historical records that the *Pendock Barry-Barry* service could have been painted by William Billingsley on Derby porcelain up to 1807 or thereabouts when he was still engaged in his contract work for Derby at Brampton-in-Torksey. It is highly unlikely, therefore, that a commissioning date later than 1807 would be possible for this service. In which case, the body of the *Barry-Barry* service porcelain should be anticipated to be that of William Duesbury II and carried over as such from the late eighteenth century. Later Derby porcelains from the Bloor Derby period involve a change in composition of the body and the applied glaze, which is very prone to exhibiting a *craquelure* effect, which is definitely not present in the *Barry-Barry* specimen. Although not every piece in a large service was marked, the observation of a Derby factory mark on those surviving pierces of the *Barry-Barry* service does not help as all marked pieces known exhibit a Derby gold mark: the puce mark commonly associated with the Duesbury period gave way to a red mark in the earlier Bloor period and then a red enamelled stamped crown within two concentric circles and a Bloor Derby stencil legend later on, as was manifest with the Lord Ongley service mentioned above. The analytical studies on the *Barry-Barry* dinner plate shown in Fig. 6.12 using stand-off Raman spectroscopy confirm that the porcelain body spectral characteristics match precisely those of a *Prince of Wales* Derby service dessert plate, also painted by William Billingsley and logged in the factory pattern books for the year 1786—this latter plate is shown here in Fig. 6.11. This provides another example of the synergistic use of scientific analysis and historical records to arrive at a conclusion for the correct attribution of a factory of origin for porcelains which may have some questionable tenets: albeit, in this case there is no doubt that the *Barry-Barry* service is a Derby service, and some but not all pieces have been marked with the cursive Derby crown and crossed batons in gold enamel. The novel research into the historical genealogical family record has facilitated a re-definition of the possible timeline for commissioning the service and has confirmed that it is quite possible that William Billingsley did paint this service after all, not whilst he was at Derby but most probably whilst he was engaged in porcelain decoration at Brampton-in-Torksey. This would also explain why the service is not mentioned at all in the Derby China Works pattern books, but it was nevertheless acknowledged as a late eighteenth century Derby service by John Haslem. This provides another example of analytical science working together with historical documentation in the holistic appraisal of the evaluation of the factory of origin of an important piece of porcelain and moreover, defining more closely the period of its manufacture. In this case, all three components of the holistic appraisal process are supportive of this item of porcelain being of Derby China Works manufacture, even though some academic discussion still reverberates around the origin of the artistic decorative execution.

8.3.4 A Swansea Trumpet-Shaped Spill Vase?

Figure 6.5 illustrates a particularly fine porcelain spill vase in a classic trumpet shape, which has a green duck-egg translucency that is so characteristic of the Swansea factory around 1817 (Fig. 8.13). The decoration depicts a dancing Chinaman with a conical straw hat in front of a tropical background with tendrils of gilt seaweed, a characteristic Swansea formulation of William Billingsley. Trumpet-shaped spill vases are more a feature of Swansea porcelain than of their Nantgarw neighbour. The Raman spectral data are matching those of fine duck-egg porelain as

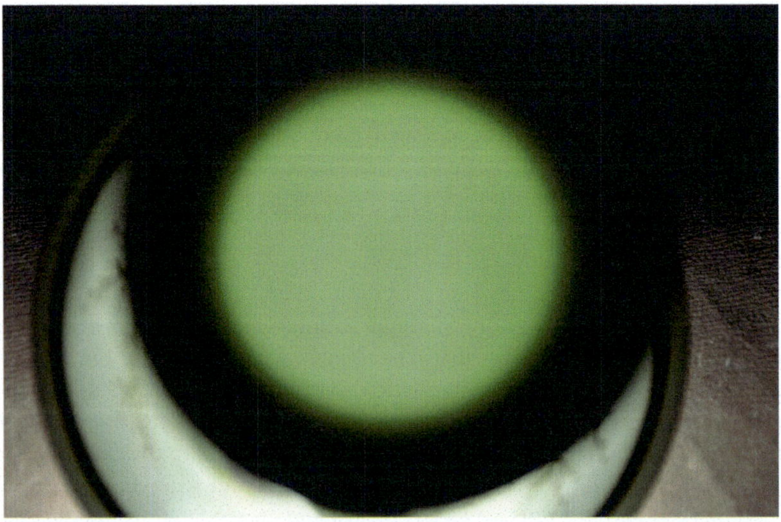

Fig. 8.13 Swansea spill vase shown in Fig. 6.5 painted with a chinoiserie scene by William Billingsley; photographed in transmitted light to show the characteristic duck-egg translucency of the finest Swansea porcelain. (*Private Collection*)

evidenced by their comparison with the spectra obtained from the deep dish marked SWANSEA, shown in Fig. 3.2, and are thus consistent with the attribution of the spill vase being assigned to the Swansea factory. There is no doubt concerning the analytical attribution to Swansea of this piece of porcelain, which is also acceptable with the unequivocally historical documentation and expert opinion and connoisseurship.

8.3.5 A Nantgarw Coffee Can?

The coffee can shown in Fig. 8.14 poses a particular problem for factory attribution from the connoisseurship standpoint in that it is very plainly decorated with only a chrome green enamel band, some 5 mm in width, is ungilded and unmarked, so no clues can be forthcoming from artistic study of the decorative style or an observable pattern. However, the translucency of the soft paste porcelain and the faint "ribbing" texture which is apparent inside the can is indicative of a Nantgarw attribution: the coffee cans at Nantgarw were all hand thrown and not moulded as they were elsewhere, and invariably demonstrate this ribbed effect sensation to the touch. The spectral analysis confirms that its assignment is consistent with a Nantgarw attribution by comparison with marked Nantgarw pieces such as those depicted in Figs. 3.1, 3.7, 6.2 and 6.4. The coffee can, therefore, is confidently assigned to the Nantgarw factory on the basis of the analytical evidence as well as from its connoisseurship expert opinion. It is interesting that this particular coffee can has the pure white glaze, known as the Billingsley/Walker glaze or the *Nantgarw No.1* glaze, so this coupled with the simple local decoration perhaps indicates that it was a remnant

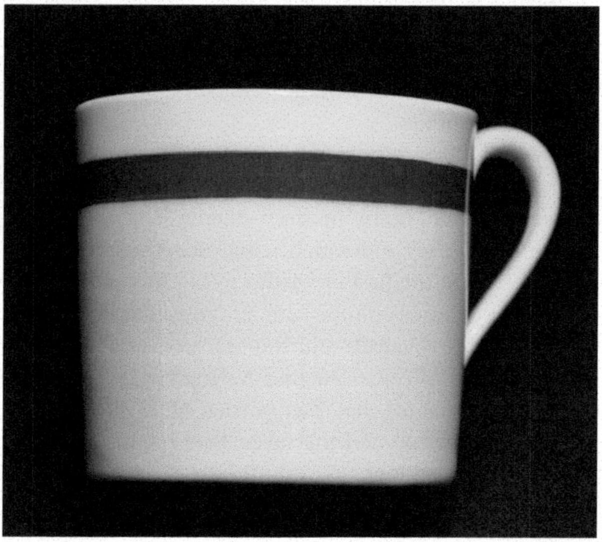

Fig. 8.14 Nantgarw porcelain coffee can very simply decorated locally with a concentric plain chrome green band and without gilding. (Courtesy of Bryan Bowden Esq)

left glazed and in the white at the factory after the departure of Billingsley and Walker in 1819/1820, or possibly was glazed using the remnant supplies of Billingsley/Walker glaze frit after William Weston Young took over responsibility for running the china works operation. The chrome green colour was a known favourite of Thomas Pardoe at this time and the absence of gilding further adds the dimension that this item was locally decorated at a time of austerity in the factory; it has been recorded that the expensive 24 carat gold gilding was prohibitive on economic grounds, as noted by William Weston Young in his notebooks and diaries (Edwards, *Porcelain to Silica Bricks: The Extreme Ceramics of William Weston Young, 1776–1847*, 2019), and if required was replaced by enamelling in green, blue and chocolate brown. Here, we have scientific analysis and expert opinion agreeing upon the attribution with some limited historical documentation in support of this conclusion.

8.4 Scientific and Artistic Expertise – The Holistic Approach and Potential for Re-interpretation?

It is clear from a consideration of the specific case studies recounted above that it does not always follow that the analytical data interpretation and suggested factory attribution are in agreement with expert connoisseurship and opinion unequivocally and that, as a result in several cases a doubtful attribution for a porcelain artefact is still prescribed. It is interesting to speculate upon the origin of the perceived impasse or disagreement that can arise between the conclusions of scientific data analysis and artistic expert opinion in the area of ceramics research. Clearly, in a holistic approach, which is deemed ideally to provide the forensically acceptable evidence for factory attribution *par excellence,* both avenues, plus the input of documentary records and accounts where available, have essential and potentially unique information which should be counterbalanced internally and considered objectively without involving the arbitrary and summary rejection of one batch of evidence in favour of the other.

For example, from the earliest days of scientific analysis dating from Sir Arthur Church in the 1880s, the analytical chemical data and results, relatively meagre though they may be by modern standards of analytical presentation, were groundbreaking and they generated some deep suspicion amongst art connoisseurs and experts. One of the first art critics to put this into words was Robert Drane, a highly respected ceramics connoisseur and museum curator, who in William Turner's classic book on the Swansea and Nantgarw ceramic factories published in 1897 (Turner, *The Ceramics of Swansea and Nantgarw*, 1987) wrote that the expert eye of the connoisseur was all that was needed to assign the factory of origin of a piece of porcelain correctly, and that even factory marks were not required to achieve this result. This statement was admittedly made on the eve of scientific investigation when relatively few pieces of porcelain had actually been analysed and only expert opinion held sway. This theme was much later taken up by the well-respected ceramics expert David Battie, who in an article entitled "*The Eyes Have It*" (Battie 1994) encouraged ceramics collectors to acquire as much experience as they could in handling pieces in which they were interested and thereby they would out-compete any instrumentation then available for ceramics characterisation and attribution, and that this would probably also apply far into the foreseeable future too. Certainly, this advice for the acquisition of knowledge through the handling of porcelain was unimpeachable, but the outright statement that instrumental analysis was definitively not required for potential ceramics factory attribution in the case of doubtful pieces surely cannot be substantiated now, although it was probably made in good faith at that time. In between Drane and Battie's revelations a rather negative criticism of the information that could be provided from the scientific analysis of ceramics was announced by Stanley Fisher, again an eminent and prolific author in the ceramics field, who attacked the whole concept behind the application of scientific analysis to porcelains:

> … *repugnant science should have little or no role to play in the study of English porcelains, which is surely an exercise in the artistic pursuit* (Fisher, *English Blue and White Porcelain of the 18th Century: An Illustrated Descriptive Account of the Early Soft Paste Porcelain Production of Bow, Chelsea, Lowestoft, Derby, Longton Hall, Bristol, Worcester, Caughley and Liverpool Potters, ca., 1740–1800,* 1947).

The particular use of the word *repugnant* here is in itself revealing as it has come to mean abhorrent, unpleasant, loathsome and detestable nowadays, but it is actually derived from the Latin verb *repugnare*, actually meaning "to oppose". Perhaps this provides a clue as to the use in this context of this particular word by Stanley Fisher – is his statement really offering an archaic basis of his mistrust in analytical science in that perhaps, in some specific cases he was associated with, the analytical data were actually found to be in opposition to other conclusions about the factory origin of a particular piece that were based solely upon his own or other well-respected artistic and expert assessments? In other words, is Fisher objecting to the analytical evidence that he sees to be in contradiction with or in opposition to the established opinion which otherwise would have been considered sacrosanct? Or is the word repugnant being used in its more modern context and thereby implying a distasteful and unpleasant usage of a means for contributing to the verification of the origin of a piece of porcelain – in other words, keep science away from artistic endeavours, and by inference a machine cannot supplant the extensive experience of an expert human being? The second point that arises here from Fisher's concern about porcelain analysis is that this statement was made in 1947, when analytical chemistry had hardly been applied to porcelain analyses: as can be seen from our survey in this current text, the only reported analyses performed on English and Welsh ceramics prior to the 1940s were those of Sir Arthur Church and of Herbert Eccles and Bernard Rackham from the 1890s and 1920s, respectively (Church, *English Porcelain*, 1894; Eccles and Rackham, *Analysed Specimens of English Porcelain in the Victoria and Albert Museum Collection*, 1922). It appears, therefore, that without further elaboration one or more of these unspecified reported analyses and their determinations could be assumed to have been at variance with the independent conclusions made by Stanley Fisher on artistic grounds alone – and that possibly a mis-identification of a particular piece perhaps was thereby made on the basis of analytical interpretation alone, which admittedly does set a rather dangerous precedent. Yet, as we saw earlier in this book, we have evaluated the results of Church and of Eccles and Rackham for a significant and representative collection of English and Welsh porcelains and compared these with the analogous results obtained using modern instrumental techniques dating from the 1990s and later, and they are found to be essentially in good agreement numerically with each other, so in this respect we cannot attribute any dissension hinted at by Fisher relating to a particular item caused by potential modern mis-attributions rendered by the earlier analyses to earlier analytical methodology. So the "repugnance" felt by Stanley Fisher at the application of analytical science to ceramics is rather inexplicable and would be better understood now if the analytical attributions of the exemplars which had caused the offence and dissension with opinions made on purely artistic grounds had been specified by him in his work. In fact, as was noted in an earlier chapter, only one piece of porcelain of doubtful origin was analysed by Eccles and Rackham and was then reported by them (Eccles and Rackham, *Analysed Specimens of English Porcelain in the Victoria & Albert Museum*, 1922) and this has been discussed in some detail hitherto, namely a putative "Bristol" mug which could well have been actually an early acquisition by the Worcester factory in its assimilation of the Bristol operation in 1751 based on their analytical evidence. It is doubly interesting, therefore, that the decision made by Eccles and Rackham on analytical grounds of the factory of attribution in this particular case was actually quite acceptable to the generation of a re-consideration of the originally expressed expert opinion at that time, although today there may be grounds for a further analytical data re-interpretation, as has been discussed earlier.

Rather than querying the implications of the analytical data themselves, it appears that the source of any potentially acrimonious debate surrounding the polarised science *versus* arts approaches to the determination of the origin of porcelains actually focusses on the *interpretation* of the data: for the analytical scientists, this strictly concerns the establishment of standard experimental values for the elemental oxide proportions in the body paste which itself requires that genuine pieces of each type of porcelain from each particular factory have also been presented at some time for analysis. Likewise, the interpretation of historical documentation at the other extreme is also not always straightforward and is occasionally open to debate: for instance, it is well known that Lewis Weston Dillwyn made a series of scientific experiments on his porcelain body compositions, aided by Samuel Walker, at the Swansea China Works between 1815 and 1817, and that these are documented in his handwritten notes and terse comments relating to the quality of the fired bodies, which have been lodged in the Victoria and Albert Museum archive (Dillwyn's *Notebooks*, see in Eccles and Rackham, *Analysed Specimens of English Porcelain in the Victoria and Albert Museum Collection*, 1922: Edwards, *Nantgarw and Swansea Porcelain: An Analytical Perspective*, 2018). From these detailed notes it is clear that production trials were undertaken on at least nine, and possibly up to thirteen, different variations in the Swansea body and involving the trialling of three associated glaze recipes during this period, yet modern writers still maintain that Swansea only ever accomplished the manufacture of three distinct bodies, namely, the well-known glassy, duck-egg phosphatic and "trident" magnesian porcelains. It could be reasonably argued, of course, that only these three variants were made in any quantity at Swansea, but we have no way of knowing this – indeed, even the contents of one kiln's experimental firing charge could easily have created several hundred pieces of porcelain or more, which could have been sold on later to recoup some finances for the operation. Hence, Professor Victor Owen and his associates (Owen et al. 1998) claim from their analyses of Swansea china shards that they can identify at least five, if not six, different porcelain compositions, which does not agree at all with the established expert artistic opinion pertaining to the Swansea China Works. A potential problem has been created which can be laid firmly at the door of the misinterpretation of historical documentation (or in some respects even a discarding or disregard of it), highlighted by the data from these recent analytical experiments.

This ongoing saga of scientific data interpretation being proposed in possible contravention with expert artistic opinion for artwork attribution cannot embrace all the scientific analytical programmes being undertaken on artworks and art-related specimens and a way forward must be addressed so that the holistic approach is adopted along the lines of much of the standard forensic science investigation in the criminological field, whose evidence and its interpretation is then rigorously tested by debate and cross-examination in an open forum. Very often, of course, it is the expressed *opinion* of the forensic scientist regarding his or her interpretation of the scientific data that is debated in the open forum of a courtroom, rather than a detailed criticism of the way in which the data themselves were gathered, although even that is open to discussion.

In a forensic analytical investigation of a ceramic artefact four major questions are generally required to be addressed and this is also true for its application here to the sourcing of porcelain specimens:

1. What elements or molecules comprise the specimen analyte – a qualitative analysis?
2. How much of each is present in the specimen analyte – a quantitative analysis?
3. Where did the specimen originate – the sourcing question?
4. What has happened to the specimen since its manufacture – have potential restoration, the insertion of false marks and the embellishment of existing decoration and gilding taken place since the original manufacture?

The answers to questions 1 and 2 are straightforward in the sense that the potential errors in the analytical data are quantifiable and can be estimated from the techniques used as described earlier in this book. However, the answers to questions 3 and 4 require a little more interpretation and deduction and this is exactly where the holistic approach is necessary to acquire a potentially meaningful result: the analyst must now seek out the historical accounts of the factory concerned to determine the source of the raw materials, which may have special and differential analytical impurity signatures (for example, the barytes impurities found in the early Bovey Tracey ceramics) and also the processing conditions used in the firing of the paste and in the preparation of preliminary mixtures of the raw materials, such as the calcined bone ash and the addition of flint glass cullet to early glassy porcelains.

The major input of analytical science to the determination of porcelain body compositions, however, has undoubtedly emanated from the accessibility of shards found in factory waste pit sites. The consideration of the work undertaken thus far on English and Welsh porcelains has revealed that the adoption of experimental porcelain trials to improve the product quality and robustness has been much more frequent than was at first envisaged by merely accessing the factory records and personal diary entries alone. In fact, several conjectural statements have been propagated and sustained by ceramic historians over many years which when questioned just cannot now be borne out in the light of hard analytical evidence obtained from the factory shards, or in some cases from finished pieces of porcelain. For example:

- The Bow porcelain factory exclusively made soft paste porcelains;
- The Swansea factory had only ever made three types of body paste in its history, namely, glassy, duck-egg and trident magnesian porcelains;
- The Nantgarw factory only ever had one composition body, and that was a soft paste porcelain which was high in a calcined bone ash component;
- The Coalport factory changed its body composition dramatically when Billingsley and Walker joined the factory in 1820;
- The Derby factory changed its porcelain body composition when Robert Bloor took over there in 1807 and that this gave rise to the characteristic crazing of the glaze noted in the Bloor Derby porcelains;
- Caughley persisted in the manufacture of a soft paste porcelain product until it closed in 1799 and its operation was then taken over by the Coalport China Works;
- Pinxton had a novel paste composition formulated by William Billingsley and John Coke compared with its neighbouring china works at Derby;
- Rockingham experimented with a variety of porcelain paste bodies between 1826 and 1842.

All of these statements originated from potential mis-interpretations of the reading of the historical documentation, which to be fair, is often incomplete and can be misleading if taken out of context. So, the revelations of analytical science and their impact upon long-held historical views about the origins and growth of the English and Welsh porcelain manufactories and their products were bound to come into collision with accepted doctrine at some time and this will increase in magnitude as more analyses are carried out. An important figure in the establishment of the English porcelain industry was William Cookworthy (Penderill-Church, *William Cookworthy, 1705–1780: A Study of the Pioneer of True Porcelain Manufacture in England,* 1972) about whom much has been written. In a recent book on the life of William Cookworthy, who first discovered a source of kaolin and Cornish stone near St Austell and thereby founded one of the first English porcelain factories to make hard paste porcelain at Plymouth in 1768, Brian Adams (Adams, *True*

Porcelain Pioneers: The Cookworthy Chimera, 1670–1782, 2016) made the following recent statement in a preface to his book:

> *Several new theories have been proferred in recent years. Some, out of the enthusiasm of their authors, are supported by proof, definite evidence, and science that are actually nothing more than the fancies of overactive imaginations, misguided sources and a lack of basic knowledge ... Much of the background information given here is based on the work of previous authors, but mainly on those who put some work into understanding what they were writing about, Their conclusions are not questioned, nor are their sources checked or specified when they are not relevant to the main theme of this book*

This statement directly challenges the application of analytical science to porcelains and further suggests that the science/arts divide is irredeemable, minimising the impact of the analytical data in implying that the involvement of analytical scientists in ceramics was rather dilettante in concept and suffered from a deficiency in knowledge. Could it be that the true reason lying behind the potential mistrust shown to analytical science by some art and ceramic connoisseurs is that the opinion expressed by art experts is normally based upon many decades of experience and study and that the analogous analytical scientist is perceived perhaps not be able to offer so much experience in the application of their science to art analysis? However, it should also be realised that the analyst has probably similarly spent several decades in acquiring the experimental expertise to undertake correctly their scientific measurements and in the objective interpretation of their observed data. In the case of the present author for example, more than five decades of personal research in analytical techniques and their application to chemical problems is counterbalanced by over five decades of personal study and collecting Welsh and English porcelains, handling them and writing about them: not every analyst is a tyro in the field of application to artworks and several have spent significant periods personally in learning about porcelains, their origins of manufacture and their histories, just as have the ceramics connoisseurs themselves, whose own knowledge of the scientific principles may also be rather desultory!

The author recently came across a quotation from the classic literature of over 500 years ago which beautifully reflects the most polarised statements of analysts and connoisseurs and their contretemps in interpretation which results from two intransigent viewpoints and an apparent inability to perceive the other side of the story. In 1516 Sir Thomas More wrote his famous book, briefly entitled "*Utopia*", whose full title in its original Latin is "*Libellus vere aureus, nec minus salutaris quam festivus, de optimo rei publicae statu deque nova insula Utopia*" (More, 1516, translated into English in 1551 by Raphe Robinson as "A little true book, both beneficial and enjoyable, about how things should be in the new island Utopia"). In this book, More deplores people who cannot see anyone else's point of view, with a most apposite quotation:

> *He thinketh himself so wise that he will not allow another man's counsel* (R. Robinson, *More's Utopia*, transl, 1551).

Which surely describes those entrenched protagonists against the adoption of the holistic forensic approach advocated here for porcelain attribution.

It has been suggested personally to the author that perhaps art historians see the interpretations of chemical analysis as a grey area: as more and more analyses are now being performed on porcelains, the two extreme approaches of *hard* and *soft* science surely need to be brought together to reinforce a healthy collaboration, which will address potentially incorrect historical assumptions and statements and thereby facilitate a true engagement in establishing the holistic forensic appreciation of the artworks. Simply stated, experts in analytical science, in historical document retrieval and in handling and judging porcelains are all needed for the holistic settlement of an attribution. Realistically, however, there will always most certainly be a failure to agree upon a source of origin for the most difficult samples and this will perhaps then require a postponement of the decision whilst awaiting further research and the potential access to more specimens of the same type.

8.5 Public Perception of the Two Cultures of Scientific Analysis and Expert Opinion

The influence of the media in the interactive role play between scientific analysis and expert opinion is claimed to be a contributory factor to the interest generated in the application of science to artwork identification. Nowhere more so is this seen but in the scientific investigation of old master paintings – and this is the parallel mantra of the holistic approach which we have been considering here.

The public fascination with antique porcelains has certainly grown in recent years and the ceramic medium is now truly appreciated as a fine work of art much more generally than it has perhaps been in the distant past, where it was the regime of a relatively small number of appreciative connoisseurs. Even a century ago, fine porcelains were being brought to a wider public awareness through trade cards accompanying chocolate and cigarettes sold over the counter in shops, advertising porcelains and the merits of different factories historically to the general public! Shown here are two examples of this advertising medium, namely, a Fry's chocolate card (No. 15) from a series of 15 entitled *China and Porcelain,* issued in 1907 (Fig. 8.15), showing a Nantgarw China Works plate alongside a young girl in a floral dress. Others in the series depicted an eclectic mixture of Coalport, Bow, Dresden, Satsuma, Plymouth, Longton Hall, Willow, Nankin, Chelsea, Bristol, Sevres, Worcester, Crown Derby and Delft. The information provided about the factory on the reverse of each card needs to be approached with some little caution despite its presumed derivation from original historical writings and documents: for example, the Nantgarw text accompanying the card is:

> *The little factory at Nantgarw was started in 1811 and was practically abandoned in 1819. This porcelain is noted for its transparency and whiteness. Many writers have described it as superior to Bow, Chelsea or early Worcester. Owing to this fact, and its scarcity, forgery is rife and the mark in red is suspicious. The mark NANT-GARW C.W. is given on the other side.*

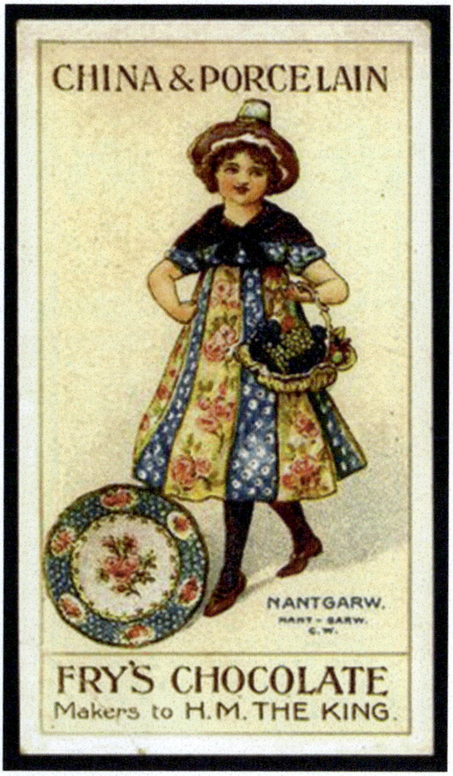

Fig. 8.15 Fry's chocolate bar gift card, No. 15, from the series *China and Porcelain*, issued in 1907, entitled *Nantgarw Porcelain*

We can infer from more recent studies that this thumbnail sketch is very accurate except for the initial sentence, where the start-up and closure date years of the factory are mentioned: otherwise we can note that already in 1907, there was an appreciation of a large number of potential forgeries and fakes in circulation on the market and the spurious red stencilled *Nantgarw* mark was even then recognised as a suspected fake mark – some 35 years before Morton Nance described it as such so eloquently in his authoritative text on Nantgarw porcelains (Morton Nance, *The Pottery and Porcelain of Swansea and Nantgarw,* 1942). Whereas some of the porcelain factories on the cards issued in this series may seem a little odd, such as Nankin, Satsuma and Willow the card selection did bring the large variety of porcelain factories to the public attention.

In 1912, R.J. Lea of Manchester issued another large series of 50 cards entitled *Old English Pottery and Porcelain* which accompanied Chairman cigarettes and these included most of the English porcelain factories we have studied here, with Swansea, Nantgarw, Dresden and Sevres and potteries such as Whieldon. For comparison purposes, the Nantgarw card (No. 14 in the series) is shown in Fig. 8.16.

The growth of public interest in scientific analysis at the end of the nineteenth century doubtlessly emanated from a Victorian inquisitiveness into the power of science to solve crimes, problems encountered in public life and in the workplace.

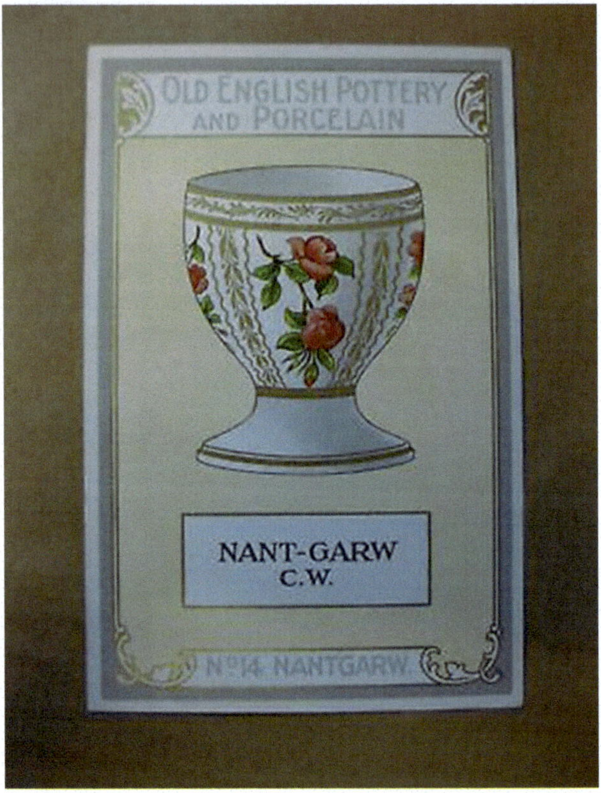

Fig. 8.16 Chairman cigarettes card, No. 14, from the series *Old English Pottery and Porcelain* issued in 1912 by R.J. Lea of Manchester, entitled *Nantgarw Porcelain*

In classic literature, for example, the great Gothic novelist Edgar Allan Poe, with his detective Auguste Dupin featuring in the *Murder in the Rue Morgue, Mystery of Marie Roget* and *The Purloined Letter,* and Wilkie Collins in *Armadale,* brought scientific principles into focus for complex murderous plots and their eventual unmasking. The snippets of chemical science brought into Charles Dickens' *A Tale of Two Cities,* the rise of the first practising "forensic detective", Sherlock Holmes, in Conan Doyle's the *Sign of Four* and *A Study in Scarlet* in the late 1880s, culminating in the novels of Agatha Christie from the 1930s, all explored the involvement of chemical poisons, the acquisition of databases and the principles of detection to expose criminal activities. Until that time, scientific analysis was the preserve of the highly trained specialist, whose prowess eventually gave birth to forensic science and the fight against crime from the 1920s onwards. It is now the application of these same forensic principles that are being advocated to assist in the attribution and identification of porcelains, their origins, their decoration and their body and glaze composition using all the information available from a holistic analytical approach in the widest sense.

References

B. Adams, *True Porcelain Pioneers: The Cookworthy Chimera, 1670–1782* (B. Adams, Dawlish, Devon, 2016)

D. Battie, The eyes have it, *Antique Collecting*, 28/9, Foreword page (March 1994)

P. Bradshaw, *18th Century English Porcelain Figures, 1745–1795* (The Antiques Collectors Club, Woodbridge, 1981)

W. Burton, *A History and Description of English Porcelain* (Cassell & Co. Ltd., 1st Edition reprinted by Wakefield EP Publishing Ltd., 1902)

W. Chaffers, *Marks and Monograms on Pottery and Porcelain, with Short Historical Notices of Each Manufactory and an Introudctory Essay on the Vassa Victilia of England* (J. Davy & Sons, London, 1863)

Sir A.H. Church, *Cantor Lectures: Lecture IV*, "Soft paste porcelains, European and Oriental", J. Soc. Arts, January 14th, pp. 126–129 (London, 1881)

Sir A.H. Church, *English Porcelain: A Handbook to the China Made in England During the 18th Century as Illustrated by Specimens Chiefly in the National Collection*, A South Kensington Museum Handbook (Chapman & Hall Ltd., London, 1885 and 1894)

V. Clewes-Salmon, "Wentworth Castle", *Antiques Magazine*, Issue 958, 12th April – 2nd May, pp. 56 and ff., 2003

P. Colomban, H.G.M. Edwards and C. Fountain, Raman Spectroscopic and SEM/EDXS analysis of Nantgarw soft paste porcelain. J. Eur. Ceram. Soc., submitted for publication (2020)

L.W. Dillwyn, *Notebooks,* reproduced in Eccles & Rackham, 1922

H. Eccles, B. Rackham, *Analysed Specimens of English Porcelain in the Victoria and Albert Museum Collection* (Victoria and Albert Museum, London, 1922)

H.G.M. Edwards, *Derby Porcelain: The Golden Years, 1780–1830* (Penrose Antiques Ltd., Thornton, 2017a)

H.G.M. Edwards, *Swansea and Nantgarw Porcelain: A Scientific Reappraisal* (Springer, Dordrecht, 2017b)

H.G.M. Edwards, *Nantgarw and Swansea Porcelain: An Analytical Perspective* (Springer, Dordrecht, 2018)

H.G.M. Edwards, *Porcelain to Silica Bricks: The Extreme Ceramics of William Weston Young, 1776–1847* (Springer, Dordrecht, 2019)

H.G.M. Edwards, P. Colomban, B. Bowden, Raman spectroscopic analysis of an English soft paste porcelain plaque-mounted table. J. Raman Spectrosc. **35**, 356–361 (2004)

H.G.M. Edwards, R. Denyer M.J. Denyer, *The Pendock-Barry Derby Porcelain Service: A Holistic Reappraisal*, to be published (2019)

S.W. Fisher, *Blue and White Porcelain of the Eighteenth Century: An Illustrated Descriptive Account of the Early Soft Paste Production of Bow Chelsea, Lowestoft, Derby, Longton Hall, Bristol, Caughley and Liverpool Potters, ca., 1740–1800* (B.T. Batsford, London, 1947)

J. Haslem, *The Old Derby China Factory* (George Bell, London, 1876)

L. Jewitt, *The Ceramic Art of Great Britain from the Prehistoric Times Down to the Present Day,* vols. I and II (Virtue & Co. Ltd., London, 1878)

W.D. John, *Nantgarw Porcelain* (The Ceramic Book Company, Newport, 1948)

W.D. John, *Swansea Porcelain* (The Ceramic Book Company, Newport, 1958)

W.D. John, G.J. Coombes, K. Coombes, *The Nantgarw Porcelain Album* (The Ceramic Book Company, Newport, 1975)

A.E. Jones, S.L. Joseph, *Swansea Porcelain: Shapes and Decoration* (D. Brown & Sons Ltd., Cowbridge, 1988)

E. Morton Nance, *The Pottery and Porcelain of Swansea and Nantgarw* (Batsford, London, 1942)

J.V. Owen, M.L. Morrison, Sagged phosphatic Nantgarw porcelain (ca. 1813–1820): Casualty of overfiring or a fertile paste? Geoarchaeology **14**, 313–332 (1999)

J.V. Owen, N.G. Panes, Bow and A-marked porcelain: A tangible link from the Stratford (East London) factory site. Trans. Engl. Ceram. Circle., 153–162 (2012)

J.V. Owen, J.O. Wilstead, R.W. Williams, T.E. Day, A tale of two cities: compositional characteristics of some Nantgarw and Swansea porcelains and their implications for kiln wastage. J. Archaeol. Sci. **25**, 359–375 (1998)

J. Penderill-Church, *William Cookworthy, 1705–1780: A Study of the Pioneer of True Porcelain Manufacture in England* (D. Bradford Barton, Truro, 1972)

E.G. Ramsay, W.R.H. Ramsay, An A-marked porcelain tea canister: Implications for early English porcelain production, World of Antiques and Art, August 2005 – January 2006 issue (Andre Jaku Publishing, New South Wales, Australia, 2005), pp, 76–79

E.G. Ramsay, W.R.H. Ramsay, Bow First Patent porcelain: New discoveries on science and art. The Magazines Antiques (Brant Publications, New York, September, 2006), pp. 122–127

E.G. Ramsay, W.R.H. Ramsay, Bow: Britain's Pioneering Porcelain Manufactory of the Eighteenth Century, *The International Ceramics Fair and Seminar, Park Lane, London,* 16pp (2007a)

W.R.H. Ramsay, E.G. Ramsay, A classification of Bow porcelain from First Patent to closure, 1743–1774. Proc. R. Soc. Vic. **119**, 1–68 (2007b)

W.R.H. Ramsay, E.G. Ramsay, A case for the production of the earliest commercial hard paste porcelains in the English-speaking world by Edward Heylyn and Thomas Frye in about 1743. Proc. R. Soc. Vic. **120**, 236–256 (2008)

W.R.H. Ramsay, A. Gabszewicz, E.G. Ramsay, *Unaker* or Cherokee clay and its relationship to the Bow porcelain manufactory. Trans. Engl. Ceram. Circle **17**, 474–499 (2001)

W.R.H. Ramsay, A. Gabszewicz, E.G. Ramsay, The chemistry of A-marked porcelains and their relationship to the Heylyn and Frye Patent of 1744. Trans. Engl. Ceram. Circle **18**, 264–283 (2003)

W.R.H. Ramsay, G.R. Hall, E.G. Ramsay, Re-creation of eth 1744 Heylyn and Frye ceramics patent wares using Cherokee clay: implications for raw materials, kiln conditions and the earliest English porcelain production. Geoarchaeology **19**, 635–655 (2004)

R. Robinson, *More's Utopia*, English Translation thereof by Raphe Robinson, 1st edition, with a foreword by William Cecil, Lord Burghley, Lord High Treasurer of England and Chief Adviser to Queen Elizabeth I, published by Abraham Veale at the sign of The Lamb (St Paul's Churchyard, London, 1551)

J. Twitchett, *Derby Porcelain: An Illustrated Guide, 1748–1848* (The Antiques Collectors Club, Woodbridge, 2002)

I.J. Williams, *The Nantgarw Pottery and its Products: An Examination of the Site* (The National Museum of Wales and the Press Board of the University of Wales, Cardiff, 1932)

Chapter 9
The Future for the Holistic Analysis of Porcelains

Abstract The necessity for a holistic approach to a porcelain factory attribution for unknown specimens is outlined, wherein the scientific analytical data and their interpretation, the connoisseurship expertise and the relevant statements contained within historical documentation are all assessed on an equal footing to arrive at a consensus conclusion. In some cases, a re-evaluation of the existing evidence will be indicated where disagreement between one or more the individual components of the original assessment is present and a balanced judgement advocated for the evidential information being considered.

Keywords Holistic analysis · Scientific analysis · Historical documentation · Connoisseurship · Expertise

9.1 The Holistic Approach

In this book the adoption of a combined holistic forensic analytical scientific, expert opinion and historical documentary approach to the factory attribution of a piece of porcelain has recommended a focus upon three contributions which should be treated with an equal weighting of importance; the scientific (chemical) analysis, the documentation (historical) and the expert (connoisseur) opinion. It is imperative to evaluate these independent assessments for all new investigations into porcelains of doubtful origins and not to dismiss one or more because they do not conform with the others and thereby comprising an "outlier". Taken individually, it should be appreciated that just one of the recommended three contributions alone and in isolation should not command the authority and credibility of a result that is formulated upon the available evidence derived from all three taken together. In some cases, we may of course expect naturally that some deviation in potential attribution could occur when, for example, two distinct contributions point to a particular assessment and the third does not; the resultant procedure should then normally involve a considered and weighted re-interpretation of all three to provide a synergistic

H. G. M. Edwards, *18th and 19th Century Porcelain Analysis*, https://doi.org/10.1007/978-3-030-42192-2_9

compromise. This would be expected to occur especially for rare and possibly unique pieces whose style, composition and shape do not conform to the standard patterns normally expected and attributed to the products of a particular factory, which may arise from an unusual ceramic artwork exemplified by a one-off commission executed to a special request by a valued client.

Although the advocacy in an ideal situation is for an equal treatment of all three contributions, in reality, the largest number of porcelain evaluations will have been undertaken primarily and necessarily by expert opinion and connoisseurship based on many years experience, and this is only right – but this also is by its very nature not infallible as we have seen here! As Archbishop Justin Welby, Primate of England, has advised in the national press: "There is *opinion* and *fact* – do not confuse the two", which in our case translates as expert *opinion* cannot be considered in the same dimensional arena as analytical scientific data or historical *facts*. Secondly, the perusal of the existing historical documentation can be accomplished from the extant literature held in museum archives and in private hands and to this can be added the new documentation which surfaces from time to time which can perhaps shed some light or clarity on ambiguous statements which have been made in the existing literature hitherto. Generally, however, the historical documentation is rather more frequently open to a re-interpretation by modern historians, who attempt to place it in more contextual surroundings, and this itself may confer different meanings upon apparently straightforward factual statements made over two centuries ago. We have already drawn attention to the potential unreliability of eye-witness statements made about factory personnel and practices which hitherto have had irrefutable reliability conferred upon them by earlier historians and authors: however, as we now realise, the reliability of eye-witness accounts in modern forensic scientific investigations can be considered extremely untrustworthy even in cases of apparently authoritative and unequivocal witness statements. This evidential unreliability of personal eye-witness accounts in modern forensic investigations has been recognised by law enforcement agencies and treated with caution by modern historians: this is acknowledged by acclaimed authors such as Laurence Rees (Rees, *Their Darkest Hour*, 2007; *The Dark Charisma of Adolf Hitler*, 2012) who realised that whereas his personal interviews of subjects who had lived under Adolf Hitler's rule offered some very significant clues pertaining to his subject matter, he also cautioned against the unverified acceptance of eye-witness testimony which sometimes proved to be at variance with accepted and established independent documentation. Laurence Rees pointed out that, in almost all instances, his acerbic and well-appreciated insight into the charisma surrounding Adolf Hitler was derived from a weighted and balanced assessment of the eye-witness statements taken with documentation which elaborated upon the complex and turbulent social and political ethos of the time. One of the that witnesses that Rees interviewed, an SS officer on Hitler's administrative staff, admitted that his initial perception of Adolf Hitler's infallible judgement was shrouded by a charismatic personality which later became manifest as megalomania – which gave two very different personal interpretations of the events he recalled some years later to his interviewer. In a parallel context, a similar balanced judgemental appraisal therefore needs to be applied to the categorical statements made by early historians about the key manufactory personnel and the reasons for their critical decisions made during the often rather short lifetimes of many early porcelain manufactories.

Finally, the relative rarity of chemical analyses performed and reported upon porcelains thus far has been commented on already and the possibility of utilising analytical data interpretation for all cases of modern porcelain attribution alongside the other two components, namely, expert opinion and historical documentation, is sometimes not always achievable. However, historically, a hierarchical degree of attributive acceptance exists in the series: expert opinion > historical documentation > chemical analysis, generated from a observed frequency of usage and this probably relates from the earliest times when so little chemical analysis had been performed on porcelains which then naturally relegates the perception and interpretation of analytical scientific data to a lower degree of acceptability than expert opinion.

The relative rarity of the reported chemical analyses on specimens of porcelain compared with the extensive contributions of the other two components historically is undoubtedly further amplified on the basis that most analytical determinations have been reported from shards and porcelain fragments and very few indeed from perfect and finished pieces – the very pieces that usually have been at the origin of expert opinion evaluations. This dichotomy is easily explained because most owners or curators of expensive porcelains are not enamoured of analysts removing small specimens for the conduct of their experiments. This applies generally to all artworks and any damage to a ceramic object is reflected negatively in its appreciation and resultant financial evaluation by collectors, who generally prize perfection in their specimens. Another result of scientific analysis is the perception that its conclusions can destroy a previously held opinion about the sourcing of a particular piece and thereby affect its value immensely: this has been noted in the scientific analyses of oil paintings which historically have sometimes been incorrectly attributed to a particular period or to a particular artist. There now exists a catalogue of paintings which have failed scientific tests carried out upon them in recent years and they have now been relegated to categories labelled as copies or fakes: their previous value based on their attribution of originality or authenticity, usually pronounced from an expert opinion, is severely compromised thereby and this makes owners rather reluctant naturally to offer their art works for analysis if therefrom a "trashing" of the formerly well-appreciated artwork results. An oil painting has the advantage in that small micro-samples of pigment can be removed acceptably during its restoration or conservation, without detriment to its appearance and this has little or no parallel for porcelains. It is also the case now that many of the major art auction houses insist upon scientific analysis being carried out on important oil paintings prior to their sale to expose any irregularities and potential litigation which may be cause for a later objection by a potential purchaser, and this is establishing the modern basis of the acceptance of scientific data in forensic art analysis.

The indicated way forward, therefore, is the use of non-destructive and non-invasive analytical instrumentation for the interrogation of porcelains which is amenable to both curators and collectors. The adoption of such techniques should facilitate the achievement of attaining the goal of porcelain evaluation using the three contributions highlighted above. The adoption of Raman spectroscopy as an analytical technique for porcelain body characterisation is subtly different from that resulting from an elemental oxide determination. Both techniques provide complementary information about porcelain bodies but the Raman molecular spectral data are rather more distantly removed from the derivation of recipe formulations. Nevertheless, in this book an attempt has been made to demonstrate the potential versatility of this analytical technique in accessing information from perfect pieces of porcelain, ranging from glaze formulation and body composition to pigment utilisation, whilst also proposing some explanations for surface blemishes on otherwise perfect ceramics. In the current analytical science climate several other modified techniques are being proposed for the non-destructive testing of art and archaeological objects, such as mobile XRF and laboratory-based environmental SEM/EDAXS.

9.2 Potential Analytical Techniques for the Dating of Ceramics and Porcelains

Several suggestions have been made in the past about the possibility of using analytical techniques to date porcelains precisely, so providing a potential way of identifying modern reproductions and fakes of ancient porcelains and facilitating their discrimination from the genuine articles: however, the main drawback to such an exercise would be that the possible inbuilt temporal errors could be large. For example, in a 200-years old piece of porcelain an experimental error of only +/−50 years would be enough to cast some doubt upon the reliability of the determination for an unequivocal attribution to a particular factory. Thermoluminescence is the accepted analytical method for the dating of ancient ceramics, such as Roman Samian wares where it has become the standard technique of choice, but here the artefact age timescale is approaching 2000 years and not 200 years as for most European porcelains, so an error in the determination of +/−50–100 years is perfectly acceptable generally for the correct placement of a period timeline.

The principle of thermoluminescence is that when materials are subjected to exposure to the environment they absorb cosmic or gamma radiation which creates trapped electrons in their body matrices – additionally, if buried, the artefacts can acquire naturally occurring radioactive elements from their depositional surroundings. For ceramic materials which have been fired to temperatures higher than 500 °C, thermoluminescence is ideally suited as a dating technique as the heating process to which they have been subjected effectively zeroes their background free electron count, so a measurement of the current photon luminescence from an object can lead directly to a determination of its age (Aitken, *Thermoluminescence Dating,* 1985; Aitken, *Introduction to Optical Dating,* 1998).

X-Ray fluorescence on the other hand depends upon the emission of characteristic secondary and fluorescent X-rays from materials upon bombardment with high-energy X-Rays and gamma radiation and it is particularly important for ceramics, glasses and minerals as each X-Ray emitted is at a characteristic wavelength of a particular element (Beckhoff et al., *Handbook and Practice of X-Ray Fluorescence Analysis*, 2006; Van Grieken and Markowitz, *Handbook of X-Ray Spectrometry*, 2002). In scanning electron microscopy (SEM) a specimen surface is scanned with a focussed beam of electrons which interact with atoms present in eth sample to give information about the surface composition: in the associated energy dispersive X-Ray spectroscopy (EDAXS) the characteristic X-Rays emitted during the scanning process are matched with the elements to provide compositional abundance data (Goldstein et al., *Scanning Electron Microscopy and X-Ray Microanalysis*, 1981).

A more recent dating technique for ceramics is that of RHX, which is based upon the reversible rehydroxylation of fired clays over time, developed from the observation of the thermal shrinkage (contraction) of fired archaeological bricks which was seen to be dependent on their age (Brazil 2019). The absorption of moisture by fired clay bricks progresses slowly according to the fourth root of the time and a logarithmic plot of moisture uptake with time has a slope of 0.25 (Wilson et al. 2003). Measurement of the moisture uptake of refired archaeological clay specimens heated to temperatures of >500 °C attributed to the rehydroxylation of bridged Si-O-Si bonds to form Si-OH bonds can thus be related to the age of the ceramic when correlated to the environmental exposure conditions of their burial environments (Clelland et al. 2015). Archaeological specimens from well-dated burial sites were analysed, including seventh century CE Anglo-Saxon and first century CE Roman Samian wares: a specimen of the latter which had been identified with a production in Gaul during 45–75 CE was dated using RHX to 59 +/−30 CE (Wilson et al. 2012). It should be noted that these analyses have all been accomplished with archaeological specimens and that this would not be immediately transposable for more modern porcelains, which would generally have been glazed: probably, the best comparator in the eighteenth and nineteenth century ceramics field would therefore be unglazed porcelain shard wasters from burial pits.

Of these cited techniques, only the thermoluminescence and rehydroxylation methods are applicable as dating procedures for ceramics and in itself these are probably not very applicable for the detection of contemporary fakes or for the factory attribution of unknown porcelains synthesised relatively recently and then perhaps only since the eighteenth century. In summary, the accelerator mass spectrometry/radiocarbon dating techniques that have been so applicable to the dating of archaeological biomaterials and organics unfortunately have no counterpart for the inorganic bodies of ceramics and porcelains.

9.3 Elemental Oxide Versus Molecular Spectroscopic Analyses

The analytical data from these two techniques are complementary in that, although both provide compositional information, the former provides quantitative elemental percentages which can be idealised as metal oxides whereas the latter yields semi-quantitative chemical structural information about the molecules and molecular ions which are present in the specimens and indicated which elements are linked with each other and bonded together. These types of information are subtly different regarding the interpretations which may be forthcoming from them in their analytical conclusions, Hence, the elemental oxide data from a porcelain specimen can be related to the raw materials used in its construction and their relative proportions whereas the molecular spectroscopic data reveal the chemical species that have been created in reactions at the high temperature in the processing kiln. The latter does not generally give direct information about the raw materials used in the synthesis of the porcelain body but an inference can be made about which of these raw materials were used or not, such as bone ash (calcium phosphate), soapstone (magnesium silicate), cullet (lead silicate) and borax (sodium borate decahydrate). The presence of derivatives of silicates, aluminates and phosphates and of polymorphs of silica such as tridymite and cristobalite are indicative of the processing conditions and especially the temperatures of the reactions which have occurred in the kiln: these are critically important for categorising correctly the type of porcelain operation being undertaken in the factory. Additionally, information about the surface blemishes and the composition of sagged specimens can be obtained from both techniques. Finally, the usage and composition of the pigments used in porcelain decoration is best achieved using the molecular spectroscopic method, which determines unequivocally which elements are bonded together: the detection of lead, mercury, sulfur and oxygen by SEM for example could be interpreted in several ways for pigment analysis but the molecular spectroscopic method can conclusively identify the presence of cinnabar, calomel, minium, galena, massicot and plattnerite as individual pigments, all formed from various chemical combinations of the same elements.

An important difference between the elemental oxide and the molecular spectroscopic techniques, however, is not so much the data interpretation but rather the methodology of its acquisition: the SEM method is potentially destructive of specimens, even using micro-sampling, whereas the RS method is non-destructive and non-invasive: the resultant sampling regime is therefore a simple one – SEM usually needs to be adopted for application to broken specimens, fragments and shards whereas RS can be used quite safely with finished and perfect pieces of specimens, even for cases of very rare or unique examples of a factory product, and this must be a very large positive advantage over the taking of even minute microspecimens from otherwise perfect artefacts, where even the mechanical procedures of extracting a sample can cause damage to highly stressed matrices and bodies, such as chipping, cracking and flaking.

9.4 The Potential of Raman Spectroscopy for Dating Porcelains: Is This a Viable Procedure?

Recently, several research papers have appeared in the literature in which the authors have claimed to use Raman spectroscopy to date ancient artefacts or relics: in all of these cases, however, the materials are essentially organic in origin and the basis for the approach is the determination of the extent of the environmental deterioration suffered by the artefact or relic and its relation to an arbitrary timescale. The major problem, of course, is that the environmental deterioration being assessed spectroscopically is a variable parameter and the combined influence of radiation insolation (exposure to sunlight), humidity, temperature, exposure to reactive chemicals (salts, acidic and alkaline environments) and biological colonisation are all different and changes are often difficult to quantify over a period of time, which could be many centuries.

The first attempt to use Raman spectroscopy as a specimen dating method was a study of antique Italian drinking glasses by Bertoluzza et al. (1995). These authors examined the Raman spectra of 28 drinking glasses which they grouped into 8 age classes between 1750 and 1940 to characterise the authenticity by age of these objects. Sampling was strictly forbidden from these brittle objects on account of the potential damage that could result from drilling or from the vibration induced during the excision of even microspecimens. Although the Raman spectra did not show any change or relationship with age in the samples studied the fluorescence emission intensity increased significantly with the age of the sample. Mathematically, the logarithm of the ratio of the Raman band intensity of the 1080 cm^{-1} stretching band of the silicate matrix and the broad fluorescent band intensity showed a linear relationship with the known age of the sample. The authors concluded that the method therefore presented a non-destructive spectroscopic method of evaluating the age of a glass specimen and thereby its authenticity. This paper does not seem to have been picked up for its potential application, not perhaps to porcelain body age discrimination, but for the dating of the glaze on finished porcelain specimens which is materially tantamount to offering the same dating experiment. It is interesting that the research of Bertoluzza et al. on their drinking glasses spanned a wider temporal range that is required for the assessment of the porcelains that we have been considering here, *viz.* the period 1750–1840. It would be quite realistic, therefore to institute a research programme to evaluate the relationship between porcelain glazes and their age determined in a similar fashion: the question of establishing "authenticity" of the porcelain from such a study is another matter, however, as age alone does not determine this absolutely. For example, we have already seen that questions have been raised relating to the attribution of some early Bow porcelains and to the products of other contemporary factories such Chelsea, Limehouse and the so-called "A-marked" porcelains. Nevertheless, the methodology, with its advantageous non-destructive capability that is so essential for the interrogation of perfect specimens of porcelains in collections could be used for selected demonstrations of authenticity for the identification of late-period interlopers and more modern fakes being passed off as genuine early porcelains from a particular factory. The only other example of such a Raman spectroscopic procedure being adopted for archaeological artworks in this context is provided by the study of mediaeval painted and stained glass shards from a monastic foundation (Edwards and Tait 1998).

In a sequel to this idea, an opportunity arose for the assessment of RS as a dating device using the Bertoluzza principle and it appears that, after conducting some trial experiments on known factory examples of porcelains which can be dated precisely, the method was judged unfortunately to have no applicability to the dating of porcelain glazes. Examples of the porcelain studied in this experiment were: Derby *Prince of Wales* service (1786), Derby *Earl Camden* service (1795), Derby *Viscount Cremorne* service (1790), Swansea *Burdett-Coutts* service (1815–1817), Nantgarw *Duke of Cambridge* service (1817), Nantgarw *Lady Seaton* service (1817–1819).

References

M.J. Aitken, *Thermoluminescence Dating* (Academic, London, 1985)

M.J. Aitken, *Introduction to Optical Dating* (Oxford University Press, Oxford, 1998)

B. Beckhoff, B. Kanngiesser, N. Langhoff, R. Wedell, H. Wolf, *Handbook and Practice of X-Ray Fluorescence Analysis* (Springer, Dordrecht, 2006)

A. Bertoluzza, S. Cacciari, G. Cristini, C. Fagnano, A. Tinti, Non-destructive *in situ* Raman study of artistic glasses. J. Raman Spectrosc. **26**, 751–755 (1995)

R. Brazil, Chemical clocks for ancient artefacts. Chem World **16**(/9), 54–57 (2019)

S.-J. Clelland, M.A. Wilson, M. Carter, C.M. Batt, RHX dating: Measurement of the activation energy and rehydroxylation for fired clay ceramics. Archaeometry **57**, 392–404 (2015)

H.G.M. Edwards, J.K. Tait, FT-Raman spectroscopic studies of decorated stained glass. Appl. Spectrosc. **52**, 679–683 (1998)

G.I. Goldstein, D.E. Newburg, P. Echlin, D.C. Joy, C. Fiori, E. Lifshin, *Scanning Electron Microscopy and X-Ray Microanalysis* (Plenum Press, New York, 1981)

L. Rees, *Their Darkest Hour* (Ebury Press/Random House, London, 2007)

L. Rees, *The Dark Charisma of Adolf Hitler: Leading Millions into the Abyss* (Ebury Press/Random House, London, 2012)

R. Van Grieken, A.A. Markowitz, *Handbook of X-Ray Spectrometry* (Marcel Dekker Inc., New York, 2002)

M.A. Wilson, W.D. Hoff, C. Hall, B.M. McKay, A. Hiley, Kinetics of moisture expansion in fired clay ceramics: A (time $^{1/4}$) law. Phys. Rev. Lett. **90**, 125503 (2003)

M.A. Wilson, A. Hamilton, C. Ince, M. Carter, C. Hall, Rehydroxylation (RHX) dating of archaeological pottery. Proc. R Soc. A **468**, 3476–3493 (2012)

Chapter 10
Summary and Conclusions

Abstract A brief summary and some conclusions resulting from this research study now follow. An appreciation of the information derived from the two types of analytical experiment surveyed here, namely the elemental oxide qualitative and quantitative determinations, and the molecular spectral species signature identifications reinforce their complementarity for the characterisation of porcelains, comprising the bodies, glazes and also the pigments used in their decoration. A feature of these analytical studies is the work carried out on shard materials excavated from the waste pits of derelict porcelain manufactories and the comparative rarity of the corresponding and complementary work carried out on perfect finished porcelain artefacts is outlined. A key requirement for future studies in this area will be the need for specialised instrumentation which can interrogate rare and perfect porcelain specimens from museum and private collections without the need for sampling, even at the microscopic level. Thereby, a definitive database of analytical information will be realised to assist in the attribution of source factory origin to unknown or doubtful pieces. Finally, the requirement for a genuine holistic approach to porcelain assignment and attribution is advocated, whereby the three areas of specialist input will be accommodated on an equal basis of interpretation for a mutually agreed decision, namely, analytical science data, expert opinion and connoisseurship, and historical documentary research. In this context the importance of the unequivocal interpretation made of the information obtained from all these three components is stressed to produce a forensically acceptable result for the attribution of unknown pieces of porcelain.

Keywords Porcelain bodies · Glazes · Pigments · Elemental analysis · Molecular species identification · Holistic approach · Connoisseurship · Analytical science · Historical factory documentation · Interpretation

H. G. M. Edwards, *18th and 19th Century Porcelain Analysis*,
https://doi.org/10.1007/978-3-030-42192-2_10

10.1 What Has Analytical Science Achieved in Porcelain Identification?

The major contribution of analytical science to the characterisation of porcelains from the first analyses carried out towards the end of the nineteenth century and continuing into the last decade or so has undoubtedly been in the quantification of the elemental oxides which comprise the porcelain body compositions and the deduction therefrom of the raw materials that were used in their manufacture. With the advent of analytical instrumental techniques used currently such as XRay diffraction, Scanning Electron Microscopy and Energy Dispersive XRay Spectrometry and XRay Fluorescence Spectrometry the determination of the elemental composition of porcelain bodies and their glazes has been advanced to provide quantification at the bulk and microdomain levels.

The identification of the mineralogical composition of porcelains is a particularly favourable aspect of analytical work and has enabled the correlation between the starting materials in the formulation recipes and the products of the high temperature chemical reactions occurring in the kiln to be accomplished. Generally, however, this has been achieved using microsampling techniques, which necessitate the use of damaged porcelain artefacts or of shards excavated from the factory waste pit sites. In contrast, the development of remote probes and stand-off imaging devices for the non-invasive interrogation of specimens has facilitated the recent adoption of molecular spectroscopic analytical instrumentation; thereby the identification of the minerals and species involved in the porcelain bodies and glazes can be carried out in a non-destructive mode which does not involve the invasive sampling of porcelain specimens or of their chemical or mechanical treatment in any way prior to the investigation. This is particularly advantageous for the analytical examination of perfect and finished porcelain items without compromising their integrity and without even incurring even minimal superficial damage.

The information available for the elemental and molecular spectral analytical techniques is complementary and self-supporting and several ceramics studies have been carried out recently using a combination of both techniques. Clearly, many more studies need to be carried out in a combinatorial way to develop and to extend further a database of characteristic porcelain body features which could be recognised as a definitive forensic indicator of the factory origin of a sample or specimen. In this respect, the other two aspects which could contribute to the unequivocal definition of the origin of a porcelain artefact must also need to be invoked, namely, the interpretation of relevant historical documentation relating to the factories and their key personnel (whose personal diaries and workbooks should not be under estimated) and the expert opinion of ceramics historians, collectors and curators who have acquired many years' experience in assessing and attributing the products of a particular factory. The agreement between the contributions made from the interpretations of all three approaches would be quite demonstrably reassuring for the correct attribution of a particular piece of porcelain to a source origin. However, in the case where all three lines of research actually fail to agree on an affirmative and cognisant result there will still be a problem in securing a proper unequivocal attribution, and then it should be necessary to review the individual cases to locate the source of the problem in which the disagreement resides. One particular approach should not then be taken as supernumerary or more important than the other two in this respect unless there is a specified reason for so doing. From the analytical case studies presented in this book, the major bone of contention that arises between the three approaches used to define the factory of origin of a porcelain artefact centres upon the interpretation being made of the key features of each individual contributory approach, whether this be the elemental oxide percentages which may be considered to be unusual for the standard factory production, a mis-interpretation of the factory records or statements made by proprietors about their experimental changes in composition, or a mis-identification of a factory specimen based upon an erroneous expert opinion occasioned by its unusual texture, style, shape or potting quality which has led to this misconception. This is expected to be particularly susceptible for specimens from the very early porcelain factories where it became necessary to amend the porcelain body composition, often arbitrarily and experimentally, and perhaps with or without establishing a written record of this procedure, to achieve a more marketable and finer or more robust product in the face of intense competition from rival manufactories.

10.2 The Forensic Requirement for the Holistic Approach to Porcelain Attribution

The acquisition of scientific data from porcelain specimens which generally would comprise the quantitative elemental oxide percentages or the qualitative molecular spectral identification of minerals, molecules and molecular ions also needs to be examined in terms of the evidential interpretation required to formulate the conclusions which will determine its viability as unimpeachable factual information in a forensic sense. Firstly, the analyst must be certain that the analytical data are truly representative of the specimens being analysed and that distortion of the chemical compositional data does not occur from interference between the biscuit porcelain body, the glaze and any applied pigments: this can normally be achieved through multiple sampling and the use of replicates from different regions or domains of the samples which is necessary to obviate or counteract any effect of heterogeneity in the specimen. Secondly, a key component of the data interpretation that then follows will be the comparison of the analytical results from the unknown specimens under analysis with the analogous data obtained ideally with similar analytical instrumentation from specimens which have been authoritatively designated through connoisseurship and historical documentation as standard products of the porcelain factories in question. This is an area which could itself generate some speculative argument if the so-called "standards" are themselves not fully accepted by expert opinion as belonging to Factory X or Y; indeed, in some cases, the attribution of a specimen to a particular factory based upon expert opinion alone is itself already open to question in the light of analytical data which do not apparently support such a conclusion and this too can be misleading for making any further advance in the subject! This is an area which encompasses shards from excavated factory waste pits and unusual specimens of a factory output which have nevertheless been accepted hitherto as genuine products of that factory but which now in reality should perhaps be re-assessed using a more holistic interpretation. In other words, the acceptability of porcelain standards as specimens which then become the gold standard for future analytical data comparative assessments must be unequivocal and unimpeachable – and this is rather more indefinitely achievable when the specimens themselves are perhaps fragmentary or unmarked. The forensic nature of the chemical analytical data interpretation that is advocated for adoption here as an integral part of the holistic and definitive attribution of an unknown piece of porcelain is *sine qua non* rigorously applied and must therefore provide unique solutions without offering avenues for debate which encompass the possibility of alternative derivations and conclusions. In a recent book, Patricia Wiltshire (Wiltshire, *Traces: The Memoir of a Forensic Scientist and Criminal Investigator,* 2019), has written a fascinating account of her decades spent as a forensic scientist, citing from the hundreds of cases on which she has worked and providing a detailed description of the critical procedures required to acquire meaningful analytical data and information from specimens, which often cannot be accessible through other means, all of which required that the forensic evidential data and their interpretations were eventually offered up for public scrutiny in the courtroom. It is interesting that Dr. Wiltshire's initial approach came from the multidisciplinary scientific analysis of archaeological specimens and that her procedural interpretations have the same sensibilities and rigour that we have found here and applied to the scientific analysis of porcelains.

10.3 Future Work

From the various arguments and case studies presented in this work it can be seen that the future adoption of the holistic approach to porcelain attribution demands an appreciation of situations where divided opinion is expressed between the three components of science, historical documentation and expert opinion which afford observable differences in perception for the attribution of porcelains. The detailed complexity of scientific data interpretation which is often difficult to comprehend by laypersons does little to contribute to the debate and it is the overall conclusion which is then relied upon. Likewise, expert opinion and connoisseurship which has been accumulated through many decades of personal study of porcelains should not be dismissed lightly: any discrepancy in outcome between these two component parts of the analysis should be investigated and explanations should not be merely dismissed to the detriment of the proper attribution of an artwork. The advent of the viability of analytical science demonstrated here in the non-destructive and non-invasive interrogation of perfect and rare pieces of porcelain could establish a marker for future work in this area as the pieces which are examined by experts and by chemical analysts are then identical and do not involve the comparative assessment stage necessarily coming between shard fragments and the finished, decorated pieces. As compositional changes were made to porcelain body recipes, sometimes unsuccessfully and often perhaps not documented properly, so stylistic changes could also have been made which then have created apparently non-standard shapes for the typical factory output and it is here that the independent inputs of scientific and expert connoisseurship especially can be brought together effectively to deduce the correct attribution for unknown specimens.

It is in this context that the categorisation of unknown pieces, including those with unusual or non-standard applied marks, as fakes comes into focus: it is realised that several large collections of porcelains may well contain some pieces of a rather questionable origin which by their association with genuine pieces from a particular factory gives them some credibility in attribution to the same source, perhaps even as "rare" items. These should obviously merit a targeted analytical scientific investigation to establish their potential authenticity as has occurred with collections of oil paintings and art works referred to earlier – where some surprising results have been obtained *vis-à-vis* their attribution to the *genuine* and *fake* categories. Again, the adoption of non-destructive and non-invasive analytical techniques such as molecular spectroscopy could be of invaluable assistance in this respect for providing ancillary information for the determination of the origin of suspect artefacts. Analytical scientists have continuously attempted to improve the accuracy of their techniques and the determination of the quantitative percentages of elemental oxides and mineral components in heterogeneous specimens attests to this: however, it is a fact that significant improvements of the highest precision or accuracy of these determinations really will have little effect upon the eventual interpretation of the data for the attribution of porcelain factory origin purposes since, as we have seen hitherto, the changes in original body composition were often made arbitrarily and small changes in the relative composition of the raw materials and their sometimes rather selective mixing at source would render the acquisition of more precise analytical data rather superfluous. What is of more importance is the comparison between reliable standards and specimens and a range of factory products to establish some realistic sigma mathematical error bars and ranges over which the acceptable percentages of elemental oxide compositions lie for that particular factory period of production.

It is to be hoped that the next few years will see more applications advanced in this direction which will further strengthen the collaboration and discourse between analytical scientists and art connoisseurs, especially in the ceramics arena, for the eventual and unequivocal attribution of factory origin to some of the finest ceramic art works. In closing, a very recent success for holistic analytical science has been the approval of a Van Gogh self – portrait, which had been purchased as genuine in 1910 by the Norwegian National Museum, but whose authenticity had been challenged from the 1970s. Eventually in 2014, the Norwegian National Museum submitted the painting for assessment to the Van Gogh Museum in Amsterdam, where it was subjected to analytical testing. At that time the Norwegian Museum owners were told categorically "You might not like the results and it might be that we will never find out". In January 2020, the pronouncement was made that the painting was a genuine work of art by Vincent van Gogh and that it represented a psychotic period of the artist's life in the late summer of 1889 during which he languished in the Saint-Remy asylum in the south of France. Several features of the painting's composition reflected this period as well as his strong use of the palette knife in its construction and Mai Britt Guleng of the Norwegian Museum summed up this definitive result appropriately as "It is reassuring to know it's genuine". The Norwegian Museum is to be applauded upon taking the decision to undertake analysis of their artwork, as otherwise there would always have been reasonable public doubt and conjecture about the authenticity of the work which has now been unequivocally classified as genuine.

Reference

P. Wiltshire, *Traces: The Memoir of a Forensic Scientist and Criminal Investigator* (Blink Publishing, Chelsea/London, 2019)

Appendices

Appendix I: What Quantities of Raw Materials Were Used in a Typical Kiln Charge?

Although there are several recipes in existence which give details of the composition of the porcelain bodies as formulated before they were fired in the biscuit kiln, very few of these were specific in terms of the actual "charge" or quantity of raw materials which were compounded for a single firing process used in the manufacture of porcelain. For example, Lewis Dillwyn in his *Notes on Experimental Porcelain Bodies* (reference to Eccles & Rackham, *Analysed Specimens of English Porcelain in the Victoria and Albert Museum Collection,* 1922, and reproduced from the originals preserved in the museum archive and lodged there in 1921) does provide examples of the relative weights of the raw materials that he used in his trial experiments at the Swansea China Works between 1815 and 1817 with the assistance of his kiln manager, Samuel Walker, but does not state the quantities used in the kiln charge. Simple addition of the weights of each component from Dillwyn's data results in an estimate of the total charge for the kiln being approximately 60 lbs (27 kilos). On reflection, this charge seems surprisingly small, when it is realised that perhaps at best only a few hundred pieces might result from an extensive and prolonged firing process which would require up to 2 days to heat the kiln to the maximum temperature of *ca.* 1300–1350 °C, then to maintain this temperature for up to another 4 days and finally to lower the temperature gradually over another period of a further 2 days or more. Hence, at least a whole week was required for the kiln firing cycle to be initiated and completed per charge of biscuit porcelain.

It is found that a typical Swansea cup and saucer in duck-egg porcelain weighs approximately 200 g, a dessert plate 320 g, a dinner plate 400 g and a bowl 300 g, so a small kiln charge of 27 kilos would only represent perhaps some 80 individual items at most, of which kiln losses would be in Dillwyn's estimation some 75%, so only some 20 pieces of porcelain would be expected to have been produced

H. G. M. Edwards, *18th and 19th Century Porcelain Analysis*,
https://doi.org/10.1007/978-3-030-42192-2

realistically per charge for each trial firing on the basis of the recipe formulation given. This must be considered to be ineffectively small when the cost of firing up the kiln and monitoring its temperature regime is taken into account. At the Pinxton China Works between 1795 and 1799, William Billingsley calculated the operational costs of wood and charcoal fuel and maintenance of the biscuit kiln and he informed the proprietor, John Coke, of his findings and concerns: Billingsley reckoned that each kiln charge should produce about 25 tea sets per week for commercial sale after factoring in his estimated kiln losses of about 15%, this being surely a conservatively low figure? Billingsley also reckoned that he would lose up to 15–20% of finished, decorated porcelain in transit to the client or agent. This is borne out by a letter from Joseph Lygo, the Derby China Works London agent, to William Duesbury, the proprietor of the Derby factory, beseeching him to increase his supplies of china per order by up to 20% to allow for damage in transit and to ensure that this additional increment of china from the works was decorated to the standard required for the other pieces in the same service.

Each basic standard tea set from a china works, for example, would comprise twelve trios (tea cup, coffee cup and saucer), a slop bowl, a sugar bowl/box and lid, a milk/cream jug, a teapot, lid and stand and possibly a hot water jug and lid, with additional extras such as small tea plates and special items such as serving dishes which could be provided as desired or commissioned by the client. A breakfast service would also include larger versions of the tea cups, known as breakfast cups and possibly additional items such as a muffin dish and cover, egg cups and stand and assorted larger plates: the *Hensol Castle* Swansea breakfast service comprised over 100 pieces of porcelain with additional items of the sort described here. The basic tea or coffee set would therefore consist of some 42–45 separate items of porcelain with a total weight of approximately 5.5 kilos. This means that only 4 or 5 sets would have been produced for a week's firing at Pinxton, based on the quantities specified in Dillwyn's detailed work notes at Swansea and his recipe manifest without kiln losses, far below the estimate quoted by William Billingsley to John Coke for the average production run of 25 sets that Billingsley wrote that they needed to make at Pinxton per week in a single kiln firing process to achieve profitability.

Clearly, a scale-up factor needs to be employed and we can get some idea of this from a handwritten recipe provided by Josiah Spode II (1755–1827), who was the son of Josiah Spode I and founding proprietor of the Spode China Works in Staffordshire. In 1820, Josiah Spode II revealed the recipe for his "*China Body No.6* as:

160 lbs blue clay; 240 lbs Cornwall clay; 360 lbs Cornish stone; 400 lbs bone; 160 lbs flint; 5 oz (0.3 lb; 150 g) blue calx for stain. All intended for one charge of the kiln. A description of *blue calx* follows later.

This recipe adds up to a total of 1320 lbs (600 kilos) and would therefore represent about 3000 pieces of porcelain per firing charge which is significantly different (about 20 X larger) from the previous estimate based on Dillwyn's figures in his own recipe formulations.

Blue calx probably refers to the "blue calx of copper" described by John Strachey FRS in his treatise of 1727 (*Observations on the Different Strata of Earths and Minerals and More Particularly of Such as are Formed in the Coal-Mines of Great Britain,* 1727). Calx is defined as a residue, usually in the form of a fine powder, which is left after the calcination of a mineral at high temperatures, its name derived from the Latin for "lime" – and calcination itself from the heating of lime. The most common blue carbonaceous mineral of copper is azurite, basic copper carbonate, $Cu_2CO_3(OH)_2$, which upon heating eliminates carbon dioxide and leaves copper(II) oxide, CuO. Its addition to an alkaline medium in the form of a fused glaze forms a blue colour which removes traces of yellow in the incipient porcelain body (which generally can be ascribed to traces of iron impurities in the raw materials), which was desirable for the Spode bone china clear white translucency. Dillwyn had set out to achieve a similar result using blue smalt, a glassy cobalt silicate (Edwards, *Nantgarw and Swansea Porcelain: An Analytical Perspective,* 2018).

Morton Nance in his well-researched book on Nantgarw and Swansea porcelains (*The Pottery and Porcelain of Swansea and Nantgarw*, 1942) has found from his studies of Dillwyn's workbooks that the Swansea China Works entered into a contract for the supply of 11 tons of high quality china clay from the Gewgraze Mine near St Austell in Cornwall per annum, which was to be delivered by sea via the Bristol Channel to Swansea docks. Simple calculation reveals that if Dillwyn's kiln charge was as little as 26 lbs. of china clay per firing then 11 tons of china clay would be sufficient for almost a thousand firing cycles, that is approximately 20 per week, an inordinately large number for a small china works! Even if Dillwyn utilised a second biscuit kiln this would not have approached the china clay raw material consumed on an annual basis. We must conclude, therefore, that the published recipes provided at the china factories were scaled down and that these would be increased according to the local kiln capacities. In summary, therefore it should be realised that the detailed analysis of errors in weighing carried out in this text would be made even worse when larger weights and less accurate industrial scale balances were employed for porcelain synthesis. In contrast, on the basis of Josiah Spode's formulation recipe for his firing charge of *China Body No. 6* given above, the 11 tons of china clay would provide about 100 runs per annum for a single kiln or one run per week for two kilns run in tandem, which a large factory such as Spode might well promote. Of course, a rather simple potential explanation for the reduced size of a kiln charge particularly for experimental studies relating to porcelain body improvements is that only a few pieces were needed to establish whether or not a sensible improvement in china quality had resulted, with consequent savings on raw materials but also introducing an unknown factor in process scale-down using full size kilns which is sometimes difficult to quantify commercially.

An entry in Cox and Cox (*Rockingham Porcelain,* 1745–1842, 2001, page 175) is relevant to this discussion: the standard porcelain biscuit recipe in Thomas Brameld's notebook, cited in Appendix II of Cox & Cox, comprised the following mixture – 1 part blue clay, 11/2 parts Cornish clay, 1 part Cornish stone and 21/2 parts calcined bones. The Cornish clay was obtained from St. Stephen's China Clay & Stone Co., St Austell, and from the Meledor China Clay Co., St. Columb, and shipped to Hull Docks, whence it was transported via the River Ouse and the Don Navigation Canal to the Swinton wharf factory site. The access to waterborne transport via rivers and canal systems in the 18th and early nineteenth centuries was vital for the operations of a ceramics factory, whereby raw materials could be shipped directly in ton quantities from seaports and even more important, the finished porcelains could be despatched safely to their agents and consumers in large cities such as London. The consequences of transporting quantities of delicate porcelains by waggons on rutted and bumpy roads or turnpikes resulted in significant damage being caused and, as indicated above, Joseph Lygo insisted upon the Derby factory sending about 20% more porcelain items than the number required on their orders to allow for this (Anderson 2000). Many successful porcelain factories, therefore, were sited with wharves at seaports or near the river and canal systems, such as Coalport, Caughley, Nantgarw, Bow, Limehouse, Bristol, Plymouth, Chelsea and Liverpool.

References

J.A. Anderson, *Derby Porcelain and the Early English Fine Ceramic Industry, ca, 1750–1830*, PhD thesis (University of Leicester, 2000)

A. Cox, A. Cox, *Rockingham Porcelain, 1745–1842* (Antiques Collectors Club Publishing, Woodbridge, 2001)

L.W. Dillwyn, Notes on experimental porcelain bodies, reproduced in H. Eccles & B. Rackham, *Analysed Specimens of English Porcelain in the Victoria and Albert Museum Collection* (V&A Museum Publications, South Kensington, London, 1922)

H.G.M. Edwards, *Nantgarw and Swansea Porcelains: An Analytical Perspective* (Springer, Dordrecht, 2018)

E. Morton Nance, *The Pottery and Porcelain of Swansea and Nantgarw* (Batsford, London, 1942)

J. Strachey, *Observations on the Different Strata of Earths and Minerals and More Particularly of Such as are Formed in the Coal-Mines of Great Britain* (John Walthoe Publisher, Royal Exchange, Cornhill, London, 1727) Appendix II, #276 *Copper with an Organic Form*, page 126

Appendix II: A Combined Analytical Study of the Nantgarw Porcelain Glaze on Shard No. NG6 and Others (Colomban et al. 2020; Edwards and Surtees 2020): Implications for Nantgarw Porcelain Attribution

The idea for this analytical study was generated by the discovery of an unusual type of porcelain composition in a Nantgarw shard, numbered NG6, which had been obtained from an archaeological excavation of the China Works site in the 1990s. As described in a case study in Chap. 8, this shard analysed as a highly siliceous hard paste porcelain body hitherto not encountered in the Nantgarw production, and, of even greater interest, it was also glazed which meant that the porcelain biscuit body had survived the initial kiln firing stage. It was decided to try and seek some further information about the composition of this glaze from a combination of Raman spectroscopy (RS) and SEM/EDAXS molecular spectroscopic and elemental oxide analytical experiments (Colomban et al. 2020). The primary objective of this work is to verify if an analytical differentiation was possible between the Nantgarw glaze used by William Billingsley and Samuel Walker between 1817 and 1819 (known historically as the "Billingsley glaze") and that used by their successors, William Weston Young and Thomas Pardoe (known historically as the "Pardoe glaze"), between 1820 and 1823, when the Nantgarw China Works finally closed. Ceramics historians and connoisseurs have noted that the "Billingsley glaze" (also called the *Nantgarw No. 1 glaze*) was much whiter and more reflective than the so-called *Nantgarw No. 2 glaze* used by Young and Pardoe (and often termed the "Pardoe glaze") which was thinner, of a creamy hue and often appeared "pitted" superficially by comparison with the whiter, smoother *Nantgarw No. 1 glaze*. Although the Nantgarw phosphatic porcelain body composition of Billingsley and Walker was a closely guarded secret and was never revealed during the lifetime of William Billingsley, John Taylor (*The Complete Practical Potter*, 1847) published a recipe for Nantgarw porcelain which he claimed came from Samuel Walker, some 19 years after the death of William Billingsley in 1828. However, the formulation of the *Nantgarw No. 1 glaze* was never formally made public anywhere. William Weston Young and Thomas Pardoe were engaged primarily in decorating the residual stock of china left "in the white" at the Nantgarw China Works after the departure of Billingsley and Walker for Coalport early in 1820 (Edwards, *Nantgarw and Swansea Porcelain: An Analytical Perspective,* 2018; Renton 2005) and to this purpose they invented the *Nantgarw No. 2 glaze* for application to the biscuit porcelain items remaining there, which would have then been glazed, decorated and sold locally. This locally decorated china is now highly prized by collectors and characteristically has the thinner *Pardoe* or *Nantgarw No. 2 glaze,* is simpler in decoration (although still beautifully decorated by Thomas Pardoe and others at the factory), has a deficiency of gilding (this being replaced in the interests of economy by coloured edging and borders, especially favouring blue, chocolate brown and green colours) and often exhibiting small firing defects and perhaps small blemishes which would have rendered the pieces unacceptable to the London agent, John Mortlock of Oxford Street (John, *Nantgarw Porcelain*, 1948).

In our study here, it was of interest to examine analytically the glazes on the Nantgarw shards provided from the archaeological excavations of the Nantgarw China Works site to see if any differentiation between the glaze specimens could be obtained. Firstly, the glaze compositional analyses in terms of their elemental oxide content are summarised in Table 1 for the five shards NG2, NG6, NG7, NG8 and NG9, from which it can be clearly seen that shard NG6 is highly unusual: with a very high silica content of 85% and an almost insignificant lime content of only 0.2%, this specimen is highly siliceous as also befits its porcelain body type attribution to a *hard paste* porcelain. The other four shards do also show some compositional glaze variation but generally group around the standard glaze composition of NG2, selected as our Nantgarw standard.

The following nine glazed specimens, which comprise seven Nantgarw glazed shards (including the five analysed by SEM/EDAXS above and two others), one Nantgarw finished saucer and an analogous contemporary specimen from the Swansea factory in the finest duck-egg porcelain of the same vintage (Edwards, *Swansea Porcelain: The Duck-Egg Translucent Vision of Lewis Dillwyn,* 2017; Jones and Joseph, *Swansea Porcelain*, 1988) were also analysed using RS for molecular spectral information.

G1. Shard NG6. (Fig. 8.2). A glazed shard which exhibits the characteristic body composition of a hard paste porcelain type; this was found to be highly siliceous and devoid of a phosphate content.

G2. Shard NG2. A glazed shard which exhibited the expected Nantgarw high phosphate/bone ash content.

G3. Unnumbered glazed Nantgarw shard, base of a bowl or tureen.

G4. Unnumbered glazed Nantgarw shard, base of a coffee cup.

G5. Three small glazed shard fragments, numbered NG7, NG8 and NG9, which were found unusually with applied enamel decoration and therefore were obviously derived from final stage firing of the enamels in the glost kiln at lower temperatures – to lose finished porcelain artefacts at this stage which demanded their destruction after firing is rather unusual and unexpected especially when the biscuit kiln firing losses in the initial stage approached 90% anyway.

G6. Nantgarw saucer, finished, perfect and complete, as shown in the coffee cup and saucer depicted in Fig. 3.7; bought by John Mortlock of Oxford Street, London agent for the Nantgarw China Works, in the white from William Billingsley, *ca.* 1817–1819, and decorated by Moses Webster in the Robins and Randall London atelier with pink roses and foliage and with dentil edge gilding. Billingsley *Nantgarw No.1. glaze.*

G7. Swansea saucer (Fig. 1), finest duck-egg porcelain of Lewis Weston Dillwyn, *ca.* 1817, set pattern 449, locally decorated at Swansea with a band of gilt berries and marked "SWANSEA" and "449" in red stencilled enamel. Glazed with Dillwyn's *Swansea No. 2* glaze.

Firstly, the RS molecular spectroscopic data can be summarised as follows:

Table 1 Glaze formulations for Nantgarw China: composition %[a]

Specimen	SiO_2	Al_2O_3	CaO	P_2O_5	K_2O	Na_2O	MgO	Fe_2O_3	PbO
Shard NG6	84.5	9.6	0.2	0.0	1.1	2.2	1.1	0.5	0.8
Shard NG2	76.7	9.7	9.0	0.0	1.2	2.3	0.3	0.2	0.5
Shard NG7[b]	75.7	13.5	6.3	0.0	1.8	1.0	0.0	0.3	1.2
Shard NG8[b]	68.5	17.4	3.0	0.0	1.4	3.6	1.5	0.8	2.5
Shard NG9[b]	86.3	4.0	0.8	0.0	1.2	4.0	1.2	0.2	0.7

[a]Colomban et al. (2020)
[b]These shards are enamelled and glazed wasters

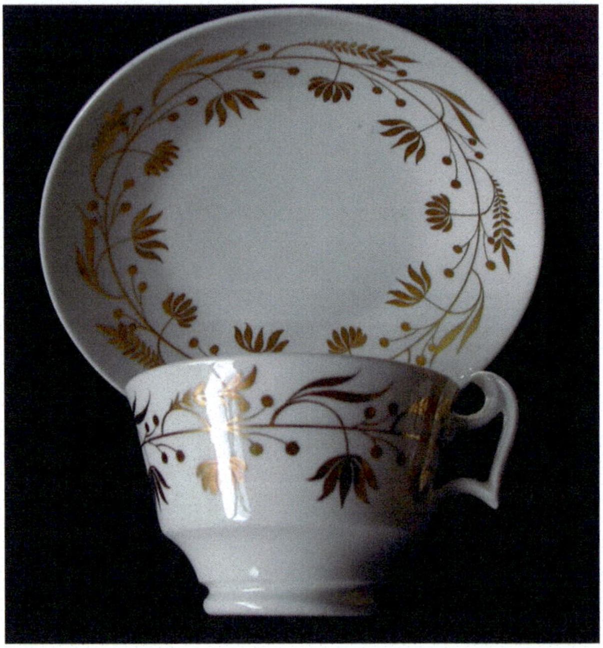

Fig. 1 Swansea duck-egg porcelain, coffee cup and saucer, London shape with triple curved ogee handle, decorated with gilt berries and foliage, *ca.* 1815–1817, marked on base of saucer with red enamel stencilled "SWANSEA" and script "449", the latter being the set pattern number. *Private Collection*

- The Raman spectra of the glaze over the wavenumber range 100–1500 cm^{-1} for the shard NG6, specimen G1, and NG9, specimen G5/3, are very different in pattern from the analogous spectrum given by the other shards (Fig. 1), such as NG2, NG7, NG8 and the unnumbered shards comprising the coffee cup footrim and the bowl/tureen footrim. The inference is that there are two distinct glazes being employed on the Nantgarw china wasters comprising the shard specimens analysed here.
- The glaze spectrum of the Nantgarw finished, perfect saucer artefact (Edwards and Surtees 2020) exactly matches that of the other "standard" Nantgarw glaze spectra (as shown in Fig. 2) and differs from that of the two shards NG6 and NG9 for which a spectral stackplot of the two types of glaze identified spectrally is shown in Fig. 3. It can be concluded from this observation that the "standard" Nantgarw glaze spectrum represents the *Nantgarw No.1 glaze* of Billingsley and Walker and that of shard NG6 is therefore a later glaze, quite possibly that of Young and Pardoe, the so-called *Nantgarw No.2 glaze*. The reasoning behind this assertion is that the finished, perfect Nantgarw china specimen has the identical glaze characteristics spectrally of the "standard" shards and that the finished specimen is unequivocally a sample of William Billingsley and Samuel Walker's best Nantgarw china, decorated in London, which means that the "standard" shards also have the Billingsley, *Nantgarw No.1 glaze*. The Raman spectral band intensity comparison between the different glazed shards is summarised in Table 2.

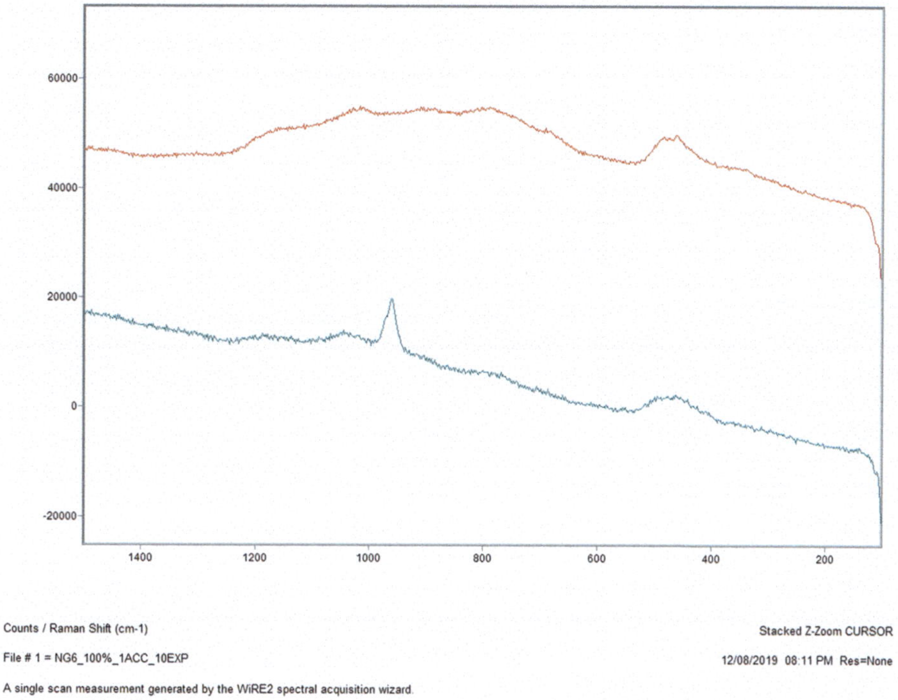

Fig. 2 Stackplotted Raman spectra of glazes, wavenumber range 100–1500 cm^{-1}, of *upper,* shard NG6 and *lower,* Nantgarw saucer

- The RS of the glazed, perfect Swansea saucer is close to that of the Nantgarw shards, with the exception of NG6, NG9, and of the Nantgarw saucer. It is concluded from this observation that, firstly, the saucer, although London-decorated by Moses Webster in the Robins and Randall atelier for a complete Nantgarw tea and coffee service commission, actually has a *Nantgarw No 1 glaze* and not a different composition glaze formulated by the London workshops. Hence, this strongly implies that the Nantgarw "in the white" porcelain that was sold to John Mortlock for his decoration commissions in London had been glazed by Billingsley and Walker locally in Nantgarw already prior to its being enamelled in London. Secondly, it transpires that the glaze composition used by Billingsley and Walker for their *Nantgarw No. 1 glaze* was perhaps close to that adopted by Lewis Weston Dillwyn at the Swansea China Works after his experimental trials which matched three types of glaze evaluated in his trials with his new duck-egg porcelain body (Dillwyn, *Notebooks,* see Eccles and Rackham 1922). Eventually, Dillwyn selected his own *No. 2 glaze* formulation in 1816/17 for his duck-egg china and, of course, he was actively assisted in these experiments by Samuel Walker, who not long afterwards left Dillwyn at Swansea to join Billingsley at Nantgarw for the Phase II start-up operations there. It is superfluous to add here that Walker would have had primary knowledge of Dillwyn's trial glaze formula-

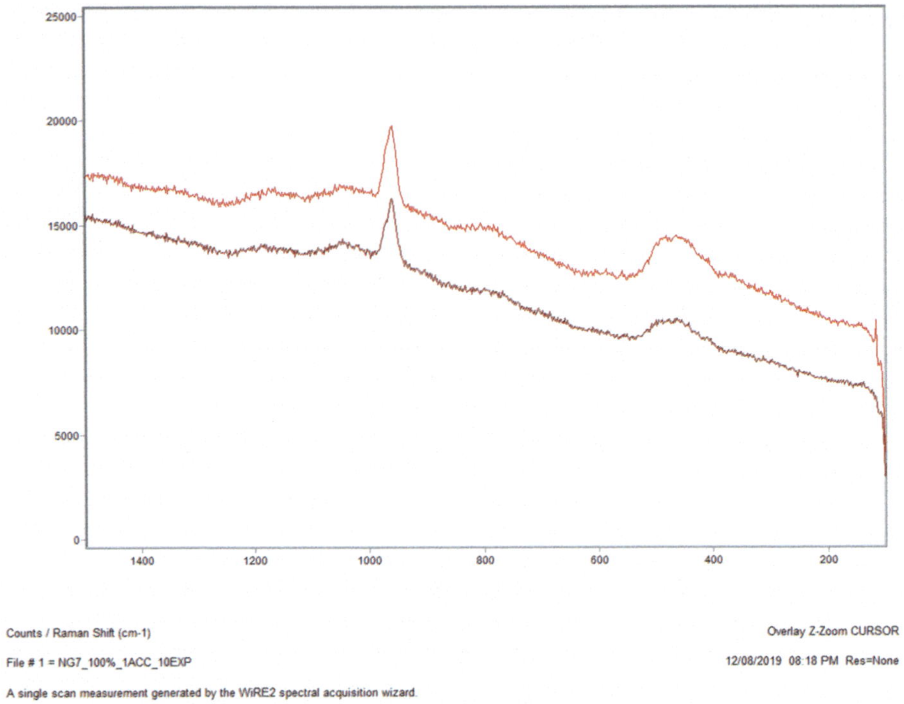

Fig. 3 Stackplotted Raman spectra of glazes, wavenumber range 100–1500 cm^{-1}, of *upper*, shard NG2 and *lower*, Nantgarw saucer

tions and experiments that he carried out at Swansea during the period 1815–1817 and it is reasonable to propose that both he and Billingsley would have been in an ideal situation to adopt a glaze recipe at Nantgarw that was perhaps quite similar to Dillwyn's *No. 2* Swansea glaze for their Nantgarw china, even though the phosphatic bodies at Swansea and Nantgarw were of sensibly different formulaic compositions.

Implications for the History of Nantgarw Porcelain

The analytical spectroscopic and elemental oxide determinations of the work undertaken on shards from the Nantgarw China Works site are quite important for our historical understanding of the china which was produced there some 200 year ago. Firstly, most authors allege and maintain that only one porcelain body was ever produced at Nantgarw and this was the esteemed, highly translucent phosphatic soft paste porcelain made by William Billingsley and Samuel Walker between 1817 and early 1820, by which time they were facing bankruptcy and had then departed for employment with John Rose at his Coalport China Works (Edwards, *Nantgarw and*

Table 2 RS glaze[a] comparison band intensity data for shards and porcelain artefacts

Specimen	Area Ratio $I^{650-1200} / I^{350-600}$	Comment
NG2 shard, G2	4.0 +/− 0.4	No. 1 Nantgarw glaze
NG6 shard, G1	6.1 +/− 0.1	No. 2 Nantgarw glaze
NG7 shard, G5/1	3.7 +/− 0.2	No. 1 Nantgarw glaze
NG8 shard, G5/2	3.6 +/− 0.2	No. 1 Nantgarw glaze
NG9 shard, G5/3	6.0 +/− 0.3	No. 2 Nantgarw glaze
Saucer, G6	3.4 +/− 0.4	No. 1 Nantgarw glaze
Bowl shard, G3	3.8 +/− 0.2	No. 1 Nantgarw glaze
Cup shard, G4	3.9 +/− 0.4	No. 1 Nantgarw glaze
Swansea saucer, G7	2.7 +/− 0.3	Dillwyn glaze

[a]Billingsley/Walker glaze: No. I Nantgarw
Young/Pardoe glaze: No.2 Nantgarw
Dillwyn glaze: No.2 Swansea

Swansea Porcelains: An Analytical Perspective, 2018). There have been several reasons proposed for this assertion and these can be summarised as follows:

- Billingsley and Walker had devised a most satisfactory porcelain body composition prior to 1817 and this was successfully put into practice, gaining many plaudits from a discerning clientele for its beauty and high-quality production (for illustrated examples, see John et al., *Nantgarw Porcelain Album,* 1975). However, kiln losses at the high temperatures required for firing the biscuit paste resulted in a commercially unacceptable loss of approximately 90% of the production at this first stage, which rendered the china manufacturing operation non-viable.
- Billingsley and Walker had neither the time nor the inclination to amend their porcelain composition by undertaking experimental trials: several possible reasons have been advanced for this conclusion, namely, they just did not have recourse to stop the production lines when all that they could successfully produce anyway was being bought up by their London agent, John Mortlock, who then sold it on for decoration to his awaiting client commissions. To stop the existing successful production whilst engaging in empirical experimentation, which may have had no fruitful outcome eventually regarding china body improvement would have been judged to have been foolhardy for cash flow purposes and this would also have consumed valuable supplies of raw materials which may not have been recouped through sales finally – an expensive exercise and undertaking. Also, Billingsley and Walker would have been aware of the precisely similar situation that faced their former colleague, Lewis Weston Dillwyn, at the neighbouring Swansea China Works, who just a year or so before had started some experimental trials to improve the robustness of his porcelain, which extended over 2 years between early 1815 and August, 1817. Although these experiments did realise Dillwyn's objective, in the meantime they also decreed that he lost his market edge with a discerning clientele, and their loss of affection for his presumed lower quality product then resulted in his having to sell his factory at a loss to Timothy and John Bevington in 1820. This situation would have been in the minds of Billingsley and Walker as they had been involved actively in Dillwyn's decision to seek an new experimental porcelain body at Swansea—both were then employed at Swansea by Dillwyn, and Billingsley departed in late 1816 in disagreement with Dillwyn's decision, and Walker a little later in 1817, having stayed on to assist Dillwyn in his final experimental practicalities. The message was clear: the extended interruption of an already successful commercial production, despite its inherent processing faults, to hopefully achieve a better product was a dangerous exercise and might probably backfire economically into a disastrous loss for the China Works.
- A major problem historically has been that Billingsley and Walker initially kept their porcelain body, glaze formulations and recipes top secret, such that even their third partner, William Weston Young, without whose financial acumen, investment opportunities and business contacts the Nantgarw China Works would not have been able to start up in 1817, was ignorant of the compositional data for the porcelain being produced there. However, in 1847, John Taylor (*The*

Complete Practical Potter, 1847) published a book in which he declared the recipe for the Nantgarw porcelain body which it is believed originated from information provided hitherto by Samuel Walker. This recipe was repeated in the *Pottery Gazette* issue of 1885, apparently provided by an anonymous contributor. Professor Mellor, a distinguished experimental inorganic and ceramics chemist was engaged by Ernest Morton Nance in the 1880s to synthesise a Nantgarw porcelain body based upon the Taylor and *Pottery Gazette* formulations, which he successfully achieved. Morton Nance (*The Pottery and Porcelain of Swansea and Nantgarw,* 1942, page 398) quotes a statement from Professor Mellor upon the results of his experimentation with the published formulations provided for the porcelain body and glaze:

Judging from the satisfactory appearance and behaviour of this glaze it can be unhesitatingly be assumed to be of the correct composition in every respect. It is very gratifying to know that this glaze does resemble that used by Billingsley and that the current recipes have been isolated. It is also interesting to note that both the body and glaze recipes have been provided by the anonymous writer, suggesting that he had a very intimate knowledge of the china in question.

- As a corollary, it should be noted that the two glaze formulations cited by John Taylor in his book were both judged to be rather unsatisfactory by Professor Mellor, the first of these was only partially successful and comprised the following recipe: 14 lb china clay, 18 lb Lynn sand, 14 lb bone, 13 ½ lb feldspar, 12 ½ lb china stone, 11 ½ lb flint, 110 lb borax and 110 lb lead glass. The second of the glaze recipes was found to be utterly unworkable by Professor Mellor, namely comprising, 8 lb Lynn sand, 6 lb chalk, 2 ½ lb china stone, 10 lb borax, 40 lb china clay, 3 ½ lb.feldspar, ½ lb flint, 6 lb soda and 3 lb nitre. Note that this second recipe is completely different from the first in that it does not contain any bone ash or lead glass cullet and that neither contain any lime flux specifically although the second recipe does contain chalk which would be converted in the kiln to lime. Dillwyn at Swansea was of the clear opinion that the use of chalk or calcite would be highly detrimental in the kilns because of its decomposition into carbon dioxide around 700 °C (Edwards, *Nantgarw and Swansea Porcelains: An Analytical Perspective,* 2018). Morton Nance does make an interesting statement in this context that "*Billingsley only used one Nantgarw body and one Nantgarw glaze for his china but there is evidence that he did experiment with other glazes*". Eventually, Professor Mellor tried the recipe and formulation for the Nantgarw glaze cited in the *Pottery Gazette* in 1885, comprising 50 lb sand, 60 lb borax, 20 lb whiting(lime), 4 lb nitre and 4 lb lead glass which formed a frit of which 50 lbs. were then added to 30 lb china stone, 4 lb china clay, 4 lb whiting and 4 lb lead glass. According to Professor Mellor this glaze proved "*correct in every way and most excellent samples were glazed with it*" – he assumed that this glaze recipe was handed on by Samuel Walker as it seemed to represent the true so-called *No 1 Nantgarw glaze* adopted by Billingsley and Walker between 1817 and 1819 as the final production in his opinion could not be differentiated from the original glazed and completed Nantgarw specimens of Billingsley and Walker.

- Professor Mellor also experimented with a formulaic glaze that was given in William Weston Young's *Diary*, which is assumed to be the so-called *Nantgarw No. 2 glaze* used by Young and Pardoe to glaze the remnant specimens of Billingsley's china left in biscuit form in storage at Nantgarw after his departure in 1820 for Coalport (Edwards, *Porcelain to Silica Bricks: the Extreme Ceramics of William Weston Young*, 2019). The composition of Young's glaze is 5 parts Lynn sand, 4 parts borax and either 1 or 2 parts lead glass: Professor Mellor confirmed that this glaze indeed gave very satisfactory results when applied to the Nantgarw china body but qualified it with the statement, "*.......but not as good as the Billingsley glaze*". Young's glaze in comparison with Billingsley's recipe is much simpler, with fewer components especially in that it contains no additives such as china clay, china stone, lime or nitre. We should therefore expect that the analysis of Young's glaze will be sensibly different from that of Billingsley's: indeed, from one's reference to the shard NG6 in Table 1, which is putatively but confidently now attributed to Young's glaze, it is seen that this has an almost zero content of lime (0.2%), and higher silica content (as expected from the greater Lynn sand content) than the "standard" *Nantgarw No.1* Billingsley glaze exemplified by shard NG2 (Colomban et al. 2020). The absence of lime in Young's glaze recipe could be a reflection of his concurrent and successful experiments on *Dinas* silica refractory bricks, which were very high in silica and extremely low in lime content (Edwards, *Porcelain to Silica Bricks: the Extreme Ceramics of William Weston Young*, 2019; Jenkins 1942).
- Morton Nance has expressed the opinion, without stating his source, that Billingsley did not experiment with a body formulation recipe but that he did nevertheless attempt experimental glaze trials at Nantgarw: this probably arose from several contemporary comments that Billingsley had difficulties in achieving a consistent glaze firing process for his biscuit porcelain which sometimes appeared after firing in the glost kiln with significant crazing, or craquelure, and this obviously detracted from the perfection of the fired porcelain artefact. Isaac Williams (I.J. Williams, *The Nantgarw Pottery and its Products: An Examination of the Site*, 1932) commented on a number of glazed shards in the waste pit which exhibited this unusual crazing effect. Morton Nance also comments that Billingsley eventually managed to achieve an effective re-glazing operation whereby he was able to re-fire and recover his damaged glazed porcelain using perhaps a new glaze formulation in the glost kiln at a higher temperature of between 1100 and 1120 °C to remove the craquelure effect. The other Nantgarw glazed shard specimens listed in Table 1 are therefore possibly evidence of Billingsley's glazing attempts with changes in composition, especially in lead glass frit and lime component content.
- The burning question remains in the realm of the forensic arena: did William Weston Young manufacture porcelain at Nantgarw after the departure of William Billingsley and Samuel Walker in early 1820? Morton Nance (*The Pottery and Porcelain of Swansea and Nantgarw*, 1942, page 397) makes a very controversial statement about William Weston Young, namely that "*Young's porcelain made at Nantgarw is of a similar quality to that of Billingsley – an unassailable fact and*

entirely accurate!". The basis for this statement seems to be the definitive comment made by Herbert Eccles in his objective evaluation of Nantgarw porcelain exhibited at the Centenary Exhibition of Swansea and Nantgarw porcelains at the Glynn Vivian Art Gallery in Swansea in 1914, that all Nantgarw porcelain had an identical body composition. This statement, taken with the records made in William Weston Young's *Diary* copies for February and March 1821 (which are now lost, but which fortunately were transcribed by Turner, *The Ceramics of Swansea and Nantgarw*, 1897) specifying his personal manufacture of the glazes and china body at the Nantgarw China Works, seems irrevocable, yet Morton Nance is perhaps unsure of his stance here in that he also alludes to Young categorically and in a contradictory way that he "*.... probably made no new porcelain at Nantgarw*" (*The Pottery and Porcelain of Swansea and Nantgarw,* 1942, page 251 and later pages 397 and ff.). In contrast, William Turner, the author of the first dedicated text on Swansea and Nantgarw china (*The Ceramics of Swansea and Nantgarw,* 1897, page 228) is definitive in his belief that Young did make his own porcelain at Nantgarw and he was able to interview personally the surviving former employees of Billingsley and Young at Nantgarw to elicit detailed information about the operations being carried on at the site, even making the intriguing statement that "*A harder porcelain was made at Nantgarw (author: compared with Billingsley's) by Young*". Turner then recounts that specimens of Young's porcelain were even marked in a similar fashion to those made by Billingsley and that it required some expertise to differentiate between them. Morton Nance was aware of Turner's prior information about Young's porcelain manufacture and this clearly caused him some confusion when it did not correlate with the Eccles' opinion. Of course, an entirely reasonable explanation would be afforded by the fact that all the Nantgarw exhibits at the Centenary Exhibition commented on by Herbert Eccles were actually of Billingsley's porcelain body and there were none of Young's artefacts there anyway!

- The consensus of opinion thus must be that it is very likely that William Weston Young did make quantities of porcelain at Nantgarw before he departed the site in 1823, as he was still making *Diary* records about his attempts to interest other porcelain manufacturers in his *Nantgarw Dry Mix* from 1832 onwards with especial focus from 1838, when he allied himself with John Wright at Bristol in an unsuccessful attempt to market the product more widely. In January and February, 1821, Young records that he was still persisting in the manufacture of a Nantgarw body frit!
- What is definitely agreed is that Young made a new glaze at Nantgarw – the *Nantgarw No.2 glaze* – which he needed to do to enable his decoration and enamelling plans for the residual china left over by Billingsley to be carried out. His *Diary* records that on February 24th, 1821, he "*fired up the glaze, our own, very good*" and on March 31st, 1821, the intriguing statement describing "*further glaze trials ...... old glaze now exhausted, new trials numbered 1,2,3*". These remarks prove that he had successfully created a satisfactory glaze by early 1821 which was necessary because by then the supplies of the glaze frit that were left behind by Billingsley and Walker were now exhausted .... up until that

time Young and Pardoe clearly were using Billingsley's glaze frit to glaze the remnant biscuit porcelain supplies, which apparently were present on the site in large quantities. It is significant that in the first Nantgarw public auction sale advertised in 1820, a large quantity of biscuit porcelain items was mentioned, but in the final sale in 1822 there is none, so confirming that all remnants left by Billingsley and Walker had by then been glazed! Young comments that even as early as January, 1821, he mentions his concern at the depletion of the remnant Billingsley glaze frit to which he is now adding further quantities of lead glass to eke out the supplies even further. This also carries over into the statement in Morton Nance (*The Pottery and Porcelain of Swansea and Nantgarw,* 1942, page 397) that Pardoe's painting is found on Nantgarw porcelain with the *Billingsley glaze* – this must surely only apply to porcelains decorated by Thomas Pardoe between his arrival in Nantgarw in the middle of 1820 and early 1821 because after March that year, the supplies of Billingsley's glaze had been exhausted, as declared by William Weston Young. It is interesting that modern collectors can now identify Pardoe's painting on Nantgarw china with the characteristic creamy and thin *Nantgarw No. 2 glaze*, which would have been perfected by Young in his glaze trial experiments alluded to above and which would therefore have probably only come into operation after March, 1821, and that these persisted until Thomas Pardoe died in July 1823. Pardoe's painting on the Billingsley glaze would therefore have been executed in the "early" period between mid-1820 and early 1821, whereas his painting on the *Young glaze* would be dateable to the "later" period from post-March 1821 to July 1823.

- A comment should be made about Lewis Weston Dillwyn's Swansea glaze for his biscuit duck-egg porcelain as this is mentioned through the appropriate saucer specimen listed in Table 1 for analytical comparison with the contemporary Nantgarw shards from the period 1817–1819. Dillwyn's *Notebook* diaries specify his use of a *No.2 glaze* which was favoured for the glazing of his high quality porcelain: this was one of three he trialled and which are recounted in his *Notebooks* (see Eccles and Rackham, *Analysed Specimens of English Porcelain in the Victoria and Albert Museum Collection,* 1922, Appendix). The composition of Dillwyn's *No. 2 glaze* was as follows: firstly a frit is made of 12 parts sand, 6 parts lime, 3 parts lead glass, 8 parts calcined borax and 1 part of pearl ash which would be fired ground and then added to 40 parts china stone, 28 parts lead glass, 6 parts lime and 4 parts St Steven's clay: analytically this reduces to the following composition, 38% china stone, 11% lime, 4% clay, 11% sand, 29% lead glass, 8% calcined borax and 1% pearl ash. The very high lead oxide/flint glass content should be noted – although of an indefinite composition, this could well approach a 5–10% lead oxide content and Dillwyn mentions that the key criterion is the fine grinding necessary to achieve a good homogeneity and the production of a creamy slip suitable for glazing.

Raman Spectroscopy of Glazes: Potential for the Dating of Porcelain Substrates?

In a paper published in 1995 Bertoluzza et al. described the non-destructive Raman spectroscopic study of a range of glass drinking vessels which dated form between 1750 to 1940. As expected, the Raman bands characteristic of the silica glass, which have been described in detail therein, do not show any significant differences based upon the age of the glass but surprisingly, when the Raman spectral data near 1080 cm^{-1} are compared with the broad band intensity of the accompanying fluorescence emission envelope between 1500 and 4000 cm^{-1} (centred on 2800 cm^{-1}) then a linear mathematical relationship is found between the sample age and the *log* ratio of the Raman and fluorescence band intensities according to the expression:

$$\log_{10}\left(I_{1080}/I_{2800}\right) = -25.384 + 0.013\left(\text{Age of sample}\right)$$

The range of glass wares encompassed in this research article actually matches quite well that of the porcelains we have been surveying here, especially over the first hundred or so years; say from 1740 through to 1840. Preliminary calculations reveal that around the selected date of 1815 appropriate to our consideration of the Nantagrw China Works it is possible to determine a date of production of the artefact to better than about 5 years, represented by a maximum of +/− 3.3% change in the logarithmic ratio of the band intensities.

It seems clear that if this principle can be translated into the field of ceramic glazes then it may be possible to "date" a piece of glazed porcelain with sufficient accuracy to establish its authenticity in terms of a time-line if not a factory of production as we have been working towards in this text. This information would be invaluable as an adjunct in the determination of the factory attribution as it would facilitate the narrowing down of periods when the most likely production would have taken place, so better defining the composition of the porcelain body in situations where much experimental variation in manufacture was employed.

As this exercise was undertaken to analytically determine the composition of glazes on Nantgarw porcelain the fluorescence emission was examined at 514.5 nm: unfortunately, however, this could not be determined as it was for drinking glasses, so it seems that it can be concluded that the method described for drinking glasses is not applicable at the wavelength of excitation used here to glazed porcelains. However, in a very recent paper, (Kirmizi et al. 2019), the Raman signatures of protonic species accumulated at the surface of porcelains and glazed ceramic artefacts as a result of corrosion processes induced by water over a course of time seem to be related to the age of the item concerned The relative intensities of the broad Raman bands arising from H_2O and OH moieties were found useful for the differentiation between modern specimens and fakes and ancient examples for lead free glasses. Glazed porcelains and celadons. It is implicit that lead silicates interfere with the analytical measurement procedure and this is unfortunate for the vast range of porcelains from the 18th and nineteenth centuries which used lead glazes, often with a significant proportion of lead oxide in their content. Earlier porcelains may, nevertheless, still be adaptable to this analytical measurement, which arose from Professor Philippe Colomban's earlier studies on the mediaeval stained glass in the St. Chapelle church, Paris, in which samples of the earliest mediaeval glass in the twelfth century Rose window could be differentiated spectroscopically from samples used in the nineteenth century restoration work (Colomban 2008).

References

A. Bertoluzza, S. Cacciari, G. Cristini, C. Fagnano, A. Tinti, Non-destructive *in situ* Raman study of artistic glasses. J. Raman Spectroscopy **26**, 751–755 (1995)

P. Colomban, On-site Raman identification and dating of ancient glasses: A review of procedures and tools. Journal of Cultural Heritage **9** (2008). https://doi.org/10.1016/j.culther.2008.06.005

P. Colomban, H.G.M. Edwards, C. Fountain, Raman spectroscopic and SEM/EDXS analysis of Nantgarw soft paste porcelain. J. Eur. Ceram. Soc., submitted for publication (2020)

L.W. Dillwyn, *Notes and Workbooks of Recipes at the Swansea China Works, 1815–1817,* reproduced in H. Eccles & B. Rackham, *Analysed Specimens of English Porcelain*, 1922, and in Edwards, *Nantgarw and Swansea Porcelains; An Analytical Perspective,* (2018)

H. Eccles, B. Rackham, *Analysed Specimens of English Porcelain in the Victoria and Albert Museum Collection* (Victorian and Albert Museum, London, 1922)

H.G.M. Edwards, *Swansea Porcelain: The Duck-Egg Translucent Vision of Lewis Dillwyn,* (Penrose Antiques Ltd. Short Guides, Thornton, 2017). ISBN:9780244325787

H.G.M. Edwards, *Nantgarw and Swansea Porcelains: An Analytical Perspective* (Springer, Dordrecht, 2018)

H.G.M. Edwards, *Porcelain to Silica Bricks: The Extreme Ceramics of William Weston Young (1776–1847)* (Springer, Dordrecht, 2019)

H.G.M. Edwards and A.P.H. Surtees, "A Raman spectroscopic study of early 19th Century porcelain glazes", to be published (2020)

R. Jenkins, The silica brick and its inventor, William Weston Young. Trans. Newcomen Soc. **22**(1), 139–147, (1942)

W.D. John, *Nantgarw Porcelain* (The Ceramic Book Company, Newport, 1948)

W.D. John, G.J. Coombes, K. Coombes, *The Nantgarw Porcelain Album* (The Ceramic Book Company, Newport, 1975)

A.E. Jones, S.L. Joseph, *Swansea Porcelain: Shapes and Decoration* (D. Brown and Sons, Ltd., Cowbridge, 1988)

B. Kirmizi, S. Chen, P. Colomban, Raman signatures of protonic species as a potential tool for the dating and authenticity of glazed pottery. J. Raman Spectrosc. **50**, 696–710 (2019)

E. Morton Nance, *The Pottery and Porcelain of Swansea and Nantgarw* (B.T. Batsford Ltd., London, 1942)
A. Renton, Thomas Pardoe and William Weston Young, in *Welsh Ceramics in Context, Part I*, ed. by J. Gray, (Royal Institution of South Wales, Swansea, 2005), p. 129
J. Taylor, *The Complete Practical Potter* (Shelton, Stoke-upon-Trent, 1847)
The Pottery Gazette, *Organ of the China & Glass Trades, Stationers' Hall*, No. 92, vol IX (Ludgate Hill, London, 1885)
W. Turner, *The Ceramics of Swansea and Nantgarw* (Bemrose & Sons, Old Bailey, London, 1897)
I.J. Williams, *The Nantgarw Pottery and its Products: An Examination of the Site, Cardiff* (The National Museum of Wales and the Press Board of the University of Wales, 1932)
W. W. Young, *The Diaries of William Weston Young, 1776–1847 (1802–1843)*, 30 volumes, West Glamorgan Archive Service, Swansea, SA1 3SN. https://arcgiveshub.jisc.sc.uk/data/gb216-d/dxch/ddxch/i/hub

Appendix III: William Billingsley and His Pursuit of Perfection in the Manufacture of Highly Translucent Porcelain

William Billingsley, arguably perhaps the greatest exponent of ceramic artistry on British porcelain in the late eighteenth and early nineteenth centuries, was born into a family of ceramic decorators at Derby, where his father was a ceramic painter. He was born in Derby in 1758 and apprenticed in the decorating workshop at the age of 16 in September, 1774, at the Derby China Works under the proprietorship of William Duesbury, working under his mentor Edward Withers, who was at that time the chief ceramic artist at Derby (Edwards, *Nantgarw and Swansea Porcelains: A Scientific Reappraisal*, 2017; Edwards and Denyer, *William Billingsley: The Enigmatic Porcelain Decorator and Manufacturer* 2016); Edwards, *Nantgarw Porcelain: The Pursuit of Perfection*, 2017)). In November, 1784 he was married at St Alkmund's Church to Sarah Rigley and they had three children, all born in Derby, Sarah in 1785, James in 1793 and Lavinia, born in 1795. James died in his infancy, Sarah died in January 1817 and Lavinia died in the September of the same year. William Billingsley died and was buried in an unmarked grave in Kembleton Church, Coalport in January, 1828, where he had been employed in the decorating workshop at the Coalport China Works along with his son-in-law, Samuel Walker, who was advising John Rose, the Coalport China Works proprietor, on the composition and firing of his porcelains.

In the interim, Billingsley had established a national reputation as a highly accomplished and much esteemed ceramics artist who specialised in the life-like depiction of garden flowers, especially pink roses: at Derby he was known as "the

Fig. 4 *"The Flower Painter"*, believed to be a portrait of the young William Billingsley who spent the years between 1774 and 1795 (aged 16–37) at the Derby China Works, where he was held in the highest esteem for his naturalistic painting of roses on china. Now in the Derby Museum

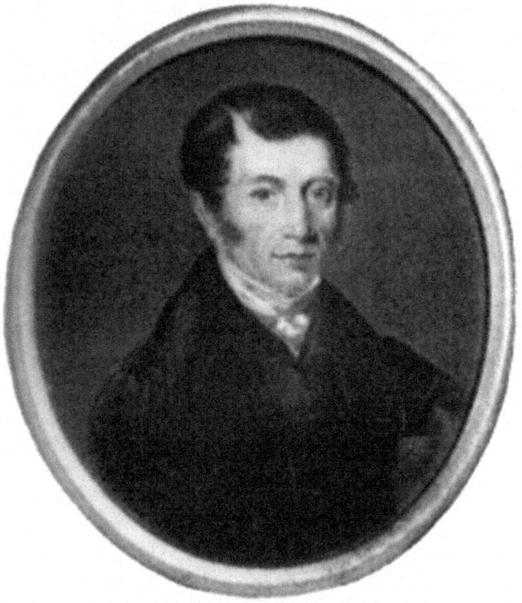

flower painter" (Fig. 4), with his characteristic artistry of "washing-out" petals with a dry brush to give a three-dimensional effect, supported by a naturalistic portrayal of the accompanying foliage and rosebuds. What could be more appropriate, therefore, in his appointment to execute his first Royal commission at Derby to decorate the *"Prince of Wales"* dessert-dinner service, ordered by Prince George the eldest son of King George III and Prince of Wales, which comprised a single pink rose and a forget-me-not spray centrally located in a circlet of gilded dots and edged with a dawn pink ground, beautifully and yet simply gilded? A dessert plate from the *Prince of Wales* Derby porcelain service, William Billingsley's first recorded prestige commission, is shown in Fig. 6.11. Each piece of this service depicts a different rose arrangement and was much admired for its cleanness of painting and beautiful accomplishment. Much admired today for its classic Georgian beauty and fine ceramic painting, pieces of the "*Prince of Wales*" Derby service are avidly collected by connoisseurs of fine porcelain and ceramic art (Edwards, *Derby Porcelain, The Golden Years, 1780–1830*, 2017). Upon the retirement of Edward Withers in 1790, William Billingsley was appointed head of the decorating workshop at Derby, which had an enviable complement of skilled artists who specialised in figures, landscapes, marine scenes, fruit, insects, animals and floral decoration, such as Zacariah Boreman, William Complin, John Brewer, Leonard Lead and William Pegg (Twitchett, *Derby Porcelain, An Illustrated Guide: 1748–1848,* 2002). A landmark at Derby was the institution of a factory pattern book in 1784, which gives much information about the commissioning of tea/coffee and dinner/dessert services and their original decorators and gilders: it should be noted that it was apparently quite a common custom to enlist several painters to execute the decoration of large services and the so-called *Rothschild* service, pattern 100, from 1790 although ascribed to Billingsley in the pattern book and verified by John Haslem (*The Old Derby China Factory*, 1876) has occasioned John Twitchett (*Derby Porcelain, An Illustrated Guide: 1748–1848,* 2002) to claim that he sees five different painting styles involved in the commissioning of this large service. However, even from an early age, the young Billingsley was involved at his parental home in Derby with his father in the decoration of porcelain purchased "in the white", which they then painted in their top floor attic studio and subsequently fired in their own muffle furnace. It is also recorded that at this time he developed his friendship with Zacariah Boreman, also apprenticed at Derby, and with whom he undertook his first trial experiments with the manufacture of porcelain in their residence at the Nottingham Arms, which Billingsley later inherited along with other properties, at 22 Bridgegate, Derby (Bambery, *William Billingsley in Derbyshire*, 2005).

It is clear from the existing documentary records at Derby that William Billingsley never made any porcelain whilst he was employed at the Derby China Works, but he obviously was germinating the manufacturing idea in the early 1790s and he must have been aware of the recipe used at Derby because of his close association with those on the synthetic processing side of the operations at the China Works. The opportunity soon arose to assist John Coke in the creation of a new porcelain manufactory at Pinxton in Derbyshire. Initially, John Coke attempted to interest William Duesbury in the undertaking of a joint venture in the manufacture of porcelain at Pinxton, but Duesbury was not enamoured of this idea. At this time, Billingsley is recorded as saying that he wanted to make the world's finest porcelain to better enhance his ceramic decoration. Billingsley departed from the Derby China Works in October, 1795 and he and John Coke started to make Pinxton soft paste porcelain with the construction of a kiln which was operational in April 1796 (Bambery, *William Billingsley in Derbyshire*, 2003). It is interesting to now make a comparison of the William Billingsley manufactured porcelains which really commenced at Pinxton between 1795 and 1799, when he left Pinxton in July, 1799, to set up a new business at Mansfield in Nottinghamshire, at premises in Belvedere Street. At Mansfield he concentrated upon the decoration of porcelain which he purchased in the white from other manufacturers, including Derby, Chamberlain's Worcester, Coalport and some imported French hard paste porcelains (Edmundson, *Billingsley and his China Artists, 1796–1808,* 2005); this being described as a "decorating establishment" in an authoritative text (Sheppard and Deverill, *Billingsley, Mansfield: The Bicentenary of William Billingsley's Mansfield Decorating Establishment, 1799–1802,* 1999) rather than a manufacturing one. Then in 1802, Henry Bankes, partner of John Coke at the Pinxton China Works encouraged Billingsley to set up a new china factory at Brampton-in-Torksey in Lincolnshire; this was a dangerous move by Bankes as it was in conflict with his legal partnership arrangement with Coke which required his undertaking not to make porcelain elsewhere (Edmundson, *Billingsley and his China Artists, 1796–1808,* 2005; Chapman, *William Billingsley at Brampton-in-Torksey,* 2003)). Nevertheless, a kiln was constructed and the manufacture of porcelain commenced at Torksey: the discovery of a large quantity of porcelain shards at excavations in Torksey (unfortunately without any stratigraphic context) confirm that the manufacturing of porcelain had recommenced there. In August 1803, Billingsley was dealt a serious blow when Henry Bankes was declared bankrupt, thereby sequestering all of the financial backing for the Torksey site. It is quite possible that the catalyst for Billingsley's start-up of porcelain manufacture in Torksey was the presence of Samuel Walker, a neighbour and future son-in-law through his later marriage to Sarah Billingsley in 1812. Walker was generally acknowledged to be a very accomplished kiln engineer and undoubtedly would have been an invaluable asset to the Billingsley operation at Brampton-in-Torksey in this regard which was not available at Mansfield. After a difficult period, during which Billingsley was still engaged in manufacturing rather precariously at Torksey, a new company, Sharpe and Co., was floated to manage the Torksey China Works: this involved five partners, namely William Sharpe, William Billingsley, Samuel Walker, James Walker and Benjamin Booth, who celebrated their factory foundation with the production of a superbly translucent jug with a monochrome landscape view of the factory, the name of the new owners and the significant date of July 1806—this is one of the very few unequivocally attributed pieces of genuine Torksey porcelain (Chapman, *William Billingsley at Brampton-in-Torksey*, 2003). It is significant that William Billingsley is described as a *decorator* at Mansfield (Sheppard and Deverill, *Billingsley, Mansfield: The Bicentenary of William Billingsley's Mansfield Decorating Establishment, 1799–1802,* 1999) whereas he is described as a *potter* at Torksey (Exley, *A History of the Torksey and Mansfield China Factories,* 1970; Gardner, *Billingsley, Brampton and Beyond: In Search of the Weston Connection – The Provenance of a Porcelain Service, Over 200 Years Old, is Investigated*, 2005). Despite this, there is evidence that Billingsley also carried out some decoration of china from other sources at Torksey, for example, the prestigious French hard paste porcelain service for Dr. Jesse Boot of Nottingham acquired in the white from the de la Courtille manufactory in Paris (1711–1841), which Billingsley painted in sepia with local landscapes and scenes (Exley, *A History of the Torksey and Mansfield China Factories*, 1970) – although there are some who believe that this was a Mansfield operation. In 1807 the firm of Sharpe and Co. was again in financial difficulties which closed down the site, so in early 1808, Billingsley and Walker, along with Sarah and Lavinia Billingsley, left Torksey and joined Martin Barr, proprietor of the Worcester China Works: here, there seems to be some confusion amongst historians about what their actual roles were --- the concensus of opinion is that William Billingsley was employed as advisor in the decorating workshop, where he undertook several commissions personally, but the roles of the two daughters and of Samuel Walker are less clear. Walker was not a decorator and probably advised on kiln firing processes, whereas Sarah and Lavinia it is believed were employed in the decorating workshops as enamellers, but perhaps also as gilders (Edwards, *Nantgarw and Swansea Porcelains: A Scientific Reappraisal*, 2017). It also transpires that Martin Barr personally encouraged Billingsley and Walker to undertake some rather clandestine trials and experiments in porcelain manufacture for him with different composition porcelain formulations and incorporating novel reverberatory kiln designs (Edwards, *Porcelain to Silica Bricks: The Extreme Ceramics of William Weston Young*, 2019).

In 1813, Billingsley and Walker decided to depart the Barr, Flight & Barr Worcester China Works upon the death of Martin Barr as it was made clear to them that the new owners, Messrs. Flight, would not be amenable to introducing modifications to their existing porcelain manufacturing operation. Prior to this they had already secured a financial incentive from Martin Barr in the amount of £200 to set up their own manufacturing operation for their new experimental porcelain, which they were free to do and did so eventually at Nantgarw in South Wales. This Phase I of the Nantgarw operation did not succeed in acquiring financial support from the British Government of the day despite the preparation of an emotive *Memorial* document submitted with examples of their decorated china presented to the Lords of the Committee of the Council for Trade and Plantations on the 5th September, 1814 (for a full transcript of this important historical document, addressed to Sir Joseph Banks and signed by William Billingsley, Samuel Walker and William Weston Young see Edwards, *Porcelain to Silica Bricks: The Extreme Ceramics of William Weston Young*, 2019). It seems credible that the recent and ongoing hostilities with the emergent United States of America and with Napoleon can be blamed for this unsuccessful action as the porcelain specimens impressed their Lordships. Despite this setback, the specimens of china submitted with their application impressed Sir Joseph Banks particularly and he brought them to the attention of Lewis Weston Dillwyn, proprietor of the Cambrian Pottery in Swansea, who wished to manufacture porcelain at Swansea and who thereby enlisted their aid. At Swansea, there is much speculation and conjecture as to whether or not Billingsley and Walker actually revealed their secret porcelain formulation to Dillwyn but it is now generally accepted that they did not, but nevertheless they assisted Dillwyn in the production of his esteemed highly translucent duck-egg porcelain in 1815 which continued thereafter until about 1817 (Edwards, *Swansea Porcelain: The Duck-Egg Translucent Vision of Lewis Dillwyn,* 2017). Billingsley decorated some of this china personally and he undoubtedly found it a good canvas for his ceramic painting skills—some of the most highly prized and finest Swansea china is painted by Billingsley. The decision of Dillwyn to move to a more robust and magnesian-rich china body in 1816/17 (See Eccles & Rackham, 1922 for a full transcript of Dillwyn's *Notebooks* containing his experiments in the systematic variation of the Swansea porcelain body) promoted the departure of Billingsley, who felt this was a retrograde step, to attempt a re-start of his Nantgarw operation, which on being joined by Samuel Walker in 1817, was financed successfully by William Weston Young and a team of local investors to generate Phase II of the Nantgarw China Works later that year. The kilns and associated potting sheds were still extant from Phase I so this required financial aid specifically to purchase raw materials and fuel for the kilns. The new porcelain, whose formulation most likely was determined in their experiments at Worcester up to 1812, was superb in every way and it was clear that Billingsley had achieved his objective, but difficulties occasioned by remarkably large and unacceptably high kiln wastage losses approaching 90% meant that the operation was commercially unsuccessful and this resulted in supplies not being compatible with the huge demands of a discerning clientele, resulting in an effective closing down of the Nantgarw China Works in 1820 and the departure of Billingsley and Walker for the

Coalport China Works. At Coalport, it is believed that they did not manufacture new porcelains or institute changes in the porcelain body being produced there but that Billingsley and Walker were again employed in advisory roles. Billingsley died in 1828 and Walker eventually went on to take up residence in the state of New Jersey, where he set up a new earthenware and perhaps porcelain manufactory too, at Temperance Hill. Michael Messenger (*Coalport Porcelain, 1795–1926,* 1995) intriguingly suggests that John Rose was eager to acquire the services of William Billingsley at his Coalport China Works to prevent his going elsewhere and starting up a competitive venture and for this he paid him an undisclosed "pension". John Randall, an accomplished decorator who later set up his own ceramics atelier as Robins and Randall, is quoted as saying that Rose really did admire the quality of the Nantgarw soft paste porcelain so much that he wanted to see if he could make a similar body at Coalport and to this end he particularly wished to engage with Samuel Walker:

> *Mr John Rose ... went over and bought the plant moulds and everything at Nantgarw and then entered into an agreement with Walker and Billingsley to make the same quality of china at Coalport for a period of seven years. However, the process was expensive and suffered from a difficulty in working the clay, with poor plasticity, and also from the loss in burning. The fret body was then abandoned and by adding pure felspar to the Cornwell stone and clay a good translucent body was obtained at a less cost than by using a fret body.*

The composition of Walker's "equivalent" Nantgarw body made at Coalport in 1825 is given as follows:

"325 lb bone ash; 175 lb sand and 25 lb potash is triturated with water, well=dispersed and fired in a biscuit kiln. This is then fritted and 2 parts mixed with I part of Cornwell clay, the whole fired in a biscuit kiln and glazed" with the formulation given below:

"56 lb raw borax; 28 lb Lynn sand; 14 lb chalk; 4 lb white lead; 7 lb nitre all fired ina biscuit kiln, then fritted. Then 84 lb of the above frit, 56 lb Cornwell stone. 28 lb white lead, 14 lb flint glass, 15 lb Cornwell clay and 14 lb chalk are all ground together very fine and ready for use".

Messenger (*Coalport Porcelain, 1795–1926,* 1995) notes the confusion about whether or not John Rose did acquire the Nantgarw moulds and equipment in the 1820 sale following the departure of Billingsley and Walker and reasons that he did not, clearly, as William Weston Young was still operating the Nantgarw China Works as a going concern albeit in survival mode after that. However, it is still possible that some equipment was purchased at the 1822 sale at the closure of Young and Pardoe's operations at Nantgarw although Messenger points out that Coalport was highly successful commercially at this time and had no need for the acquisition of material from a then defunct factory. It is still undeniable that some consideration needs to be given to Randall's observation of the manufacture at Coalport of a new porcelain soft paste phosphatic felspar body based on a Nantgarw formulation supplied by Samuel Walker, the quality of which impressed John Rose so much.

It is relevant to compare the porcelain bodies that have been worked on and synthesised by William Billingsley during his career: their compositions are summarised in Table 3 and it is also pertinent to qualitatively assess the relative translucency of each one. Unfortunately, chemical data are currently not available

Table 3 Compositional elemental oxide percentages/% for factory porcelains associated with William Billingsley, their types, periods of production and relative translucency

Factory	Period	Type of body	SiO_2	Al_2O_3	CaO	P_2O_5	K_2O	Na_2O	MgO	Fe_2O_3	PbO	Relative translucency[a]
Derby	1774–1795	Phosphatic	46.7	10.0	21.8	16.5	1.0	0.8	0.6	0.4	1.3	Fair: 4
Pinxton	1795–1799	Phosphatic	41.9	13.8	24.8	14.1	0.6	0.6	0.6	0.0	0.8	Good: 6
Mansfield	1799–1803	–										
Torksey	1803–1808	–										
Worcester	1808–1813	Siliceous	73.9	6.6	0.5	0.3	4.2	1.1	7.0	0.2	6.0	Very good: 7
Swansea	1814–1816	Phosphatic	45.2	20.0	16.7	13.0	2.8	1.4	0.4	0.3	0	Very good: 8
Nantgarw	1817–1819	Phosphatic	43.7	12.8	22.0	17.4	2.2	0.5	0.5	0.5	0.5	Excellent: 10
Coalport	1820–1828	Phosphatic	45.6	12.6	21.2	17.3	1.6	1.0	0.5	0.2	0	Very good: 8

[a]Based on a scale of 0–10 with Nantgarw being assigned the value of 10

for shards and wasters found at the Brampton-in-Torksey site, which would perhaps provide a clue as to Billingsley's thoughts immediately prior to his sojourn and experimental porcelain paste trials at the Worcester China Works, which were carried out between his arrival there in 1808 and his departure in 1813. Therefore the time-line proceeds as follows, with approximate dates of arrival and departure defining the period of production: Derby (1774–1795), Pinxton* (1795–1799), Mansfield (1799–1803), Brampton-in-Torksey* (1803–1808), Worcester (1808–1813), Swansea* (1813–1816), Nantgarw* (1816–1820) and Coalport (1820–1828). Of course, it is realised that only at the factories marked * can we definitively say that porcelain was made by Billingsley that can be identified as such and uniquely so. Although shards have been excavated at Torksey which apparently confirm that porcelain was manufactured there in the William Billingsley era, these have not been subjected to rigorous chemical compositional analysis as has been undertaken for the other sites listed in Table 3 and, as stated earlier, at Mansfield he is believed to have adopted a purely decorating role for ceramic pieces which he purchased for that purpose from elsewhere. A statement made by the signatories, namely Billingsley, Walker and Young, to the *Memorial* of September 1814 (see above and transcript in Edwards, *Porcelain to Silica Brick: The Extreme Ceramics of William Weston Young, 1776–1847*, 2019), submitted for financial support to enable the Nantgarw operation to continue declares that:

> *"Your Memorialists have to state to your Lordships that they have either separately or together been engaged for some years (near Twenty on the whole, but the last Seven very closely) in trials for the improvement of British porcelain"*

Simple back-calculation reveals that this time frame is extremely relevant for our thesis here in that it starts with William Billingsley at Derby in 1794/5, then Pinxton, Mansfield, Torksey and Worcester, with the Billingsley /Walker alliance at Torksey in 1803/4 and Worcester 1808/12 and concluding with the Billingsley/Walker/ Young alliance at Nantgarw Phase I 1813/14. The implication is that the "very close" and serious experimental trials on their new porcelain bodies culminating in the birth of the Nantgarw China Works therefore began just 7 years before, i.e. in 1807, whilst they were in the act of imminently moving to join Martin Barr at Worcester!

The analytical data comprising the list of factories in Table 3 are derived from the following published research work: Derby (Owen and Barkla 1997), Pinxton (Eccles and Rackham 1922), Worcester (Owen 1997), Swansea (Owen et al. 1998), Nantgarw (Owen and Morrison 1999) and Coalport (Owen and Sandon 2003). Several conclusions can be made about these data, namely:

Considering the porcelains actually made by Billingsley, or at least where he was known to have had some significant input into their formulation, and for which we currently have robust analytical data, these we need consider in chronological order and hence only the factories at Derby, Pinxton, Swansea and Nantgarw qualify in this respect. It is recognised from historical documentation that Billingsley and Walker did undertake experiments and trials in a new porcelain body composition and were engaged in the development of a new kiln at the Worcester China Works, with the active cognisance and encouragement of Martin Barr. However, as noted above, Barr's successors at Worcester after his death in 1813 refused to undertake any modification of their already successful siliceous porcelain which was esteemed in its own right and which had its devoted aristocratic and royal clientele. Nevertheless, it seems very plausible that whilst they were at Worcester between 1808 and 1813, William Billingsley and Samuel Walker further developed their porcelain body which had been trialled first at Pinxton and then at Brampton-in -Torksey (where a large number of porcelain shards have been unearthed to demonstrate the synthetic operations being undertaken there), as immediately upon leaving Worcester in early 1814 they were engaged upon setting up their new china works with William Weston Young on an undeveloped site at Nantgarw, which involved the building of a biscuit kiln and glost kiln along with associated potting sheds, decorating and glazing studios and storage facilities. It seems clear, therefore, that Billingsley and Walker whilst at Worcester had reached there their final Nantgarw formulation, which enabled them to go into a limited production in mid-1814, to synthesise, glaze and decorate their selected specimens of the new Nantgarw China Works porcelains for submission along with their application for government funding with their *Memorial* document in September, 1814, signed jointly by Billingsley, Walker and Young. Although unsuccessful in receipt of funding, the eminent receiving committee chairman, Sir Joseph Banks, was so impressed by the quality of the porcelain items submitted that he alerted his friend, Lewis Weston Dillwyn, who was keen to establish a porcelain manufacturing facility at his Cambrian Pottery in Swansea, to their needs. It is also clear, rejecting assertions otherwise made by earlier authors, that Billingsley and Walker jealously preserved their secret Nantgarw formulation recipe and did not reveal it to Dillwyn, whose initial efforts at the Swansea China Works produced an acceptable but rather heavily potted glassy porcelain which he them improved in experimental trials carried out between 1815 and 1817 and produced his excellent duck-egg porcelain body, for which Swansea rapidly became famous (Edwards, *Nantgarw and Swansea Porcelains: An Analytical Perspective,* 2018);

- Inspection of the available analytical data for the four relevant factories of manufacture associated with Billingsley's porcelain, namely, Derby, Pinxton, Swansea and Nantgarw, shows that all have a highly phosphatic soft paste body. The P_2O_5 content in each of these porcelains is listed as Derby 16.5%, Pinxton 14.1%, Swansea 13.0% and finally Nantgarw 17.4%, all of which embrace a bone ash component amounting to 40–46%.
- From Derby–Pinxton-Swansea -Nantgarw, the respective amounts of SiO_2, Al_2O_3, CaO, K_2O and Na_2O can be compared and some significant differences can now be evaluated which indicates that some experimental application has undoubtedly occurred. Although the silica component remains quite reasonably constant between the four porcelains in the range 41.9–46.7%, with a mean of 44.3 +/−2.4%, the situation for the others is rather differently described. The alumina content varies between 10.0 and 20.0%, with the final Nantgarw component reading being 12.8%: the Swansea component percentage is 100% more

than that of Derby. Likewise, the lime component percentage has a range of between 12.8% at Nantgarw to 24.8% at Pinxton, again a significant difference of 94% in lime content. In all of these, the pattern is that the final Nantgarw porcelain body has the highest phosphate, the lowest lime and the second lowest alumina content of all factories studied. The alkali metal oxide flux component is also widely different, adopting the range from 1.8% at Derby through to 4.2% at Swansea, a percentage difference of 133%: of even greater significance is the relative ratios of the potash to soda in the alkali metal oxide flux component, being 1.3 for Derby, 1.0 at Pinxton, 2.0 at Swansea and 4.0 at Nantgarw. Combination of these two data analyses reveals that the porcelain manufacture at each site consciously used significantly different proportions of terrestrial plant alkali (calcined soda ash) and marine seaweed kelp (calcined potash, or pearl ash). The higher contents of potash in the raw materials components at Swansea and Nantgarw can be correlated with their location near the coast (Swansea on the coast of the Bristol Channel, at Swansea Bay, and Nantgarw some 7 miles from Cardiff Bay) which would ensure the ready availability of marine supplies of local seaweed and kelp for calcination.

- The lead oxide percentages in the analysed porcelains again vary strongly between factories, this reducing from Derby at 1.3%, through Pinxton at 0.8% to Nantgarw at 0.5% and Swansea at 0%. The lead oxide can be ascribed to the addition of flint glass cullet to the porcelain paste mixture before firing but because of the variable percentage of lead in the glass, which can be in the range 10–60%, it is not possible to quantify the amount of lead glass that has been added in the recipe. Whereas the Derby China Works is on record as practising the addition of lead glass cullet to its porcelain pastes (Anderson 2000), the situation at Swansea is rather more mysterious as Lewis Dillwyn's *Notebooks* (see Appendix in Eccles and Rackham 1922) refer positively to the glass frit additive in his list of components initially but this is not specifically mentioned anywhere in his 13 trial recipes thereafter. According to the analytical data in Table 3, Dillwyn did not add any lead glass to his porcelain paste in the sample(s) analysed. Similarly, Billingsley and Walker's Nantgarw recipe as revealed to John Taylor (*The Complete Practical Potter*, 1847) in 1847 does not mention the addition of glass cullet as a raw material. The nil percentage of lead oxide in the Swansea paste and very small amount in the Nantgarw analogue cannot therefore be easily explained from the data cited here as these data do not correlate with given historical recipes.
- The percentage levels of MgO and Fe_2O_3, which can be averaged at 0.5 +/− 0.1% and 0.3 +/− 0.3%, respectively, must surely reflect impurities in the raw materials such as sand and china clay: the deliberate addition of steatite would result in a magnesia percentage as seen to be evident in magnesian or soapstone/soaprock porcelains and the use of the purest sand deposits such as Lynn sand was designed to minimise the incorporation of ferric (Fe^{3+}) iron. Dillwyn commented that he used to add small quantities of blue smalt to offset the yellow colouration afforded by increased amounts of iron impurity in his raw materials.

- It seems reasonable to presume that the final Nantgarw porcelain body was developed by Walker and Billingsley whilst they were based in the Royal Worcester China Works and in this context it is equally obvious that whilst their employers had no intention of marketing and producing this new china they were not averse to Billingsley and Walker manufacturing it themselves elsewhere as long as they did not divulge the formulation and recipe to other competitors. It is intriguing to speculate upon the potential outcome if Messrs. Flight and Barr had decided to manufacture this novel porcelain body at Worcester in 1814 as an experimental phosphatic alternative to their successful siliceous porcelain body. They would, of course, have probably had the immediate downturn of exceptionally high kiln losses (as exemplified by 90% at Nantgarw) but perhaps with the much improved financial support then existing at Worcester compared with the relative paucity of support available to the fledgling Nantgarw Phase I operation, which was personally financed at that time by William Weston Young, they may have been more able financially to overcome this setback and have been able to refine the process, if given the time to ameliorate the wastage in the biscuit firing stage.

The last column in Table 3 is a qualitative visual assessment of the translucency of each of the porcelains and several of these are discussed below to support the judgement. The details of the specimens are as follows:

Derby porcelain: a *Prince of Wales* service dessert plate, William Duesbury's factory, *ca.* 1786, representing the decoration of William Billingsley on the finest Derby porcelain of the period: pattern number 65, marked with the Derby crown and crossed batons in puce enamel and with the gilder's mark "8" for William Longden. Recorded in the Derby plate pattern book and illustrated in Dr. W.D. John (*William Billingsley*, 1948) and in John Twitchett (*Derby Porcelain: An Illustrated Guide, 1748–1848, 2002).* Illustrated here in Fig. 6.11.

Pinxton porcelain: a coffee can, *ca.* 1795–7,decorated by William Billingsley with landscape scene in sepia enamel. Unmarked, but service illustrated in Chapman, *William Billingsley at Brampton-in-Torksey,* 2003, and there attributed to Billingsley. Illustrated here in Fig. 5.

Torksey porcelain: no example was available for analysis but the two known specimens are said to have a delightful and clear translucency that is similar to that of Swansea porcelain.

Worcester porcelain: a deep saucer dish decorated by William Billingsley with a group of four sprays of pink roses, foliage and rosebuds on an attractive cerulean blue ground, carrying the backstamp of the Royal warrant accord and the BFB with crown impressed of the Barr, Flight & Barr period, *ca.* 1808–1812. Illustrated in Fig. 6.

Swansea porcelain: a spill vase in the finest duck-egg porcelain decorated by William Billingsley with an Oriental figure in a tropical landscape with swags of golden seaweed, *ca.* 1815–1816. Shown in Fig. 6.5.

Nantgarw porcelain: a very fine spill vase, London decorated in the London workshops of John Mortlock with a spray of fresh garden flowers and finely gilded, *ca.* 1817–1820. This specimen spill vase was selected for exhibition at the Special 200th Anniversary Exhibition of Nantgarw Porcelain, July-September 2019, entitled "*Coming Home*" and held at Tyla Gwyn, Nantgarw, on the actual Nantgarw China Works Site. Shown in Fig. 6.2.

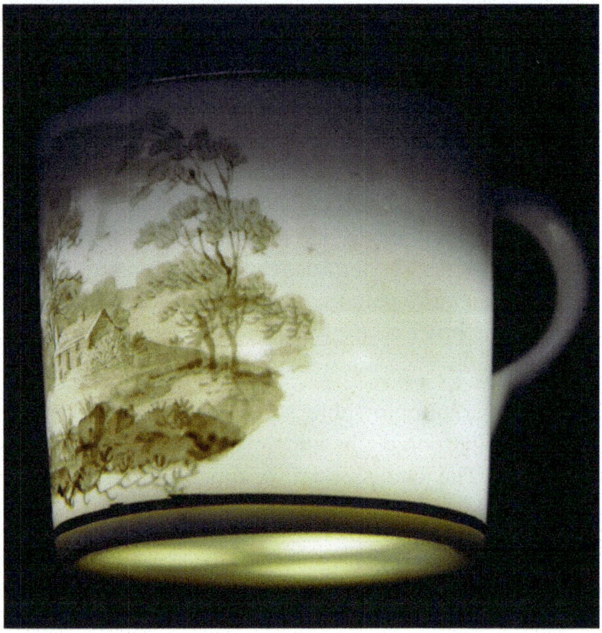

Fig. 5 Pinxton porcelain coffee can, *ca.* 1796–1799, photographed in transmitted light. *Private Collection*

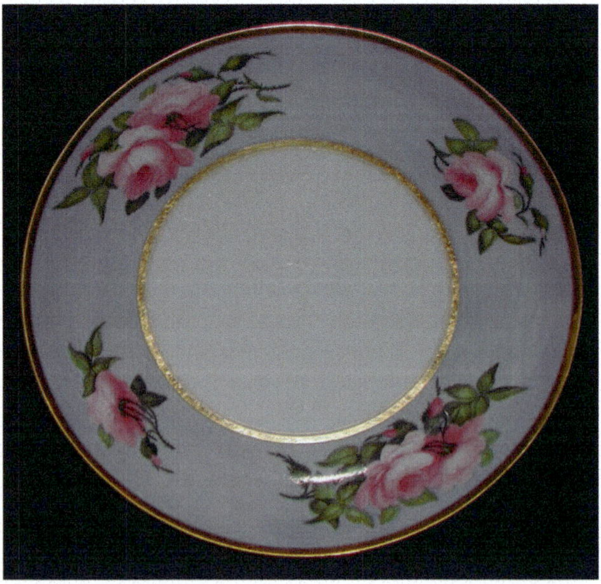

Fig. 6 Worcester porcelain, deep saucer dish impressed with BFB and crown for Barr, Flight & Barr, *ca.* 1808–1812, decorated by William Billingsley with a group of four sprays of pink roses and rosebuds on a cerulean blue ground colour. *Private Collection*

References

J.A. Anderson, "*Derby Porcelain and the Early English Fine Ceramic Industry*", PhD thesis (University of Leicester, UK, October, 2000)

A. Bambery, William Billingsley in Derbyshire, chapter 9, in *Welsh Ceramics in Context, Volume I*, ed. by J. Gray (Royal Institution of South Wales, Swansea, 2003), pp. 159–176

R.E. Chapman, *William Billingsley at Brampton-in-Torksey,* chapter 9, in *Welsh Ceramics in Context Volume I*, ed. by J. Gray (Royal Institution of South Wales, Swansea, 2003), pp. 179–192

H. Eccles, B. Rackham, *Analysed Specimens of English Porcelain in the Victoria and Albert Museum Collection, London* (Victorian and Albert Museum, 1922)

R.S. Edmundson, Billingsley and his China artists, 1796–1808, in *Welsh Ceramics in Context, Volume II, Chapter 8*, ed. by J. Gray, (Royal Institution of South Wales, Swansea, 2005), pp. 151–168

H.G.M. Edwards, M.C.T. Denyer, *William Billingsley the Enigmatic Porcelain Artist, Decorator and Manufacturer*, Penrose Antiques Ltd. Short Guides (Neopubli, Berlin, 2016). ISBN:978-3-7418-6

H.G.M. Edwards, *Nantgarw Porcelain: The Pursuit of Perfection,* Penrose Antiques Ltd. Short Guides, Series Editor: M.D. Denyer (Penrose Antiques Ltd., Thornton, West Yorkshire, 2017). ISBN:978-0-244-90654-2

H.G.M. Edwards, *Swansea and Nantgarw Porcelains: A Scientific Reappraisal* (Springer, Dordrecht, 2017)

H.G.M. Edwards, *Swansea Porcelain: The Duck-Egg Translucent Vision of Lewis Dillwyn,* Penrose Antiques Ltd. Short Guides (Thornton, West Yorkshire, UK, 2017). ISBN:9780244325787

H.G.M. Edwards, D. Porcelain, *The Golden Years, 1780–1830*, Penrose Antiques Ltd. Short Guides (Thornton, West Yorkshire, 2017)

H.G.M. Edwards, *Nantgarw and Swansea Porcelains: An Analytical Perspective* (Springer, Dordrecht, 2018)

H.G.M. Edwards, *Porcelain to Silica Bricks: The Extreme Ceramics of William Weston Young (1776–1847)* (Springer, Dordrecht, 2019)

H.G.M. Edwards, R. Denyer and M.J. Denyer, *The Pendock-Barry Derby Porcelain Service: A Holistic Reappraisal*, to be published (2019)

C.L. Exley, *A History of the Torksey and Mansfield China Factories* (Lincoln, Keyworth and Fry, 1970)

P.T. Gardner, *Billingsley, Brampton and beyond: In Search of the Weston Connection–The Provenance of a Porcelain Service, over 200 Years Old, Is Investigated* (Troubadour/Matador Publications, Leicester, 2005)

J. Haslem, *The Old Derby China Factory* (George Bell, London, 1876)

W.D. John, *William Billingsley* (Ceramic Book Company, Newport, 1948)

M.F. Messenger, *Coalport, 1795–1926: An Introduction to the History and Porcelain of John Rose and Company* (Antiques Collectors Club, Woodbridge, 1995)

J.V. Owen, Quantification of early Worcester porcelain recipes and the distinction between Dr Wall- and flight-period wares. J. Archaeol. Sci. **24**, 301–310 (1997)
J.V. Owen, R. Barkla, Compositional characteristics of 18th century Derby porcelains: Recipe changes, phase transformations and melt fertility. J. Archaeol. Sci. **24**, 127–140 (1997)
J.V. Owen, J. Sandon, A rose by another name: A geochemical comparison of Caughley (c.1772–99), Coalport (John Rose & co., c.1799–1837), and rival porcelains based on Sherds from the factory sites. Post-Mediaeval Archaeology **37**, 79–89 (2003)
J.V. Owen, J.O. Wilstead, R.W. Williams, T.E. Day, A tale of two cities: Compositional characteristics of some Nantgarw and Swansea porcelains and their implications for kiln wastage. J. Archaeol. Sci. **25**, 359–375 (1998)
J.V. Owen, M.L. Morrison, Sagged Phosphatic Nantgarw Porcelain (ca. 1813–1820): Casualty of Overfiring or a fertile paste? Geoarchaeology **14**, 313–332 (1999)
C.B. Sheppard, I. Deverill, *Billingsley, Mansfield: The Bicentenary of William Billingsley's Mansfield Decorating Establishment, 1799–1802. An Exhibition to Celebrate the Millennium and the Bicentenary of William Billingsley's Porcelain Decorating Establishment* (Pinxton Porcelain Society, Pinxton, 1999)
J. Taylor, *The Complete Practical Potter* (Shelton, Stoke-upon-Trent, 1847)
J. Twitchett, *Derby Porcelain, an Illustrated Guide: 1748–1848* (The Antiques Collectors club, Woodbridge, 2002)

Appendix IV: Analysis of Pigments on Porcelain

Most of the analytical studies of porcelains that have been carried out hitherto have been concerned with the composition of the bodies and glazes and the identification of the pigments used in the enamelling decoration has received relatively little attention (Ricciardi et al. 2009; Colomban 2013). The reason for this is quite simple in that the artistic requirement of pigment usage and adoption for ceramic decoration was that the minerals and synthetic pigments used should retain their colours at the temperatures to which they were exposed in the firing kilns: usually, for soft paste porcelains the secondary firing processes were carried out in glost kilns at significantly lower temperatures than those which were used for biscuit porcelain firing. Nevertheless, several coloured minerals that were utilised in other artworks such as oil paintings and which were normally applied "cold" and at room temperature, often suspended in organic resins and carriers such as resins, gums, linseed oil and egg white tempera, would not be suitable for use at elevated temperatures because of their chemical instability towards thermal decomposition especially in a reducing or oxidative kiln environment. Examples of these unsuitable pigments for ceramics decoration involving firing include several copper and lead based minerals which can be oxidised to the metal oxides at elevated temperatures: an example would include the blue mineral azurite, basic copper carbonate, $CuCO_3.Cu(OH)_2$, which decomposes to the brown copper oxide, tenorite, CuO, at a temperature below

300 °C and the green compound *verdigris*, a basic copper ethanoate, $Cu(OH)_2.2Cu(CH_3CO_2)_2$, of variable composition which decomposes thermally at 240 °C. Neither of these pigments, therefore, would be suitable for underglaze ceramic decoration involving firing in a glost kiln.

Decoration of a fired ceramic frequently involved gilding, using the application of gold leaf suspended in honey or as a mercury amalgam which deposited the metal upon firing. A range of naturally occurring and synthetic coloured minerals such as haematite, Naples yellow (pyrochlore), Egyptian blue and smalt were found to be favourable for ceramics decoration. Metallic carbonates and sulfides which were much used as favoured pigments in other arenas of decorative art work were prone to decomposition and oxidation at elevated temperatures and examples such as malachite, verdigris and cinnabar were not recommended thereby for ceramic work unless cold-painting overglaze was envisaged. For most porcelain decoration, the enamels were applied to the biscuit porcelain after the first firing process and then a silicate-based glazing slip was used which set the enamels firmly and protected them against future deterioration in usage as an underglaze painting. Occasionally, some enamelling was applied over the glaze but this decoration would not be robust in normal usage – a type of decoration known as "clobbering" used this effectively to increase the applied ancillary decoration particularly on simply executed transfer work and this often incorporated subsidiary gilding as well.

Dr. William John (*Nantgarw Porcelain*, 1948) was the first observer to note the presence of a peculiar but distinctive iridescence exclusively found on London-decorated Nantgarw porcelains from the workshops of John Mortlock, John Sims and Robins and Randall, which he attributed to a thermal migration of the metallic colours into the surface applied glaze which was exacerbated in the reducing atmospheres of the glost kilns used. This did not occur in the output of locally decorated porcelains from the Nantgarw China Works. This iridescence, seen as a spectral halo in the vicinity of the enamelling, was considered further by the current author (Edwards, *Nantgarw and Swansea Porcelains: An Analytical Perspective*, 2018) and discussed in terms of charge transfer colours formed between the heavy metal ions and impurities in the silicaceous glazes. What this means analytically is that a differentiation can be made between the locally decorated Nantgarw porcelains and their London-decorated analogues, which otherwise may be of identical shapes and forms. Much of the output of the Nantgarw China Works was despatched in the white to their agent in the capital, John Mortlock, in Oxford Street, London, where it was decorated to the commission of their clients by noted ceramic enamellers such as Moses Webster and James Plant in ateliers such as those of Robins and Randall and John Sims (Morton Nance, *The Pottery and Porcelain of Swansea and Nantgarw*, 1942). Analytically, therefore, the presence of this iridescence can be seen as an unequivocal confirmation of London-decorated Nantgarw porcelain and serves to direct the observer towards possible source painters in the various ateliers.

One could perhaps seek a reason for this perceived lack of interest in analysing the pigments applied to porcelains, which is in complete contrast to the analytical work being performed on oil paintings, where the identification of the pigments and the artistic palette is so vital to the detection of potential fakes and the analytical interpretation can inform the attribution of the correct period of execution for the artwork. The author believes that in the ceramics field the relatively narrow range of potential pigments used for porcelain decoration and the similarity of palettes adopted by the major manufactories in the decoration of their porcelains limits the analytical differentiation protocols for their characterisation and identification purposes for the potential attribution of a possible factory. However, this is not always so, and in several instances the discovery of an unusual mineral or synthetic equivalent on a piece of porcelain has afforded another piece of useful analytical information that could potentially be useful for attribution purposes in addition to the analytical information coming forth from the body and glaze compositional studies. For example, in this book, we have discussed earlier the body compositional analysis of an unique, unmarked Rockingham porcelain table-top (Fig. 4.1) which was compared analytically with a Royal Rockingham period dessert plate adopted as a "standard", marked with the red griffin mark characteristic of the 1835–1842 period of manufacture (Edwards et al. 2004). In both cases, molecular spectroscopy indicated the presence of an unusual bright blue pigment, described as the Rockingham *bleu de roi*, which proved to be assignable to a cobalt arsenosilicate. Likewise, the underglaze blue decoration of several Ming porcelain shards dating from the early part of the seventeenth century also showed an unusual blue cobalt aluminate pigment that was different from the more usual cobalt silicates known in the West as *Bristol blue* (de Waal 2004a, b; Kock and de Waal 2007). Perhaps a major factor in the deficiency of analytical information derived from applied porcelain pigment analyses is that most work hitherto has been undertaken on shards acquired from waste pits and in very few cases have these been found to be accompanied by remains of enamelling decoration: the reason for this is clear, in that most failures in the porcelain kilns occurred at the biscuit firing stage which involved the highest kiln temperatures which resulted in an increased likelihood of distortion, fracture and sagging of the porcelain bodies undergoing firing at these temperatures. In very few instances have the porcelain shards also been found with applied enamelling, which implicitly concludes that a failure has occurred during the relatively more gentle glazing stage which operated in the lower temperature glost kiln. The only other option for most destructive elemental analytical techniques, particularly SEM/EDAXS, XRF and XRay, is the use of fragmented and broken china specimens, which are not always accessible for analysis and for which further damage is generally deemed to be inconceivable or undesirable for their museum collection curation.

In this work we have described the discovery of just three decorated, enamelled shards obtained from an archive at the Nantgarw China Works site which was first exacavated archaeologically by Williams (1932), dating from 1817–1820, for which both elemental and molecular analytical information has been derived for the first time and from which the palette used by the local artists (William Billingsley, Thomas Pardoe, William Weston Young *et alia*) at Nantgarw has been forthcoming (Colomban et al. 2020). The most interesting part of this discovery relates to a vivid purple colour used at Nantgarw on two of these shards, which is clearly indicative of the presence of the *Purple of Cassius* pigment (Edwards, *Nantgarw and Swansea Porcelains: An Analytical Perspective*, 2018; Colomban et al. 2020). This pigment was known in the sixteenth century and its discovery and synthesis has been described in detail elsewhere (Edwards, *Porcelain to Silica Bricks: The Extreme Ceramics of William Weston Young, 1776–1847*, 2019; Hunt 1976). This result was completely unexpected for the applied local decoration at the Nantgarw China Works, since its financial operations were in a parlous state for the majority of its operational period because of the exceptionally high kiln losses approaching 90% on first firing of the porcelains, which meant that only 10% of the porcelain produced could be sent for sale to the London agents. This was a disastrous business proposition and contributed to the eventual premature closure of the Nantgarw China Works in 1820 under the stewardship of William Billingsley, William Weston Young and Samuel Walker. Also, in the latter stages of local operation, and certainly during William Weston Young's brief tenure immediately after the departure of Billingsley and Walker, a conscious effort was made to minimise the application of gilding on the local porcelain decoration (John, *Nantgarw Porcelain*, 1948; Morton Nance, *The Pottery and Porcelain of Swansea and Nantgarw*, 1942). It was extremely fortunate analytically that some porcelain pieces did not survive the secondary glazing operation after enamelling decoration had been applied as this afforded the opportunity to assess for the first time the pigment palette used by the local Nantgarw artists. The *Purple of Cassius* pigment was synthesised from pure gold according to a defined chemical procedure and this would most certainly not have been a cheap pigment to manufacture – the loss of several almost finished articles at the last stage of firing would then have been doubly disastrous at Nantgarw, therefore, for the financial recovery of the manufacturing processes and the subsequent sales of the finished china.

Table 4 comprises a list of pigments which have thus far been discovered analytically from decorated porcelains, which also includes the analytical techniques used in their derivation. In most cases reported, some pigments could not be specifically identified because of interference effects from the overlying and surrounding glaze. It should be noted that although gilding is detectable analytically using the elemental techniques such as SEM/EDAXS, it is not detectable using molecular spectroscopic techniques as metallic gold is a monatomic species and contains no chemical bonds – elemental gold is therefore said to be *silent* for the latter determinations. From the research contributions relating to pigment analysis on decorated porcelains made in Table 4 the following conclusions can be drawn:

- The range of mineral pigments used by porcelain artists is rather limited, presumably because of their stability towards decomposition in the high temperature kilns used for preliminary or secondary firing: even glost kilns could be operated at temperatures as high as 1000 °C, below which temperature mineral carbonates are susceptible to the elimination of carbon dioxide with the formation of metallic oxides of different colours.
- Comprehensive studies by Professor Philippe Colomban and his team on body compositions, glazes and pigments have hitherto revealed that the use of underglaze colours in the firing of decorated porcelains in glost kilns frequently results in observable chemical and physical interactions between the mineral pigments

Table 4 Pigments identified analytically on porcelains

Porcelain	Date	Pigment	Formulation	Technique	References
Meissen	1710	Haematite	Fe_2O_3	Raman	Colomban and Milande (2006b)
		Naples yellow	$Pb_xSn_xSb_{2-x}O_{7-8}$		
		Cassiterite	SnO_2		
		Lazurite	$Na_8[Al_6Si_6O_{24}]S_8$		
Sevres	1780	Haematite	Fe_2O_3		
		Pyrochlore	$Pb_xSn_xSb_{2-x}O_{7-8}$	Raman	Colomban et al. (2018)
		Carbon	C		
Burghley	1690	Haematite	Fe_2O_3	SEM/Raman	Spataro et al. (2009)
Vietnamese	Fifteenth century				Colomban et al. (2004c)
European	16th–20th century	Cassiterite	SnO_2	SEM/Raman	Colomban (2013)
		Bone Ash	$Ca_3(PO_4)_2$		
		Egyptian blue	$CaCuSi_{14}O_{10}$		
		Han blue	$BaCuSi_4O_{12}$		
Ming	1600	Cobalt blue	$CoAl_2O_4$	Raman	Kock and de Waal (2007)
		Haematite	Fe_2O_3		Carter et al. (2017)
		Anatase	TiO_2		de Waal (2004a, b)
		Carbon	C		
Ming	1550	Cobalt blue	$CoAl_2O_4$	Raman	de Waal (2004 a, b)
		Rutile	TiO_2		
		Gypsum	$CaSO_4.2H_2O$		
		Carbon	C		
Yingqing	Fourteenth century	Carbon	C	SEM/Raman	Widjaja et al. (2011)
		Anatase	TiO_2		
		Cobalt oxide Co_2O_3			
Limehouse	1740	Smalt		Raman	Jay and Cashion (2013)

Bow	1746		Raman		Jay and Orwa (2012)
Medici	1570	Anatase	TiO_2	Raman	Colomban et al. (2004b)
		Bone ash	$Ca_3(PO_4)_2$		
Chantilly	1725	Haematite	Fe_2O_3	Raman	Colomban et al. (2004a)
		Cobalt blue	$CoAl_2O_4$		
		Naples yellow	$Pb_2Sb_2O_7$		
		Chrome green			
		Magnetite	Fe_3O_4		
Mennecy	1740	Haematite	Fe_2O_3	Raman	Colomban et al. (2004a)
		Carbon	C		
		Cassius purple	Au		
Nantgarw	1820	Haematite	Fe_2O_3	SEM/Raman	Colomban et al. (2020)
		Naples yellow	$Pb_xSn_xSb_{2-x}O_{7-8}$		
		Cassius purple	Au		
Worcester	1754	Haematite	Fe_2O_3	Raman	Jay and Orwa (2012)
		Magnetite	Fe_3O_4		
		Goethite	FeOOH		
		Pyrochlore	$Pb_xSn_xSb_{2-x}O_{7-8}$		
		Cassiterite	SnO_2		
		Cobalt blue	$CoAl_2O_4$		
		Egyptian blue	$CaCuSi_4O_{10}$		
		Yellow ochre	Fe_2O_3.Clay		
		Carbon	C		
		Orpiment	As_2S_3		
		Smalt	CoO.nSiO_2		
		Azurite	$CuCO_3$.$Cu(OH)_2$		

and the glaze components (Colomban and Treppoz 2001; Colomban et al. 2004a, b, 2006a, b, 2018). This was utilised to great effect by the Chinese and Vietnamese ceramics manufactories and their Islamic colleagues between the eleventh and the fourteenth centuries (Colomban et al. 2004) to produce the delicate glaze colours for their celadons and porcelains through the deposition of finely divided particles of metallic oxides and compounds such as cassiterite, SnO_2, and calcined bone ash, $Ca_3(PO_4)_2$, or very small bubbles of gaseous products from the decomposition: a wide range of delicate colours from pale green, light blue through to deep red could be produced effectively in this way. Strictly, of course, this does not compare favourably with the use of pigments for depiction of specific scenes and patterns on porcelains but a similar result is given arising from the significant interaction noted between the applied glaze slip and the underlying paint which can define differences observed in the colour hues in the finished artefact.

- The Raman spectroscopic analysis of early underglaze blue Meissen porcelains (Colomban and Milande 2006b) has revealed a surprising result in that the presence of lapis lazuli (lazurite), a sodium aluminosilicate of formula $Na_8[Al_6Si_6O_{24}]S_8$, was detected: this was most unusual and has not been observed elsewhere. The reason for this probably lies in the observation that lapis lazuli decomposes at a temperature approaching 1100 °C, which is within range of the glost kiln temperatures – this means of course that Johann Bottger and Ehrenfried von Tschirnhaus, who were undertaking their manufacturing trials at Meissen in the first decade of the eighteenth century, were single-stage firing their porcelains at lower temperatures. The Raman spectroscopic detection of lazurite depends upon the presence of the free radicals S_2^{2-}, whence the blue colour in the pigment originates, and the inference must be therefore that these reactive free radical ions of elemental sulfur are still preserved as chemical entities during the firing at the glost kiln temperatures.
- A listing of the pigments detected thus far in analytical experiments on decorated porcelains comprises: haematite, Egyptian blue, Han blue, smalt, cobalt silicate (cobalt blue), cobalt aluminate, orpiment, Naples yellow, pyrochlore, terre vert, carbon, purple of Cassius, chrome green, magnetite, lazurite, cassiterite, anatase, rutile and gypsum. This may not be an exhaustive compilation but several distinct omissions can be noted when compared with the much wider ranging palette of typical eighteenth century oil paintings on canvas or panels. Examples of mineral pigments that have not yet been identified on decorated porcelains but which are quite commonly found in their eighteenth century oil painting counterparts are: goethite, realgar, azurite, malachite, calcite, aragonite, plattnerite, cinnabar, *minium*, brochantite and galena. These are all quite stable at room temperature but they would be expected to decompose at the elevated temperatures to which soft paste porcelains are subjected in the two-stage firing process, and also for most single-stage hard paste porcelains too, and applies to the subsidiary glazing operations outside of any secondary chemical reactions that may occur by their exposure to the glaze components in the applied slip. Calcite loses carbon dioxide at 650 °C and cinnabar, HgS, loses sulfur and gives volatile mer-

cury vapour around 500 °C, both temperatures being well below the operating regime of a glost kiln.

- In the light of such a narrow mineral pigment palette available for porcelain decoration it is clear that it is very unlikely that the identification of pigments *per se* will provide a definitive differentiation analytically between factories and this has been stated by Professor Philippe Colomban (Colomban 2013) in a review of his analytical work on porcelains: "*Technology changes can be recognised from the combination of body and glaze characterisation*". It is equally significant that pigment characterisation is not referred to at all in this statement. However, despite this, it appears that in several cases the discovery of unusual pigments in a particular porcelain factory production can assist potentially in the attribution of specimens. One example is the characterisation of pigments in the earliest Meissen porcelains, *ca.* 1710–1730, where in addition to the usual haematite, Naples yellow and cassiterite found for the red, yellow and white pigments, respectively, lazurite (lapis lazuli) was uniquely identified in the underglaze blue pigment colours, alone and in admixture with smalt (Colomban and Milande 2006b). Thus far, this appears to be a unique discovery and might serve to characterise very early Meissen wares if it proved to be more ubiquitous in their factory specimens of the same period. Likewise, the variation in composition of the solid solution called *pyrochlore*, which is best represented by the formulation $PbSn_xSb_xO_3$, alternatively given as Pb_O-SnO_2-Sb_2O_3/Sb_2O_5 and $PbSn_xSb_{2-x}O_{7-8}$ with extreme members represented by the chemical formulae Pb_2SnO_4 and $PbSb_2O_7$, is well recognised: several factories prepared their own versions of this yellow material or realistically obtained the yellow pigment from different sources and the elemental ratios of Pb, Sn and Sb reflect this compositional difference (Colomban et al. 2018) in the resultant pigments used.
- Inspection of the pigments identified analytically in Table 4 reveals that the range available to ceramic artists was indeed limited in comparison with their opposite numbers working with oils: the major reason for this is clearly the requirement for thermal stability of the pigments in the firing kilns and, of course, the whole range of organic pigments and plant extracts that were being used outside of ceramic art such as Indian yellow, indigo, turnsole, orchil, alizarin, gamboge, bistre, carmine, dragon's blood, madder, and Tyrian purple were also not usable by ceramic artists for the same reason. This required ceramic artists to be rather inventive in their preparation and use of colours for decoration and this is borne out by the analytical evidence from pigment analysis: hence, a darker colour or hue could be achieved by admixture of a pigment with carbon, pyrolusite or magnetite whereas a lightening of the colour could be achieved with the admixture of gypsum, anhydrite, calcite or china clay (containing anatase, or rutile). The green colour found on Nantgarw porcelain shards is a mixture of yellow pyrochlore (a lead, tin, and antimony oxide, exemplified at one extreme by Naples yellow $Pb_2Sb_2O_7$) and a blue glass pigment such as cobalt blue (Egyptian blue is often used as an alternative colour for this admixture). Orange is formed from the admixture of yellow pyrochlore and red haematite and a pink colour created by mixing white china clay with red haematite. Black pigments could be

variously amorphous carbon (C), pyrolusite (MnO_2) or magnetite (Fe_3O_4) and a grey pigment used gypsum, anhydrite (anhydrous $CaSO_4$) or china clay (kaolin) in admixture. The purple or violet colours can achieved either by use of *Purple of Cassius,* perhaps modified by admixture with haematite or a blue glass such as cobalt blue, or alternatively by mixing directly haematite and blue glass with or without the addition of carbon to darken the colour.

- In the light of the comments above about the thermal stability of the mineral pigments used for porcelain decoration it is surprising to find evidence for goethite and azurite determined analytically (Jay and Orwa 2012) for early Worcester and Bow porcelain specimens, dating approximately to 1750. It seems that there is some evidence here for the application of pigments in an overglaze decoration, which was then merely warmed or superficially dried and did not involve the elevated temperatures of the glost kilns. It is suggested that this actually occurred here as it is clear that goethite and azurite would not have survived thermally in a glost kiln: yellow goethite (FeOOH) transforms chemically to red haematite (Fe_2O_3) around 300 °C dependent upon the grain size, and azurite decomposes to brown copper (II) oxide around 400 °C and then further decomposes to black copper(I) oxide around 600 °C, and finally to metallic copper at 850 °C. Support for this assertion is provided by the analysis of a Bow porcelain specimen with underglaze blue enamelling in cobalt blue and an applied turquoise pigment overglaze which still has remnants of organic adhesives such as oil of turpentine and gum arabic resin that are identifiable analytically in the pigmented porcelain (Jay and Orwa 2012).
- The analytical study of Jay and Cashion (2013) on Limehouse porcelain shards reports the first application of Mossbauer spectroscopy in conjunction with Raman spectroscopy and SEM elemental analyses for an early porcelain: here, Mossbauer spectroscopy particularly addresses the presence of iron and its oxides in different structural arrangements, which is particularly relevant for the detection of haematite, which is ubiquitous in red pigments.

So, it may be concluded that analytically much still remains to be discovered about the combination of pigments used to achieve colour tones by ceramic artists which will provide an insight into the strategies and specific technologies used to derive the wonderful artworks that we now see on decorated eighteenth and nineteenth century porcelains and that this is entirely aside from the analytical studies which have been carried out on the porcelain bodies and their accompanying glazes: the former analyses show the capabilities of the decorators of polychrome porcelains whereas the latter analyses provide quantitative information about the experimental variations made to the porcelain bodies themselves during the necessary improvements demanded by their manufacture for a critical client base. The use of remote Raman sensing and portable spectrometers for the non-destructive and non-invasive analysis of porcelains has been reviewed by Colomban (2013) and it is clear that this methodology will further increase the capability of accessing rare and valuable pieces of ceramic artwork in the near future, especially for pigment analysis. Spataro et al. (2009) have reported the non-destructive micro-Raman spectroscopic analysis of the red haematite pigment on the *Burghley Jars*, two important and also rather controversial early pieces of English porcelain believed to date from the late seventeenth century, probably made around 1675, using a non-contact probe and interrogation of the body and pigment adjacent to the existing cracks – which is a novel approach affording access to the interior body composition without interfering with the surface pigment and glaze.

References

E.A. Carter, M.L. Wood, D. de Waal, H.G.M. Edwards, Porcelain shards from Portuguese wrecks: Raman spectroscopic analysis of marine archaeological ceramics. Herit. Sci. **5**, 1–8 (2017). https://doi.org/10.1186/840494-017-0130-9.

P. Colomban, The on-site/remote Raman analysis with mobile instruments: A review of drawbacks and success in cultural heritage studies and other associated fields. J. Raman Spectrosc. **43**, 1529–1535 (2012)

P. Colomban, The destructive/non-destructive identification of enamelled pottery, glass artifacts and associated pigments- a brief overview. Arts **2**, 77–110 (2013)

P. Colomban, F. Treppoz, Identification and differentiation of ancient and modern European porcelains by Raman micro-spectroscopy. J. Raman Spectrosc. **32**, 93–102 (2001)

P. Colomban, I. Robert, C. Roche, G. Sagon, V. Milande, Identification des porcelains "tendres" du 18eme siècle par spectroscopie Raman: Saint-Cloud, Chantilly, Mennecy et Vincennes/Sevres. Revue d'Archaeometrie **28**, 153–167 (2004a)

P. Colomban, V. Milande, H. Lucas, On-site Raman analysis of Medici porcelains. J. Raman Spectrosc. **35**, 68–72 (2004b)

P. Colomban, G. Sagon, L.Q. Huy, N.Q. Liem, L. Mazerolles, Vietnamese (15th century) blue-and-white, *Tam Thai* and lustre porcelains/stonewares:Glaze composition and decoration techniques. Archaeometry **46**, 125–136 (2004c)

P. Colomban, V. Milande, On site Raman analysis of the earliest known Meissen porcelain and stoneware. J. Raman Spectrosc. **37**, 606–613 (2006)

P. Colomban, M. Maggetti, A. d'Albis, Non-invasive Raman identification of crystalline and glassy phases in a 1781 Sevres Royal Factory soft paste porcelain plate. J. Eur. Ceram. Soc. **38**, 5228–5233 (2018)

P. Colomban, A. Tournie, L. Bellot-Gurlet, Raman identification of glassy silicates used in ceramics, glass and jewellery: A tentative differentiation guide. J. Raman Spectrosc. **37**, 841–852 (2006)

P. Colomban, H.G.M. Edwards and C. Fountain, "Raman Spectroscopic and SEM/EDXS analysis of Nantgarw soft paste porcelain", *Journal of the European Ceramic Society,* submitted for publication (2020)

D. de Waal, Raman investigation of ceramics from 16th and 17th century Portuguese shipwrecks. J. Raman Spectrosc. **35**, 646–649 (2004a)

D. de Waal, Raman identification of the pigment in blue and white Ming porcelain shards. Asian Chemistry Letters **8**, 57–65 (2004b)

H.G.M. Edwards, P. Colomban, B. Bowden, Raman spectroscopic analysis of an English soft paste porcelain plaque-mounted table. J. Raman Spectroscopy **35**, 656–661 (2004)

H.G.M. Edwards, *Nantgarw and Swansea Porcelains: An Analytical Perspective* (Springer, Dordrecht, 2018)

H.G.M. Edwards, *Porcelain to Silica Bricks: The Extreme Ceramics of William Weston Young, 1776–1847* (Springer, Dordrecht, 2019)

L.B. Hunt, The true story of the purple of Cassius: The birth of gold-based glass and enamel colours. Gold Bull. **9**, 134–139 (1976)

W.H. Jay, J.O. Orwa, Raman spectroscopy applied to early (ca. 1746–1754) English steatitic porcelains: A tentative study of compositions. J. Raman Spectrosc. **43**, 307–316 (2012)

W.H. Jay, J.D. Cashion, Raman spectroscopy of Limehouse porcelain sherds supported by Mossbauer spectroscopy and scanning electron microscopy. J. Raman Spectrosc. **44**, 1718–1732 (2013)

L.D. Kock, D. de Waal, Raman studies of the underglaze blue pigment on ceramic artefacts of the Ming dynasty and of unknown origins. J. Raman Spectrosc. **38**, 1480–1487 (2007)

E. Morton Nance, *The Pottery and Porcelain of Swansea and Nantgarw* (B.T. Batsford Ltd., London, 1942)

P. Ricciardi, P. Colomban, B. Fabbri, V. Milande, Towards the establishment of a Raman database of early European porcelain. e-Preservation Science **6**, 22–26 (2009)

M. Spataro, N. Meeks, M. Bimson, A. Dawson, J. Ambers, Early porcelain in seventeenth century England: Non-destructive examination of two jars from Burghley house. The British Museum Technical Research Bulletin **3**, 37–46 (2009)

E. Widjaja, G.H. Lim, Q. Lim, A.B. Mashadi, M. Garland, Pure component Raman spectral reconstruction from glazed and unglazed Yuan, Ming, and Qing shards: A combined Raman microscopy and BTEM study. J. Raman Spectrosc. **42**, 377–382 (2011)

I.J. Williams, *The Nantgarw Pottery and its Products: An Examination of the Site, Cardiff* (The National Museum of Wales and the Press Board of the University of Wales, 1932)

Glossary

Alabaster A type of white, semi-translucent gypsum, calcium sulfate dihydrate, $CaSO_4.2H_2O$, which in cleaved or thin sections was used in place of glass or in ornaments and statuary. Care needs to be taken with the nomenclature in some earlier literature since the term has been applied to include calcite, calcium carbonate, and even a silicaceous onyx. Aragonite, a polymorph of calcite, $CaCO_3$, is also found naturally as semi-transparent crystals and can often be confused with alabaster, although it is chemically very different in composition.

Biscuit porcelain The unglazed product of the first high-temperature firing process in the manufacture of porcelain using a "biscuit kiln" which operated typically in the region of 1200–1400 °C. The resultant porcelain is of a pure creamy-white or ivory colour and texture, which if blemish-free and perfectly shaped, was very highly prized by ceramic artists and modellers particularly for the construction of ornamental figurines. In the late eighteenth century, the finest biscuit porcelain articles were commonly placed on dining tables for admiration and as conversation pieces. Generally, the biscuit porcelain after cooling from the kiln was painted, glazed and gilded – during which any small defects could often be masked by strategically placed enamelling: hence painted and enamelled figurines are found much more commonly than their biscuit analogues.

Body paste the wet mixture of raw material components which comprise the formulaic recipes for porcelain bodies. Texturally moist to the touch it can be moulded or formed into shapes preparatory to the firing process in the biscuit kiln. This term is also applied to the fired porcelain body prior to analysis.

Burnishing The gentle polishing of gilt decoration on a glazed ceramics surface to a highly polished reflective coating. Early gilding was accomplished using "honey gilding" whereby 24-carat gold leaf was applied to the surface in a medium of honey or resins such as gum arabic, which was replaced in the late eighteenth century by mercury gilding using an amalgam of mercury triturated with gold leaf. During the final firing of the ceramic piece in the kiln at low temperatures the organic carrier component or mercury was volatilised leaving a dull

H. G. M. Edwards, *18th and 19th Century Porcelain Analysis*,
https://doi.org/10.1007/978-3-030-42192-2

golden finish, which was then hand-polished using finely powdered jeweller's rouge (iron oxide) to a brilliant finish.

Calcination The process of heating a raw material to high temperatures, and often red heat, to destroy organic and volatile components which might otherwise promote the evolution of gaseous products that could generate voids or blemishes in a viscous porcelain body paste under the firing process in the kiln. For example, limestone breaks down thermally to lime, calcium oxide, CaO, between 650–750 °C and the gaseous carbon dioxide evolved can be trapped in the matrix as unsightly bubbles. Powdered bones were generally heated to destroy the keratotic organic component in the hydroxyapatite matrix, $Ca_5(OH)(PO_4)_3$, and to eliminate coordinated water: the carbon formed initially during the heating process was converted to carbon dioxide at higher temperatures to leave a pure white material. At lower temperatures, artists favoured the heating of ivory or bone to lower temperatures but stopped the process at the formation of carbon, to harvest the "ivory black" or "bone black" pigment which was much appreciated for its depth of colour.

Coefficient of expansion and contraction A parameter which is based upon the ratio of linear change in size of a ceramic material upon the increase or decrease of temperature, usually expressed as a unit of length per degree. It is critically important for porcelain ceramics and as it results in shrinkage of the initial article upon firing and also upon its subsequent use in kiln construction where the integrity of the structure could be compromised by a flexing of the structure or movement during large ranges of temperature and through the heating and cooling stages of operational thermal cycles. It is of relevance too for the situation where a porcelain item has been copied at another factory as the copied item made from such a mould would be expected to be dimensionally smaller than the original cast because of shrinkage.

Cullet a glass additive to porcelain body recipes to increase the transparency of the fired body. Generally purchased by porcelain manufactory proprietors from glassworks, then added as a finely ground material to the second stage firing process of biscuit porcelain synthesis. A rather indefinite term scientifically as cullet could contain either flint glass (a highly refractive glass containing up to 60% lead oxide) or crown glass (also known as soda glass, which contained either soda or potash or both as an alkaline flux, but lead-free).

Digestion A first-stage chemical process during which the components of a ceramic material are rendered soluble in water through their dissolution thermal in strong acids or alkalis, such as hydrofluoric acid, sulphuric acid and sodium hydroxide. This is usually necessary to convert components such as silica, phosphates and silicates into water-soluble salts. After dissolution, the acidic or basic solution is neutralised and then divided into aliquot parts for wet chemical analysis.

Electromagnetic spectrum This describes the range of wavelengths which define the characteristics of electromagnetic radiation from the high-energy, low wavelength cosmic and X rays through the ultraviolet and then the visible spectrum from the violet to the red and into the infrared, thence through to the low-energy, high wavelength microwave radiation. Analytical techniques probe different ranges of the electromagnetic spectrum to interrogate the molecular and elemental compositions of specimens and derive quantitative and qualitative data which can be correlated with the materials used in the ceramic bodies, glazes and pigments.

Elemental oxides Usually determined from an SEM/EDAXS experiment which detects the elements such as silicon and aluminium, a conversion factor needs to be applied to estimate the percentage of silica or alumina in the sample. From the ratio of the elemental relative atomic masses in the formula to the molecular weights of the oxides the conversion factors can be listed as follows: Si to SiO_2 2.1, Al to Al_2O_3 1.9, Mg to MgO 1.7, Ca to CaO 1.4, Pb to PbO 1.1, Na to Na_2O 1.35, K to K_2O 1.2, Ti to TiO_2 1.7, Fe to Fe_2O_3 1.4 and P to P_2O_5 2.3. This means that an analytical determination of 10% silicon in the experiment would correlate with 21% silicon dioxide (silica) in the specimen.

Frit The result of a fine grinding process which is applied to the components of a ceramic paste and especially used to describe compositional mixtures of bone ash, china clay, flint glass (cullet), and soaprock. The necessity of very fine grinding to produce a homogeneous mixture of components which may have very tangible differences in hardness was appreciated by the earliest porcelain manufacturers and a fine frit usually then required just the addition of a single component or water to effect a suitable mixture for firing in a biscuit kiln. The term frit is also applied to a first-stage preparatory firing process to a component mixture, which is then fired at high temperature then ground, mixed with new components and then re-fired to produce biscuit porcelain in a two-stage process that is typical of most soft paste porcelain syntheses.

Fusibility The process of melting a ceramic material at elevated temperatures to produce a molten phase with no residual solid residues is affected by the presence of alkaline components which are called fluxes, typically soda, lime, potash and borax. The achievement and stabilisation of this molten phase at high temperatures is essential for the structural integrity of the porcelain body as the appearance of stress cracks can be initiated at lower temperatures at interfacial domains between fusible and infusible materials. The translucency of the final fired biscuit porcelain body is critically dependent upon the achievement of a single molten phase at the firing temperature used.

Gilding The application of a gilt decoration to fired porcelain involving 24-carat gold leaf in a carrier such as mercury, honey or an organic resin followed by burnishing, see under burnishing.

Glost kiln Used in the final stage of porcelain preparation after applying enamelled decoration and involving the application of an alkaline "slip" containing china clay, soda, potash and a lead oxide component usually in the form of a powdered flint glass, which at lower temperatures will form a hard, transparent glaze coating. Occasionally a glost kiln was used for calcination of components at lower temperatures and for the drying of components in which variable amounts of water were found to occur. Initially, glazes were lead-based but in the first decade of the nineteenth century when the toxicity of lead compounds was appreciated, reversion to a tin-based glaze was effected (as patented by John Rose at the Coalport China Works in 1820).

Grog The practice of grinding up damaged porcelain items (grog) discarded from the kiln firing of biscuit porcelains and re-incorporating them in the paste of new porcelain syntheses was adopted in the eighteenth century and frequently involved the purchase of grog from other factories often by the ton. Although at first sight this might seem a reasonable procedure it does create a potential difficulty in the analytical interpretation of such "mixed" pastes where the presence of unusual and exogenous interlopers can be registered in what otherwise analyses as a standard paste for a particular factory.

Instrumental Chemical Analysis the use of instrumentation to detect qualitatively and to estimate quantitatively the presence of chemical elements and their compounds in ceramic bodies and glazes. The basis of instrumental analysis is the use of special signatures or properties of chemical species or chemical bonds such as silica, SiO_2, or the Si = O bonds, in complex macromolecular structures such as silicaceous ceramic bodies from their behaviour towards the scattering or absorption of electromagnetic radiation in the electromagnetic spectrum from the low wavelength (high wavenumber) Xray region to the high wavelength (low wavenumber) microwave region. When coupled with a microscope the instrument is capable of interrogating very small specimens of the order of cubic microns or cubic nanometres.

Minor Additives Porcelain recipes or formulations frequently listed small percentages or amounts of additives such as cobalt blue, or smalt, which was added to remove vestigial traces of yellow colouration in the incipient china clays occurring from small amounts of iron oxides, or ochres. Small amounts of borax, sodium tetraborate decahydrate, to increase fusibility and the plasticity of the body at higher temperatures and arsenic oxide to assist in the generation of a uniform silicaceous phase.

Moulding The shape and size of porcelain items can be a reliable indicator of a particular factory's output and a clue to the attribution of unmarked pieces. Mouldings were generally unique to each manufactory and comprised individual and idiosyncratic cross-sections of flatwares such as saucers and plates, the shapes of cups, the pattern of footrims, the shape fo spill vases (e.g. cylindrical, fluted, tulip-shape, trumpet-shaped flared), the impressed and embossed verges of plates and the number and type of indentations at the rim.

Pigments In ceramics decoration these are coloured minerals which are thermally stable at temperatures in the glost kiln after application of the glaze slip and firing up to 600 °C. Mineral pigments are usually metal oxides, sulfates and sulfides such as haematite, gypsum and orpiment. Care needs to be taken in the interpretation of old recipes for the decoration of ceramics as minerals were often confusingly assigned an incorrect nomenclature, such as *minium*, which is used to describe both red lead, trilead tetroxide, and cinnabar, mercury sulfide.

Raw Materials The raw materials are components comprising the body and glaze recipes used in porcelain syntheses. Mostly these were natural minerals and rocks which were sourced from precise mines and locations which gave rise to several alternative names. Natural raw materials include feldspars, china clays, ball clays and soapstone. Synthetic raw materials include bone ash, pearl ash, soda ash, smalt, lime and magnesia.

Rocks Whereas minerals have a precise and assigned chemical formula and established crystal structures with recognised polymorphs (different structures for the same chemical formulation, e.g. calcite and aragonite, both formulated as $CaCO_3$), rocks are mixtures or aggregates of minerals with ill-defined chemical compositions, such as soapstone, which is often represented in its purest form as steatite, a hydrated magnesium silicate in its mineral form of talc $Mg_3Si_4O_{10}(OH)_2$.

Sagged porcelain The distortion of the shapes of porcelain items upon firing in the biscuit kiln ascribed to an excessively high temperature: it is believed that an over-reaching of the temperature by as little as 20 °C would be sufficient to cause sagging and result in unusable fired porcelains. This does not seem to arise from further chemical transformations at high temperature but rather can be attributed to physical and structural changes in the melt viscosity.

Slaked lime The product of the calcination of limestone (calcium carbonate) or dolostone (dolomite or dolomitised limestone) which results in the gaseous evolution of carbon dioxide and residual lime (calcium oxide) and/or magnesia. Reaction of this product with water gives calcium and/or magnesium hydroxide, the so-called slaked lime or limewash, used as a white base coat for buildings or limewash putty used as a filler and mortar. On reacting with moist air, the lime or magnesia gives the respective carbonate, namely calcite or dolomite, which forms as a hard surface skin so preventing further substrate reaction.

Smalt although a precise chemical terminology for a blue glass and referred to as a cobalt silicate (*Kaiser blaue, bleu d'email, azzuro di smalto*) it is actually a cobalt aluminosilicate, unlike another blue glass which is a cobalt aluminate (*cobalt blue, Thenard's blue, Vienna blue*). Frequently transposed in early ceramics literature and often confused with another *Bristol blue (cobalt blue)* which was manufactured in Bristol in the late eighteenth century, which is a cobalt lead silicate. All contain a cobalt (II) oxide component in the range 10–15%. Other blue glasses used in ceramic technology contain copper (II) oxide silicates such as *Egyptian blue (blue frit, cuprorivaite), Han blue (hedenbergite, Chinese blue) and Han purple.* Both of the latter blue glasses also contain barium (II) oxide.

Translucency This is perhaps the greatest achievable asset of porcelain manufacture to which every manufacturer subscribed and hoped to attain in emulation of the Chinese "eggshell" wares which were imported to Europe in the mid-eighteenth century. It describes the transmission of visible radiation, or light, through a solid object: at one end of the scale is glass, which is *transparent* and at the other is earthenware which is *opaque*. Porcelain is measured by its translucency, which is the clarity for transmission when viewed with background lighting. An intermediate descriptor is "semi-opaque", which was applied to some china during the mid-nineteenth century, which is rather indefinite and conveys little information about their true category.

Wet chemical analysis The earliest type of chemical analysis for ceramics subdivided into *qualitative*, which determined the chemical composition of the specimens and *quantitative*, which determined the proportions and relative percentages of each in the specimen. The analytical determinations involved the complete chemical digestion of the ceramic specimen and therefore its destruction. For ceramics analysis, the determination of silica, alumina, magnesia, lime, soda, potash, iron oxide and phosphorus pentoxide were normally undertaken, which could then be related back to the raw materials used in the formulation.

Index

Note: Entries in bold type indicate the major descriptions of porcelain factories, personnel and their initial foundation.

H. G. M. Edwards, *18th and 19th Century Porcelain Analysis*,
https://doi.org/10.1007/978-3-030-42192-2

The manufacturer's authorised representative in the EU is Springer Nature Customer Service Centre GmbH, Europaplatz 3, 69115 Heidelberg, Germany. If you have any concerns regarding our products, please contact ProductSafety@springernature.com

Printed and bound by CPI Group (UK) Ltd, Croydon, CR0 4YY
30/09/2026
02225247-0007